FROZEN IN TIME

FROZEN IN TIME

PREHISTORIC LIFE IN ANTARCTICA

JEFFREY D STILWELL AND JOHN A LONG

CSIRO PUBLISHING

National Library of Australia Cataloguing-in-Publication entry

Stilwell, Jeffrey D.

Frozen in time : prehistoric life in Antarctica / by
Jeffrey D. Stilwell and John A. Long.

9780643096356 (hbk.)
9780643104013 (epdf)
9780643104020 (epub)

Includes bibliographical references and index.
Fossils – Antarctica.
Paleontology – Antarctica.
Environmental geology – Antarctica.
Climatic changes – Antarctica.
Antarctica – Discovery and exploration

Long, John A., 1957–

560.989

Published by
CSIRO PUBLISHING
36 Gardiner Road, Clayton VIC 3168
Private Bag 10, Clayton South VIC 3169
Australia

Telephone: [+613] 9545 8555
Local call: 1300 788 000 (Australia only)
Fax: +61 3 9662 7555
Email: csiropublishing@csiro.au
Web site: www.publishing.csiro.au

Front cover image: courtesy Jeffrey Stilwell.
Back cover images: courtesy Cara Fritz, David Cantrill, Rodney Feldmann, Jeffrey Stilwell.
Title page image: An extremely windy day at Minna Bluff, McMurdo Sound, East Antarctica. Jeffrey Stilwell examines blocks of fossiliferous sandstone of Eocene age containing wood – probably southern beech. Photo: Rich Levy.

Set in ITC New Baskerville 9.75/13
Cover design by Alicia Freile, Tango Media
Text design by James Kelly
Typeset by Oryx Publishing
Index by Indexicana
Printed by Ingram Lightning Source

Mar26_RP_ILS

CONTENTS

The Transantarctic Mountains. In the background is Mt Erebus and Mt Terror on Ross island. Photo: Jeffrey Stilwell.

PREFACE

No other continent on Earth has undergone such radical environmental changes as Antarctica. In its transition from rich biodiversity to the barren, cold land of blizzards we see today, Antarctica provides a dramatic case study of how subtle changes in continental positioning can affect living communities, and how rapidly the catastrophic changes can come about. Antarctica has gone from paradise to polar ice in just a few million years of time, a geological blink of an eye when we consider the real age of Earth.

In this book we have endeavoured to present an overview of the fossil record of Antarctica framed within its changing environmental settings, keeping in mind that some parts of the Antarctic geologic column are better represented than others – we are at the mercy of a small percentage of rock exposure (not all containing fossils) in a deeply frozen land.

In telling the story of Antarctica's prehistoric past, we have chosen to group sequences of sedimentary rocks that contain fossils representative of similar past environments, with each chapter providing a window into a past time and environment on the continent. We have tried to reconstruct Antarctica's evolving animal and plant communities as accurately as its fossil record permits.

In order to demonstrate the significance of some of Antarctica's fossils, there is also some discussion of similar faunas from adjoining continents, as these were each an integral sector of the wider ranging, more extensive faunal and floral provinces. Comparisons with the fossil faunas of South America, Australia, 'Zealandia', India, and South Africa are critical to elucidating the big picture of Antarctica's evolving biodiversity.

The story of how fossils were first discovered in Antarctica is a triumph of human endeavour. It continues today with modern expeditions going out to remote sites every year to fill in more of the missing parts of the continent's great jigsaw of life. The heroic nature of the early exploration of Antarctica gives us an appreciation of how our knowledge of the prehistory of the continent was achieved. Many brave people endured harsh conditions to collect fossils. Some, like Robert Falcon Scott and his polar party, paid the ultimate price with their lives.

For many decades the inhospitable landscape of Antarctica prevented scientific parties from reaching its most remote localities, so the collection of fossil and rock specimens was often an afterthought to exploration. Yet, although some of the early explorers were instrumental in unveiling the first glimpses of the palaeontology of Antarctica, only in the last few decades have advances in polar transportation allowed scientists to undertake systematic collections of fossils from regions well outside the helicopter ranges of the base stations. Because of these advances in logistics, palaeontologists are able to penetrate deeper into Antarctica's frozen secrets. We know now that some isolated regions of the icy continent are literally paved with fossils, such as horizons with shells composed of mother-of-pearl that shimmer and glisten in the sun like jewels.

As the story of Antarctica's prehistoric past is revealed chapter by chapter, there are also a number of boxes that give background information about subjects raised in the text. We hope these additional windows of scientific information enable most questions about the general principles of palaeontology to be easily understood. There is also a glossary of scientific terms at the end of the book, along with a comprehensive list of both scientific and general references for further reading.

Dr Jeffrey Stilwell
Professor John Long

The 'Ice Pinnacles' of McMurdo Sound with exposed subfossil sea floor remains, many thousands of years old. Photo: Jeffrey Stilwell.

ACKNOWLEDGEMENTS

Our current state of knowledge of the evolutionary history of Antarctic biotas reflects the hard work and dedication of scores of researchers from the 19th century to more recent times. More often than not the researchers have faced harsh conditions during the many expeditions that have taken place since the first fossil – a fragment of wood – was discovered by American researcher James Eights in 1830.

This book was made possible by the gracious support of many world governments and scientists, who are eager to know what ancient life forms existed on the frozen continent from the Cambrian to the present. With persistence and luck, we will undoubtedly fill in more gaps in the fossil record of Antarctica, and perhaps even extend it to the Precambrian sometime this century.

We thank and appreciate the support of the following individuals and their respective institutions: renowned artists Bill Stout (Pasadena, California) and Tony Windberg (Perth, Australia); Bill and Virginia Zinsmeister (Purdue University, Indiana), who made it possible for Jeffrey Stilwell as a student to research Cretaceous and Cenozoic Antarctic biotas; Pat Quilty (University of Tasmania, Hobart); Betty Thompson, Mary Walters, Ben Doherty, Steve Morton, Draga Gelt, Marion Anderson, Jenny Kain, Ray Cas, Patricia Komarower and Pat Vickers-Rich (Monash University, Melbourne); Brian Pump (formerly of James Cook University, Townsville); David Cantrill (Royal Botanic Gardens, Melbourne); Sean Jago (Abbotsford, Melbourne); Gavin Young (ANU, Canberra); Alex Ritchie (Canberra, Australia); Bert Rowell (University of Kansas, USA); Rod Feldmann (Kent State University, Kent, Ohio); Mike Woodburne (Flagstaff, Arizona); Christopher Consoli and Phil O'Brien (Geoscience Australia, Canberra); Erich Fitzgerald, Tom Rich and Brian Choo (Museum Victoria, Melbourne); Eduardo Olivero (CADIC-CONICET, Ushuaia, Argentina); Roberto Scasso (Universidad de Buenos Aires, Argentina); Sergio Marenssi (Instituto Antarctica, Argentina); Julia Clarke (University of Texas at Austin); Anton Oleinik (Florida Atlantic University); Brian Huber, Mary Parrish and Thomas Jorstad (Smithsonian Institution, Washington DC); Michael Skrepnick (Okotoks, Alberta); Christian Sidor (University of Washington, Seattle); Margaret Bradshaw (University of Canterbury, Christchurch, New Zealand); Sylvia Hope (California Academy of Sciences, San Francisco); David Harwood (University of Nebraska-Lincoln) and former students, Steve Bohaty and John Kaser; Richard Levy (GNS Science, Lower Hutt, New Zealand); the late Andrew Grebneff, Mark Hughes and Daphne Lee (University of Otago, Dunedin, New Zealand); Margaret (Peg) Rees (University of Nevada, Los Angeles); Sue Turner (Brisbane); Paul Willis ('Catalyst' presenter, ABC-TV, Sydney) and the Albany Institute of History and Art, New York.

The text and layout of this book was enhanced greatly by the talents of Nick Alexander, Tracey Millen and the team at CSIRO Publishing, Melbourne.

This book is dedicated to polar researchers Professors Bill and Virginia Zinsmeister, Professor Pat Quilty and Sir David Attenborough, who have collectively brought much new knowledge and spirit in our quest to learn more of Antarctica's 'deep time' biotas and environments.

Cockburn Island, viewed from Seymour Island. Photo: Jeffrey Stilwell.

OPPOSITE: Searching for glacial striations in Paleozoic marble by University of Nebraska-Lincoln researchers along the Skelton Glacier. Photo: Jeffrey Stilwell.

1
THE COLD, BARREN LAND WE CALL ANTARCTICA

It's not getting to the pole that counts. It's what you learn of scientific value on the way. Plus the fact that you get there and back without being killed.

Admiral Richard Byrd, *Alone*, 1938.

Antarctica – a word that conjures up feelings of bitter cold, sounds of howling gales, and visions of alien, barren landscapes – the desolate continent at the very bottom of the world. It has been uninhabited and unknown for most of human civilisation – a *terra incognita* until the 19th century. Poignant stories of human survival, bravery and death punctuate its early history of exploration. Sir Douglas Mawson dubbed it the 'home of the blizzard', a title well derived from his meticulous weather recordings which demonstrated that his base camp at Cape Denison had the highest average wind speeds of anywhere on Earth.

Antarctica has the coldest climate recorded on Earth, with extremes as low as –89°C. Today, almost nothing at all lives on the vast expanses of its ice-bound surface. Nearly all the continent is permanently covered by ice and snow, with mean annual surface temperatures hostile to life of any kind apart from a few organisms that can survive in just about any conditions imaginable. Near the edges of the continent life abounds in and around its biomass-rich, icy ocean waters, yet much of this life is ephemeral. The penguins and seals that come to feed during the warmer, summer season are mostly gone by winter, except for some well-adapted, extremophile species, such as the emperor penguin.

Yet Antarctica today is no reflection of what it once was. The evidence for this story is written by

Rock outcrops adjacent to the Skelton Glacier, East Antarctica. Photo: Jeffrey Stilwell.

the fossils in its rich geological history. The plot is mysterious and unexpected; there is murder on a cosmic scale as global mass extinctions are caused by killer meteorites, and sudden shifts in climate that wipe out high proportions of life. Integral to all of this is the movement of continents and the formation of supercontinents. Antarctica has, for 99 per cent of its geological life spanning 3.8 billion years, been located within the hub of the various configurations of the ancient supercontinents known as Rodinia, Pangaea and Gondwana.

The story told by Antarctica's diverse fossil record reveals that it was not always a lifeless continent locked in ice, but has been home to a great pageant of life through time. Its ancient, fossil-rich rocks reveal evidence of previous warmer climates – from the 250-million-year-old coal seams that run through the Transantarctic Mountains – part of the largest coalfields in the world – to the relatively recent non-fossilised wood from forests that grew on the flanks of the Transantarctic Mountains, only a few million years ago. Indeed, Antarctica was once rich in biodiversity, with ancient forests, huge dinosaurs, strange mammals and killer birds. The continent played a pivotal role in the migration and distribution of nearly all life in today's southern continents.

The astounding fossil record shows that Antarctica – once the central landmass of the supercontinent of Gondwana – is really the key to understanding the evolution and biogeography of most of the living fauna and flora on all the major southern hemisphere landmasses today: South America, Africa, Australia, and the New Zealand microcontinent ('Zealandia'). As we search further back in time, other landmasses, now subsumed within greater Asia, such as India, parts of China, South-East Asia and the Middle East, even part of the North American continent, were also in contact with Gondwana. Some of these regions still bear the distant biological imprints from this ancient geological alliance.

The fossil record of Antarctica shows us why we have southern beech forests in Australia and New Zealand, how dinosaurs were able to walk to every continent on Earth, and where and how the largest animals on the planet, the great baleen whales, may have evolved. Antarctica's prehistoric record is also a tale of great tragedy. All that we have discovered from the decades of fossil hunting comes from a very small percentage of accessible rock around the periphery of the continent and in the mountainous regions that poke up through kilometres of ice. We see the demise of the expansive forests and the growth of the immense ice sheets that currently lock up nearly 70 per cent of the planet's fresh water (and 90 per cent of its ice). The implications of these events continue to have a direct bearing on the global climate and the fate of most of Earth's living ecosystems.

§

Antarctica is the fifth largest continent on Earth, covering approximately 14 million square km (5.4 million square miles), an area that is one-third larger than Western Europe, about 1.5 times the size of the United States of America, and comparable in size to all of South America. More than 98 per cent of Antarctica is covered by an ice sheet, which is about 5 km thick in some places. This makes Antarctica the world's highest continent, with an average relief of about 2.3 km above sea level. Many fossil secrets inaccessible below the ice will remain that way, until (or if) in some distant time global warming unlocks long-frozen fossil treasures.

The continent is divided into the vast, flat polar plateau covered by the ice sheet, and the various mountain ranges, the largest being the Transantarctic Mountains which run right across the continent, dividing it into two greatly contrasting parts, the East and West. The Transantarctic Mountains extend from the north-western corner of the Ross Sea to the south-western corner of the Weddell Sea. The tallest mountain range is the Vinson Massif in West Antarctica, reaching nearly 5 km in height.

Less than two per cent of the area of Antarctica has rock exposures available for the geologist to study. The

The Wright Valley glacier, Dry Valleys, Transantarctic Mountains. Photo: Jeffrey Stilwell.

Antarctica from space. Courtesy: NASA and Dave Pape.

ice sheets bury nearly all of the topographic features of the continent, apart from the coastal outcrops and mountain peaks – known as nunataks – that poke up through the ice sheet in mostly inaccessible areas.

Most of the world's fresh water, approximately 24.5 million cubic kilometres, is currently locked up as ice in Antarctica. The East Antarctic ice sheet, which overlies a stable cratonic block of continental crust, covers 75 per cent of the area of the continent and it contains 80 per cent of Antarctica's ice by volume. The West Antarctic ice sheet is far less stable as it covers part of the Antarctic Peninsula and a series of islands and embayments, and unites four land units each with a complex geologic history. There are also two large frozen fields of permanent ice floating atop the inlets of the Ross Sea and Weddell Sea on either side of West Antarctica – the Ross Sea Ice Shelf and the Weddell Sea Ice Shelf. There is also the Amery Ice Shelf in Prydz Bay.

Deep waters flow between these land units and the West Antarctic ice sheet, which is a floating ice mass and is held in place by great mountainous peaks. While the current glaciated state of Antarctica commenced some 34 million years ago, the ice in

Deep blue and blinding white, the Ross Sea Ice Shelf breaks up in the austral summer. Photo: Jeffrey Stilwell.

the ice sheets is not that old either, the most ancient being only one million years old – a mere blink of a geologic eye.

Sea ice covers much of the waters surrounding the continent for most of the year. Ice flows bring sheets down from the high ground of the continent to sea level. This moves across the sea ice, before calving off into icebergs when it meets warmer water. The ice flows away from the highest points of the continent towards the sea on all sides. One would not think that ice itself flows, but it does due to its plastic nature. At the bottom of the thickest ice sheets plastic deformation takes place in the ice due to immense pressure from above. This makes the ice flow more easily over the bedrock of the continent – hence the reason for the unstable West Antarctic ice sheet.

The ice that is lost to the sea becomes replenished over time by snow inland. If the Antarctic ice were to thaw out entirely, the seas of the world would rise approximately 67 metres. The removal of the ice sheets would reveal a smaller Antarctic continent of around seven million square kilometres.

The break-up of sea ice during the austral summer enables ships to deliver supplies and allows tourist vessels to enjoy the splendour of the deep south. Photo: Jeffrey Stilwell.

Mt Erebus (3794 m) the world's southern-most active volcano, is the largest of three major volcanoes forming the roughly triangular Ross Island. On the far left is Fang Ridge, representing an earlier crater wall, and behind is Mt Terror. Mt Erebus was erupting when first sighted by Captain James Ross in 1841. Since 1972 there has been continuous lava-lake activity, punctuated by occasional strombolian explosions that eject bombs of lava onto the crater rim. Photo: Jeffrey Stilwell.

The emperor penguin, *Aptenodytes forsteri*, the world's largest and heaviest penguin, can grow to more than a metre in height and lives more than 40 years. Photo: Jeffrey Stilwell.

In the past, Antarctica was a hub of volcanic activity, and even today there are still a few active volcanic centres, such as Mt Erebus on Ross Island, Deception Island in the South Shetland Islands Group, and Mt Melbourne on the west coast of the Ross Sea in Northern Victoria Land (which has not erupted for a few hundred years) and several others, which have probably erupted in pre-historical times.[1]

Antarctica is technically the world's largest desert. It almost never rains, apart from the fringing islands near the Antarctic Circle, as most precipitation falls as snow. The atmosphere is very dry, continually desiccating the landscape and any living organisms on it with savage winds. Visitors to Antarctica must drink water constantly to keep dehydration at bay. The climate is not very predictable, and random storms called 'blizzards', which can last as long as ten days, happen throughout the year, fuelled by the gravity-fed, intensely bitter, katabatic winds blowing out and rolling off the elevated polar plateau. The seasons fall neatly into the continuous 24-hour daylight of summer, the perpetually dark night of winter, and the relatively brief autumn and spring twilights. The mean monthly temperature at the South Pole in summer is –30°C and in winter it is –54°C. At Vostok Station, the temperature registered an inhospitable –89.6°C in August 1983. As far as humans are concerned this may as well be absolute zero! During summer in the Dry Valleys, the temperature can rise as high as 10°C.

For four months of the year, Antarctica is in total darkness. Due to the absence of pollution, nights in Antarctica are breathtakingly clear with innumerable twinkling stars, along with bright planets and shooting stars or asteroids. These events remind us about the vastness of time and space. If one is lucky, one can witness a natural light show called the *aurora australis* (or 'Southern Lights'), which can shimmer away in the Antarctic night

Icicles connecting boulders in the Dry Valleys, Transantarctic Mountains. Photo: John Long.

This sand-blasted rock, called a 'ventifact', at Bull Pass, Dry Valleys, has been ceated by violent katabatic winds. Photo: John Long.

Ripples in a dune of sand eroded from the Beacon Sandstone, lower Victoria Valley, Dry Valleys, Transantarctic Mountains.
Photo: Jeffrey Stilwell.

sky, mesmerising and amazing those who view this spectacular phenomenon. Even during the short summer season the water under the ice receives less than one per cent of the surface sunlight. The temperature of the water is between –2°C and –4°C and is prevented from freezing by its salt content. Fish living in the icy Antarctic waters have evolved a clever way of adapting to the sub-zero temperatures by incorporating an 'anti-freeze' in their blood.

The geographic configurations of Antarctica and other southern lands are an historic legacy relating to the ever-changing jigsaw of continental pieces throughout the history of the planet. For much of its existence Antarctica was joined to Africa, Madagascar, South America, India, Australia, 'Zealandia', and other smaller continental fragments in the South Pacific. This supercontinent, Gondwana, remained largely in the southern hemisphere throughout the 500 million years of its existence. During the Cambrian Period it was probably situated on the equator or just inside the northern hemisphere. At that time, away from the South Pole, Gondwana was free of ice sheets for much of its history.

The isolated configuration of Antarctica today has led to cooling at the poles and an area of high atmospheric pressure over the continent, which keeps it cold and also prevents precipitation. If warm air or water could find its way into the deep freezer that currently exists today on the continent, the climate of Australia could change very rapidly. It is not surprising that the frozen continent affects its northern neighbours on an almost daily basis – for example, it is the source of the cold fronts that move across continental Austrtalia.

§

Although the discovery of Antarctica by humans dates back a mere 200 years, its geological history extends back to nearly four billion years. Rocks of this age have been identified from Antarctica, yet we know nothing of Earth's continental configurations at this early time. Around 540 million years ago, we find Antarctica as part of a giant southern supercontinent called Gondwana (meaning 'land of the Gonds', a native tribe from India). Originally the name 'Gondwanaland' was coined by the Austrian geologist Eduard Suess in his book published in 1883. Because of similarities he noted between the geology of the southern continents and peninsular India, he suggested that these landmasses were joined by land bridges that had since sunk beneath the seas. Today we know this is not true as our recent knowledge of Earth's dynamic processes tells us that the continents, which are merely slabs of Earth's crust, have been slowly moving, pushed possibly by convection currents within Earth's mantle – the molten rock layer below the crust. Australia, for example, is currently moving at about six centimetres a year northwards (the rate which human hair grows) and is destined to become another part of the great Asian landmass in about 50 million years time.

A team from the University of Nebraska-Lincoln looks for fossils in glacial moraine material in McMurdo Sound. Mt Erebus is in the background. Photo: Jeffrey Stilwell.

Stormy weather at Mt Discovery during summer. Massive snow storms can last for many days with no hope of doing fieldwork. Because of gale-force winds, it is unsafe (and unwise!) to spend too much time taking photographs outside. Bare flesh gets blasted and quickly freezes. Photo: Jeffrey Stilwell.

Rocks

There are three main kinds of rocks: igneous, sedimentary and metamorphic.

Igneous rocks are those which erupt from volcanoes or seep out below the ground from molten magma. Rapidly cooled magma forms rocks with a fine crystalline structure (as in basalt) or more coarsely crystalline if slowly cooled (as in granite).

Sedimentary rocks are formed by erosion of other rocks, broken down into grains or particles, then bound together by chemical precipitates, or by compaction through pressure and heat.

Metamorphic rocks are those produced deep under the ground when sedimentary or volcanic rocks are subjected to great heat and pressure. In extreme cases this forms melting and recrystallisation of the original rocks to make new mineral assemblages. Examples of metamorphic rocks are schists formed from intense compaction and high temperature alteration of sedimentary rocks, or gneisses formed by deformation of granitic rocks.

All fossils are preserved in sedimentary rocks – either 'clastic' or 'carbonate' rocks. Clastic rocks are made up of separate grains called 'clasts'. Fine-grained particles, such as mud or silt, form mudstones, siltstones or flat-layered fissile rocks like shale once they are buried and compacted, whereas coarser-grained rocks form sandstones or conglomerates.

Carbonate rocks are bound together by calcium carbonate ($CaCO_3$). Limestones are carbonate rocks composed mostly of calcium carbonate yet may also include a proportion of mud, silt or sand. Quite often they are made up of the accumulated remains of many small organisms which have calcitic or aragonitic shells, such as clams, corals or microscopic fossils such as foraminiferans.

The term 'sandstone' applies to rocks formed principally of sand grains, and these are defined only by the size of grains, being larger than 0.2 mm and smaller than 2 mm in diameter. Grains larger than 2 mm can form 'coarse-grained' sediments like grit and conglomerate. These are formed in high-energy environments where fossils are often broken up before final burial. Sandstones can be made up of grains of various minerals, but quartz and feldspar are the commonest varieties. The finer the sediment particle size, the quieter the environment in which the rock was formed, and the better the chance of a complete, well-preserved fossil skeleton.

Very fine-grained muds accumulate in deep still waters, whereas active high-energy environments such as river beds and tidal zones tend to be dominated by sandstones or coarser-grained rocks. Very coarse sedimentary rocks, such as conglomerates, generally lack good fossils because they were deposited in turbulent conditions where delicate organic remains were broken and not preserved.

Lower Paleozoic metamorphic rocks bordering the Skelton Glacier. Photo: Jeffrey Stilwell.

This mummified seal is at least hundreds if not a few thousand years old. Like many others found in the Dry Valleys coastal regions, it probably died from illness and confusion as it travelled too far inland from the sea. Bacterial decay is very much inhibited in the Antarctic desert environment. Photo: Jeffrey Stilwell.

Antarctica started off as being a part of a very ancient supercontinent called Rodinia. It has been suggested that eastern Gondwana was connected with the northern supercontinent of Laurentia (including much of continental North America) along the Transantarctic Mountains–East Australian margin from as far back as 1.9 billion years ago until around 560 million years ago. Most of the other known continental fragments (including the ancient Baltic regions 'Baltica', Siberia, northern China, southern China and 'Tarim') were in close proximity to this landmass, so they formed the first supercontinent, named Rodinia after the Russian 'rodit' meaning to beget or grow.[2] The palaeomagnetic record of the rocks is useful here as Earth's magnetic field will align minute magnetic particles in sedimentary rocks before they are petrified to give us an indication of which way north was at the time the rocks were deposited. By about 600 million years ago continental rifting started to drive the eastern and western sides of Rodinia apart. Gondwana began to rift away from Laurentia, and South China drifted northwards, forming the nascent Palaeo-Pacific Ocean between them.

The break-up of Rodinia at this time had the effect of rotating the Gondwana continents inside-out, forming the larger Gondwana with Africa and South America joined to Antarctica and Australia. This started at the onset of the Cambrian with the coming together of eastern and western Antarctica. By the end of the Cambrian Period, around 500 million years ago, Gondwana had formed and would remain in that configuration until the break-up of major landmasses some 300 million years later, at about the same time dinosaurs started walking the land.

Over the last 500 million years large chunks of continental crust rifted away from the hub of Gondwana to become the individual continents we recognise today. North America, although somewhat controversial as to its early placement as part of

The International Geophysical Year and the Antarctic Treaty

The International Geophysical Year (IGY) was the brainchild of US scientist Dr Lloyd Berkner. The year 1957 was suggested as sun spot activity would be at a peak so data could be compared with the previously highest year of similar activity in 1932–33. To plan for the forthcoming IGY, the International Council of Scientific Unions (ICSU) was set up in order to forge a scientific unity among all countries working in Antarctica. It aimed to coordinate the research and to oversee the sharing of the scientific results.

Twelve countries took part in the IGY program, resulting in over 40 stations set up in mainland and peninsula Antarctica, and another 20 stations on islands within the Antarctic Circle. The USA set up bases at McMurdo and the South Pole, while Russia set up a base at Mirny, at what was then dubbed the 'Pole of Inaccessibility' because it was located at the inland point furthest from all Antarctic coasts. Scott Base was set up as New Zealand's main station in Antarctica, and the smaller Vanda Station in the Dry Valleys region. A highlight of the International Geophysical Year was the joint British–New Zealand crossing of Antarctica from coast to coast by Sir Vivian Fuchs and Sir Edmund Hillary.

After the success of the IGY a special committee was set up to maintain the scientific cooperation between participating nations, and the first Antarctic Treaty was formally ratified in 1961. After the Treaty ended in 1991, a more extensive set of guidelines for working in Antarctica was established as the Madrid Protocol, which came into effect in 1998. Such protocols effectively keep commercial exploitation such as mining out of Antarctica, provide stringent regulations for the disposal of all waste materials, protect its wildlife and encourage international cooperation on scientific research.

Since the establishment of the Antarctic Treaty and the founding of the main bases around Antarctica's periphery and the few remote interior stations, scientific expeditions to its remote inland mountain ranges have been regular events almost every year. Our knowledge of Antarctica's geology and palaeontology has been steadily increasing ever since, as new material is prepared, studied and described. But the geologists who work there face daily dangers from the world's worst weather condition and other deadly hazards, such as crevasses and avalanches.

Preparing to set up camp on Seymour Island, January 1984. Cockburn Island is in the background. Photo: Rodney Feldmann.

The edge of the Barwick Valley in the Taylor Valley region of the Dry Valleys is a 'scientific site of special significance' and a restricted area.
Photo: John Long.

Gondwana, was probably the first to drift away about 500 million years ago, followed soon after by parts of Europe, the Middle East and Arabia, then South America, Africa and India. During the Devonian to Carboniferous periods, some 400 to 300 million years ago, changing continental configurations caused a geologic dance of sorts (often called the 'Waltz of Gondwana'), when the supercontinent progressed southwards with Antarctica poised over the south polar region for the time, and Gondwana eventually plunged into a mass glaciation.

Despite its wandering nature, with the palaeo-South Pole sometimes being over Australia or other parts of the South Pacific, Antarctica has mostly had a home in the high southern latitudes. A series of maps based on the latest geological data showing the formation of Rodinia, and the formation and subsequent break-up of Gondwana is shown on page 14.

Around 130–128 million years ago, only Antarctica, Australia and 'Zealandia' were left as the now much reduced landmass of Gondwana. Australia began its rifting with Antarctica about 110 million years ago but the continents did not break their crustal link from each other until about 65 million years ago, at the end of the Cretaceous Period, when the non-avian dinosaurs were uttering their final death cries. However, because of long trailing pieces of Australia's crust, like the Lord Howe Rise, the complete break away from Antarctica was not properly established until the Eocene or earliest Oligocene epochs, perhaps as recent as 35 million years ago. Just before this, South America had completed its rift away from Antarctica, opening up the Drake Passage. These two events enabled the circumpolar current around Antarctica to fully circulate, starting the great cooling of the southern continent.

The two main geographic regions which make up Antarctica today are quite different in their geological histories from each other. East Antarctica consists of a stable shield of very ancient Precambrian rocks, older than 570 million years,

Plate tectonics

The continents we know today all once had different shapes and positions on the globe, and many were joined as supercontinents, such as Laurasia, the great northern landmass (containing North America and most of Asia), and Gondwana, the great southern landmass. The continents and oceans are part of tectonic plates, which are in motion. Over millions of years the movement of continents resulted in tectonic plate collisions that formed new mountain belts; allowed seas to inundate previously dry lands; joined one landmass with another; and created chains of islands within the seas. Thus, species could invade new lands and compete with the existing plants and animals for food resources. Many of the great extinction events in past geological times may have been caused by 'armies' of invading species going into a new territory, or major fluctuations in sea level and changes in climate causing a landmass to become suddenly flooded or a lush, forested area turn rapidly to desert.

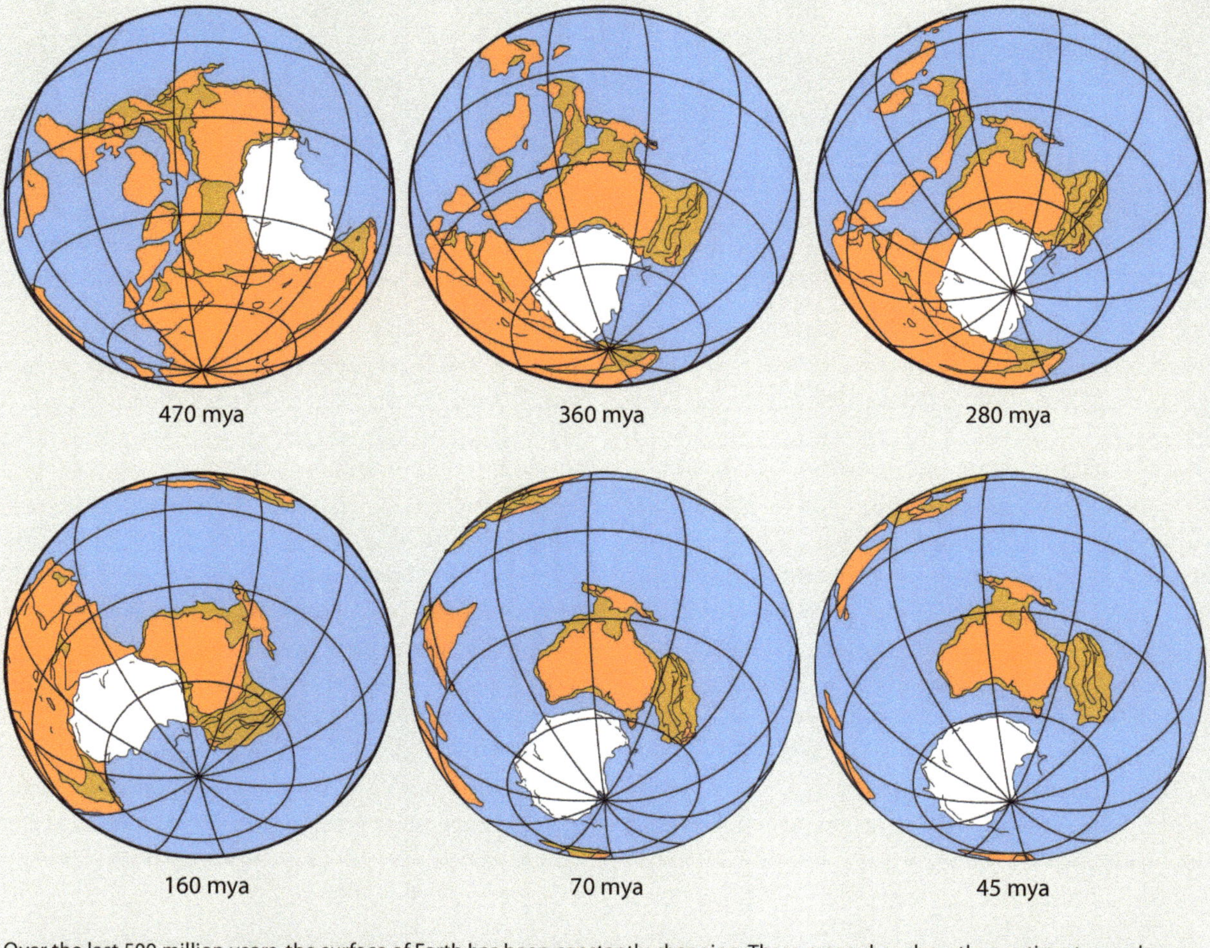

Over the last 500 million years, the surface of Earth has been constantly changing. These maps show how the continents may have been positioned 470, 360, 280, 160, 70 and 45 million years ago. Source: Geoscience Australia.

The southern-most extent of Quaternary glacial moraine cover, south of Mt Discovery campsite, during a University of Nebraska-Lincoln expedition. Mt Erebus is in background. Photo: Jeffrey Stilwell.

and mostly above sea level. Some Antarctic rocks have been dated as far back as 3.8 billion years. West Antarctica would, if the ice were removed, be simply a string of islands. The two regions are separated by the Transantarctic Mountains whose peaks protrude well above the ice cap. The ice sheets, which now cover more than 98 per cent of Antarctica's landmass, started forming about 34.5 million years ago, but did not extinguish the last forests on the land there until a mere seven million years ago, and possibly, the last trees only a few million years ago. Waxing and waning of the ice sheets has allowed re-forestation at various times throughout the last 30 million years.

Less than 100 years ago, the concept of 'continental drift' was indeed a controversial and much misunderstood theory that few scientists held seriously. The German geographer Alfred Wegener published a book on the subject in 1915, but he never gained acceptance in the scientific circles of his day. The concept of 'plate tectonics' came about later, during the last 50 years, and today, the movement of Earth's crustal plates is a well-accepted scientific fact that can be accurately measured. It explains not only the gradual movement of continents over time, but also the reason why we have earthquakes, volcanic eruptions and tidal waves. The palaeontological studies of the continents show past distributions of animals and plants that enable us, in conjunction with other lines of geological evidence, to reconstruct where the continents were in the past. However, despite many recent advances in the discovery of fossils from many localities around the world, it is the linking of these observations within the context of Gondwana that makes the prehistory of Antarctica so fascinating.

The 'Ice Pinnacles' of McMurdo Sound are pressure ridges of ancient blue ice. Photo: Jeffrey Stilwell.

Today Antarctica still holds much mystery and demands respect due to its extreme and unpredictable climate, and the fact that it is a large continent, most of which has never been explored on the ground by humans. While most of it consists of a boundless polar plateau of ice and snow, even the Transantarctic Mountains have been explored only in select regions – mostly those areas easily reached from the scientific bases which fringe the continent.

In this book we have reconstructed the story of Antarctica's evolving landscapes and animal and plant communities as accurately as its fossil record permits. No other continent on Earth has undergone such radical environmental changes as Antarctica. In its transition to the barren, cold land of blizzards we see today, Antarctica is the most dramatic case study of how subtle changes in continental positioning dramatically affect living communities, and how rapidly catastrophic changes can come about. It provides us with a salutary reminder of how a nuclear winter could wipe out life in the civilised world today if a nuclear war broke out. Antarctica has gone from paradise to polar ice in a just a few million years, a geological blink of an eye when we consider the real age of Earth.

OPPOSITE: A first glimpse of sea ice in the Southern Ocean from a Hercules C-130 cargo plane, approaching the Antarctic continent. Photo: Jeffrey Stilwell.

2
A CONTINENT DISCOVERED

[I would] rather go back to Antarctica and find a fossil marsupial than three gold mines …

Dr Laurence M Gould, *Cold: The Record of an Antarctic Sledge Journey*, 1931.

Antarctica is the youngest continent to be discovered, explored and inhabited by us humans. More than 2000 years ago the Greek philosopher Aristotle hinted at the existence of an unseen southern continent because he saw a need to balance the mass of the large northern hemisphere continents. As the northern landmasses were under the star *Arktos*, so Aristotle postulated that a great southern land must exist, which he dubbed *Antarktos.*

In the second century, the Egyptian geographer Ptolemy went further by agreeing that this southern land which he called *Terra Australis Incognita* must exist, and that it would be a fertile and populated land. He postulated that it was cut off from the rest of the world by a region of fire and, some believed, inhabited by fearful monsters. Such ideas naturally discouraged further exploration to the Antarctic region for many centuries to come.

A classic text of the Middle Ages, *Itinerarium* … by Pomponius Mela and Julius Solinus, published in Rome in 1519, expounded the argument that the known landmasses in the northern hemisphere must have antipodean counterparts to balance them. Mela and Solinus's view of the shape of the world incorporated a huge landmass surrounding the South Pole. This early notion of a fifth continent was a significant historical development.

The phantom of *Terra Australis,* an unknown southern land, haunted the minds and maps of cosmographers for more than two millennia. These speculations were still quite influential in the 18th century until the time of Captain James Cook.

This mythological Antarctica as a prosperous and populated land was properly dispelled by Cook, whose ship *Endeavour* was the first to sail into Antarctic waters. His diary entry for 17 January 1773 notes:

> I continued to stand to the south, and on the seventeenth, between eleven and twelve o'clock, we crossed the Antarctic Circle in the longitude of 66 degrees 36 minutes 30 seconds south.

During the course of his second voyage, James Cook in the *Resolution* and Tobias Furneaux in the *Adventure* became the first to circumnavigate the globe at latitudes within the Antarctic Circle. Cook achieved a new farthest south record of 71°10'S and proved that no previous land sightings constituted an Antarctic continent. He landed on South Georgia and discovered all but three northerly islands in the South Sandwich Islands. Cook eventually crossed the Antarctic Circle three times in circumnavigating the polar seas. He could not confirm that there was a continent within these icy seas, but believed that land must have been nearby.

It would be almost another 50 years before the first human laid eyes on Antarctica. The first person to do so was probably Admiral Fabian Gottlieb von Bellingshausen during the Russian Antarctic expedition of 1819–1821 in the *Vostok* and *Mirny*. Bellingshausen charted the south coast of South Georgia, discovered the three remaining northerly islands in the South Sandwich Islands, probably sighted ice attached to the mainland continent (the Fimbul Ice Shelf) on 27 January 1820, and discovered Peter I Island and Alexander Island, the first land discoveries south of the Antarctic Circle.

In February of 1821, American sealer John Davis became, arguably, the first person to land on the continent, after sailing from Connecticut, USA, to the South Shetland Islands. This landing was followed by another in December, when the American Nathaniel Palmer, along with a British sealer named George Powell, discovered the South Orkney Islands.

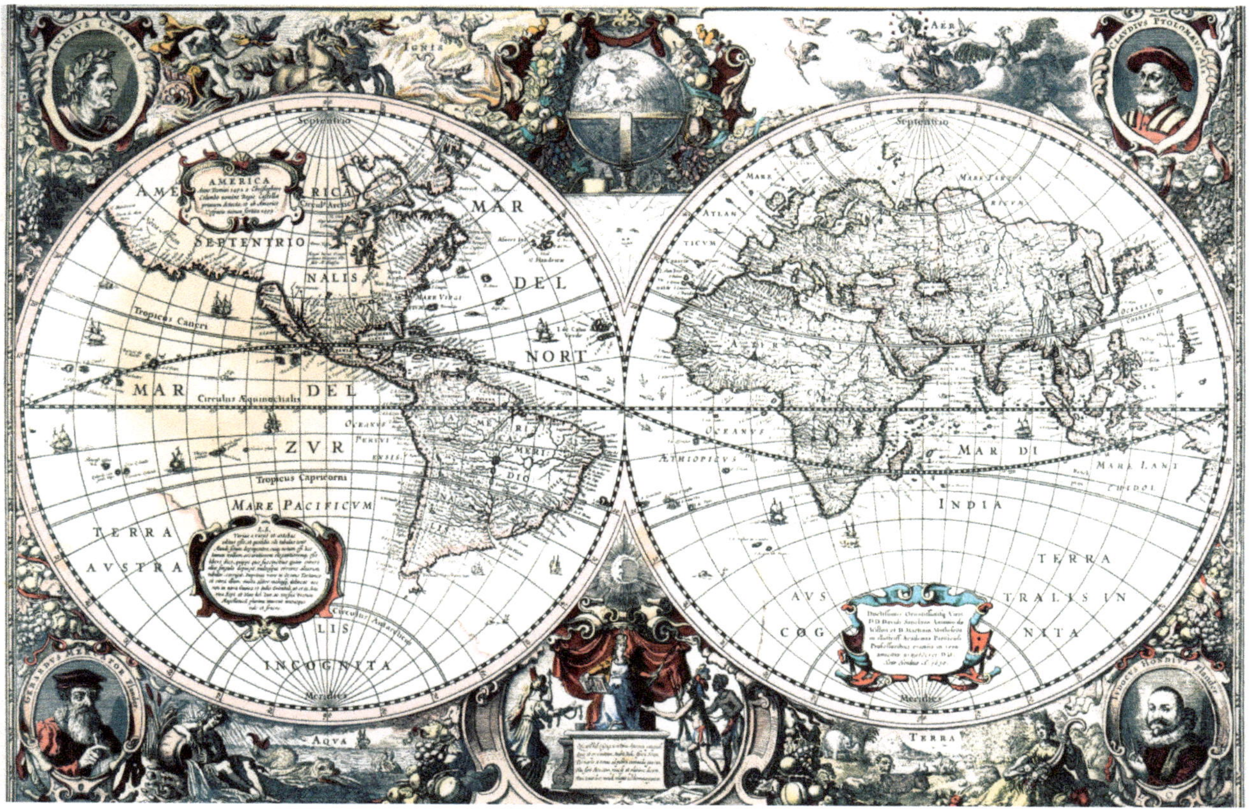

An early 17th century world map depicting the world geography as known at the time. Most of the southern hemisphere was uncharted and recognised as a *Terra Australis Incognita*. Photo: Steve Morton.

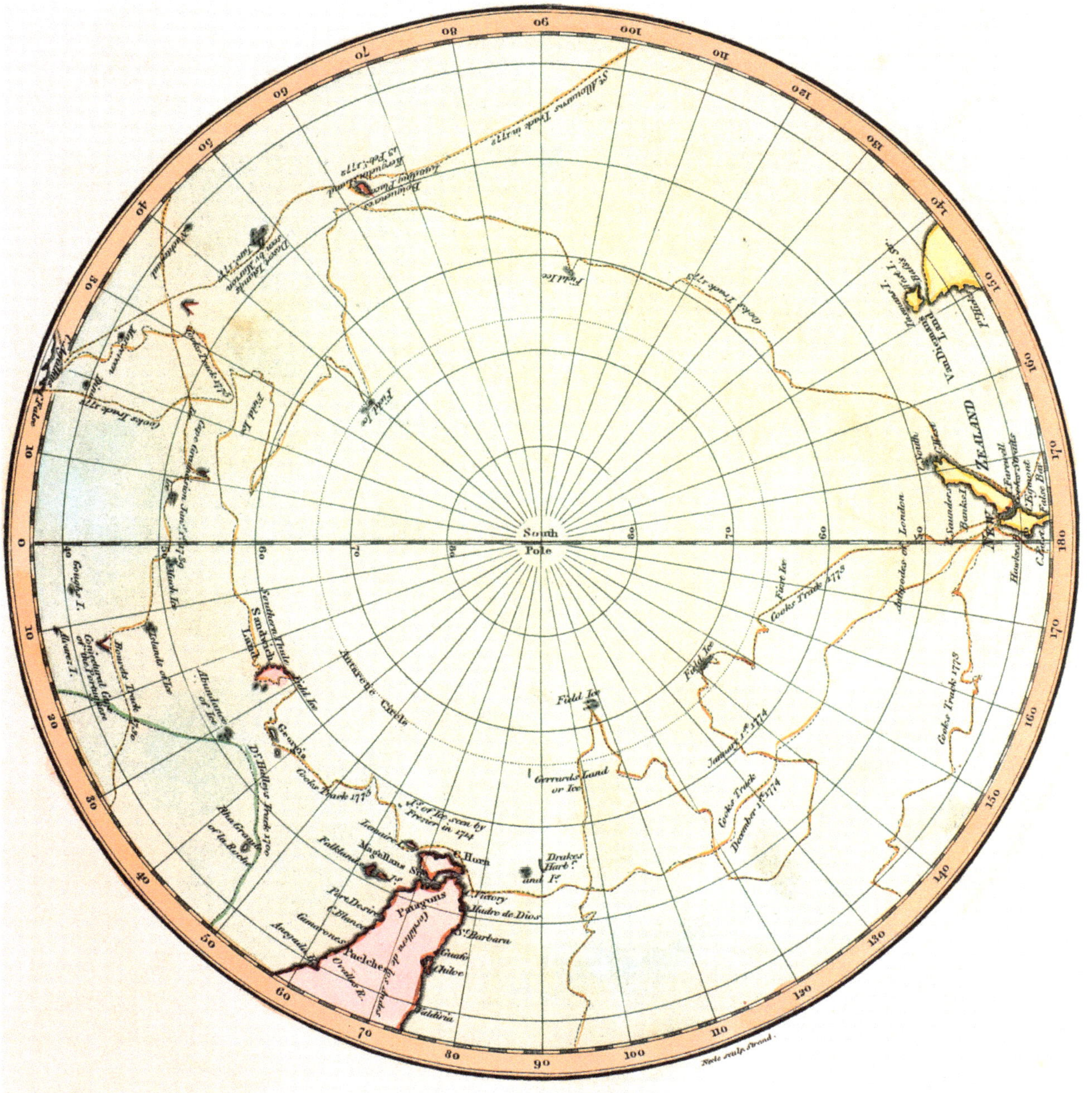

As astounding as it may seem, early 19th century explorers were still none the wiser about the existence of a continent in the far south. This early, hand-coloured engraving by J Wilkes, published in London in 1807, shows a large blank region of the South Pole.

Another celebrated Antarctic voyage was that of British sealing captain James Weddell, who visited the South Shetlands and South Orkneys in 1820–21, followed by another voyage in 1822–24 in the *Jane* and in the company of Matthew Brisbane in the *Beaufoy*. Finding the sealing poor, the two ships and crew sailed south into what was eventually to be named the Weddell Sea in search of new land and sealing opportunities. In a year of exceptionally ice-free conditions, the sea south of the main pack was wide open. Although they did not sight land, they established a new farthest south record of 74°15'S, shattering Cook's record by 185 nautical miles.

The lure of biologically rich seas and the temptation of unlimited numbers of whales, seals and penguins for skins and oil soon attracted bold mariners to risk death by venturing into its icy realm. The voyage of Captain James Clark Ross in 1839–42 was probably the most important since the circumnavigation of Cook and remains pre-eminent among Antarctic voyages.

In the first season, his two ships *Erebus* and *Terror* became the first to penetrate the pack ice of the Ross Sea. He discovered the Transantarctic Mountains as well as Possession, Coulman, Franklin, Beaufort, and Ross islands. He sighted the Ross Ice Shelf and landed on Possession and Franklin islands. During his second season, Ross bettered his own farthest south of the previous season by reaching 78°9.5'S at the face of the Ross Ice Shelf. In the third season, in the vicinity of the Antarctic Peninsula, he discovered Paulet, Snow Hill, Cockburn, and Danger islands, along with Cape Seymour – later to be named Seymour Island and containing one of the best fossil records on the continent.

An early 19th century portrait of Antarctic explorer, Sir James Clarke Ross (1800–62), when a young English naval officer.

Ross and his men made extensive and important magnetic, botanical, zoological and other scientific observations. Cockburn Island is one of the most historically significant places on the Antarctic continent; yet, despite landing there, these early explorers failed to recognise its geological and palaeontological significance. If they had visited Seymour Island it is most likely that they would have discovered the rich ancient life present there, especially as this small island is literally paved with fossils and is now referred to as the 'Rosetta Stone' of the Antarctic continent.

The account of the discovery and visit to Cockburn Island is fascinating. At 2 am on 1 January 1843, Captain Ross and his crew encountered a strong westerly wind while moving south in dangerous ice conditions near latitude 64°14'S, longitude 55°54'W, along the tip of the Antarctic Peninsula:

> … we found the ice so close, that to prevent from getting beset, we were compelled to stand back to the northward, to await a more favourable opportunity, and with the hope that the westerly breeze would drive the ice away from the shore, and leave a clear passage between them.[1]

Being New Year's Day, it was customary to present the crews with a complete suit of warm clothing and additional allowance of provisions. In the afternoon, Captain Ross, accompanied by Commander Bird, went on board the *Terror* to visit Captain Crozier and to wish him a Happy New Year. The weather was pleasingly calm and the surrounding views along the Antarctic Peninsula were spectacular. During the next few days unfavourable weather and thick ice conditions, including an immense iceberg that was an 'inconveniently close companion', prevented any landing on Cockburn Island until 9 am on 6 January, when the weather cleared and the shores of the island were free of ice. Captain Ross made a signal to Captain Crozier and both parties landed together and took possession of the island for the British Empire, an historic moment indeed as this was the only place where Ross set foot in West Antarctica. During their three-hour foray on Cockburn Island, the party, which included the botanist J D Hooker, collected some rock samples. They erroneously surmised that the island was entirely volcanic and

HMS *Erebus* and HMS *Terror* pushing through the pack ice along the Antarctic coast, with a whale and emperor penguin nearby. Jeffrey Stilwell collection, Monash University.

This scene depicting a storm on 20 January 1842 was somewhat exaggerated to show the general public just how hostile the Antarctic continent could be.

concentrated largely on the interesting botany of the island.[2]

> As we expected, we found it [Cockburn Island] to be entirely of volcanic formation; but the most interesting feature of our visit to this barren rock is that here the last vestiges of vegetation are to be found ...[3]

James Clark Ross had two surgeon/naturalists with him during the voyage – the designated expedition geologist Robert McCormick as well as the botanist Hooker. McCormick did not accompany the shore party, as Ross had a policy that one of the surgeons had to remain on board the ship whenever the other went ashore. It was Hooker's turn to go ashore.[4] Had McCormick gone ashore instead of Hooker, he would most likely have discovered fossils, as macroinvertebrates (particularly shells) abound on the small island that are Late Cretaceous, Early Paleogene (Eocene) and Neogene (Pliocene) in age.

Although it was the American sealer John Davis who is believed to have gone ashore on the Antarctic Peninsula in 1821 and recognised Antarctica as a continent, it was to be more than seven decades before anyone was able to land again on the mainland of Antarctica.

Henryk Bull, a Norwegian mercantile businessman and resident of Melbourne, Australia, was so enraptured by the idea that Australia should be the first country to set up a whaling industry in the Antarctic region that he persuaded a retired whaling master, Svend Foyn, to finance an expedition down south in search of whales. They set sail in September 1894 from Melbourne with a small crew in the refurbished whaling ship *Antarctic*. Surveyor and amateur naturalist Carsten Borchgrevink was with them. Born in Norway, Borchgrevink had earlier migrated to Australia where he taught sciences and languages. He would later become one of the first men to spend a winter in Antarctica, in 1899.

Cockburn Island is the only place where Captain Ross set foot along the Antarctic Peninsula during the expedition of 1839–43. Deeming it to be of only volcanic nature with rare plant life, Ross and his crew failed to recognise the highly fossiliferous nature of this small island.

Majestic Cockburn Island in the James Ross Island group of the Antarctic Peninsula is the only place where Captain James Clarke Ross set foot in West Antarctica in January 1843. Photo: Jeffrey Stilwell.

Example of very rare and large 19th century sperm whale scrimshaw depicting the *Erebus* and *Terror*, together with detailed portraits of Sir John and Lady Franklin. The Franklins lived in Tasmania in the 1830s and 40s, until the disappearance of Sir John in 1845 in the far north. The *Erebus* and *Terror* were used by Sir James Clarke Ross during the 1839–43 Antarctic expedition. This scrimshaw may be a commemoration of the loss of the ships and Sir John Franklin in the 1840s.

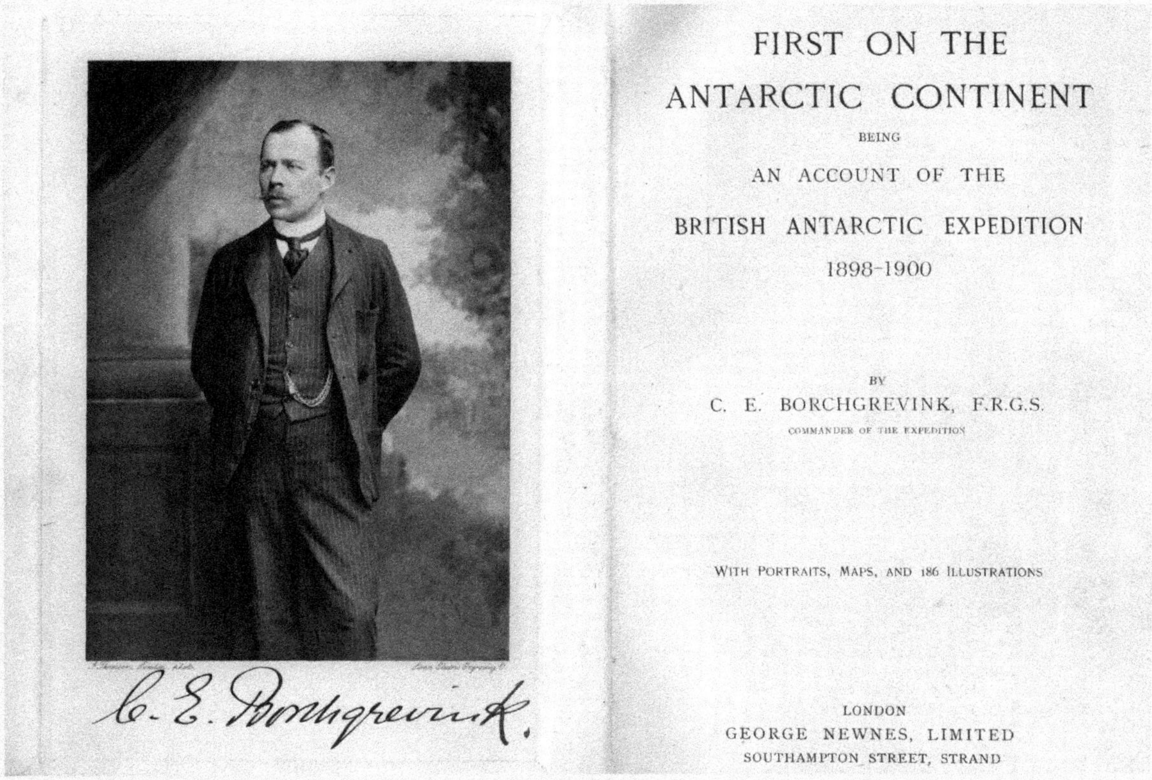

FIRST ON THE
ANTARCTIC CONTINENT
BEING
AN ACCOUNT OF THE
BRITISH ANTARCTIC EXPEDITION
1898–1900

BY
C. E. BORCHGREVINK, F.R.G.S.
COMMANDER OF THE EXPEDITION

WITH PORTRAITS, MAPS, AND 186 ILLUSTRATIONS

LONDON
GEORGE NEWNES, LIMITED
SOUTHAMPTON STREET, STRAND

One of the most famous books ever published on Antarctica by Carsten Borchgrevink (1901). This particular copy was owned by Chatham Islands sealer and whaler, H Odman, and has been in the possession of Jeffrey Stilwell since 1995.

On 24 January 1895, the *Antarctic* landed at Cape Adare on the northern Ross Sea. The landing party consisted of Captain Leonard Kristensen, second mate Carsten Borchgrevink, and Henryk J Bull, who later documented their adventure in his book *The Cruise of the 'Antarctic' to the South Polar Regions.* Bull wrote:

> ... January 24th, 1895. Cape Adare was made at midnight. The weather is now favourable for a landing, and at 1 am, a party including captain, second mate, Mr Borchgrevink, and the writer, set off, landing on a pebbly beach of easy access, after an hours rowing through loose ice, negotiated without difficulty.[5]

Exactly who stepped ashore first became the subject of a ludicrous squabble; however, Bull described the experience of being the first people on the Antarctic mainland as 'both strange and pleasurable', although he thought the crew would have preferred to find a right whale 'even of small dimensions'. But both the

Svend Foyn (1809–1894) brought whaling into the modern age with his invention of the exploding harpoon gun, which increased the efficiency of killing both large and small whales.

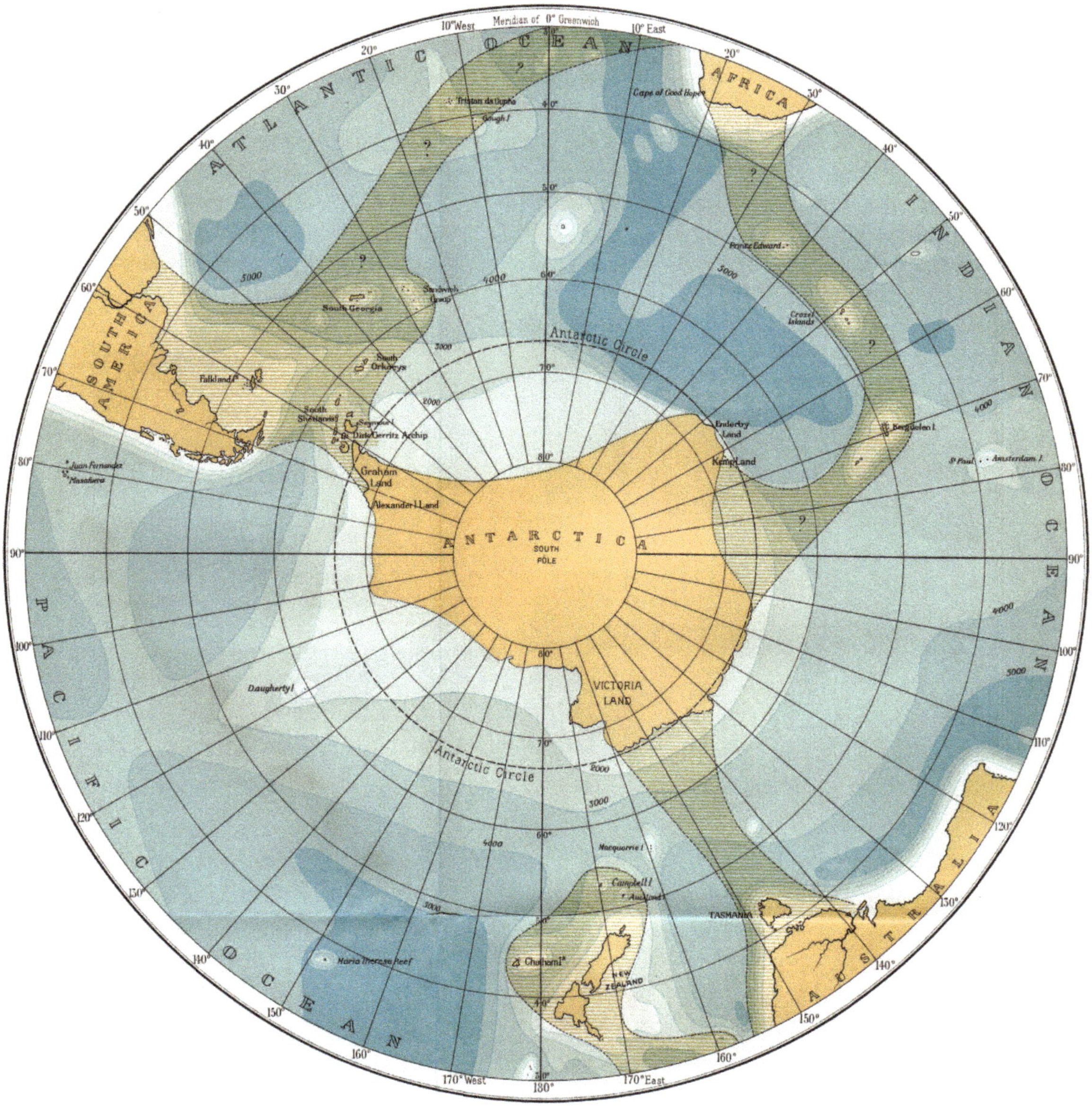

This virtually unknown 1902 map by Arnold Edward Ortmann illustrates the prevailing notions at the beginning of the 20th century. A series of land bridges were inferred to explain faunal and floral distributions. The outline of Antarctica scarcely represents the continent we know today![6]

sealing and whaling efforts during this voyage were largely a failure and apparently Henryk Bull was repulsed by the killing operations. Although Borchgrevink was a surveyor and amateur naturalist, it seems that collecting fossils at Cape Adare may have been furthest from their minds. They were unaware that the first fossils had already been collected on Seymour Island, on the other side of the continent.

OPPOSITE: Sledding journey along Snow Hill Island, Antarctic Peninsula on 30 September 1902, during Otto Nordenskjöld's Swedish South Polar expedition of 1901–03. Many famous photographs were taken on this day, including one of the expedition members in front of Nordenskjöld's Hut.

Photo: Gösta Bodman, reproduced from Andersson (1944)..

3
THE FOSSIL PIONEERS

Captain Larsen landed on the South Orkneys and Seymour Island, and in the latter he found some fossils which had fallen from a decomposing cliff. These are the first fossils ever brought from Antarctica … probably of the Tertiary age, and indicate a warmer climate than now prevails in these high southern latitudes …

William S Bruce, naturalist, *From Edinburgh to the Antarctic …*, 1894.

In 1829, naturalist James Eights departed the USA on a sealing and exploring expedition led by Captain Benjamin Pendleton to the South Shetland Islands. They sailed in the two brigs, *Seraph* and *Annawan*, in company with the schooner *Penguin*. Considered to be the first US scientist to visit the Antarctic, Eights landed, in all likelihood, on King George Island and discovered a fossil log with preserved growth rings.[1] Eights referred to this fossil fragment in an 1833 report published by the New York Albany Institute on the natural history of the South Shetland Islands along the tip of the Antarctic Peninsula:

> The only appearance of an organized remain that I anywhere saw, was a fragment of carbonized wood embedded in this conglomerate. It was in a vertical position, about two and half feet in length and four inches in diameter ; its color is black, exhibiting a fine ligneous structure, the concentric circles are distinctly visible on its superior end, it occasionally gives sparks with steel, and effervesces slightly in nitric acid …[2]

Eights apparently did not see the fossil wood as particularly significant and it would appear that he

James Eights (1798–1882) was the first US scientist to visit the Antarctic and the first to discover a fossil – a piece of wood on King George Island. Original watercolour portrait c. 1830 by Philadelphia artist Ebenezor Emmons.
Image courtesy Albany Institute of History and Art, New York.

An Antarctic specimen collected by James Eights, and illustrated by him in a paper published by the Albany Institute in 1833. These extant isopods, which are highly reminiscent of fossil trilobites from the Paleozoic, confounded Eights and other researchers at the time, who called them 'trilobitoids'.

left it where it was and didn't collect it. Most of his collected natural history specimens are housed in the the New York State Museum. In his 1833 report, Eights did a beautiful job of defining modern arthropods such as *Glyptonotus* (a large isopod reminiscent of ancient trilobites, hence his comparisons to these fossils in the paper) and one of the sea spiders called decalopods. More than 60 years would lapse before the first documented collection of fossils from the Antarctic continent.

The first fossils to be described formally from the Antarctic region were pieces of fossilised shells and wood along with a few round stones with a concentric arrangement of layers that reminded one of an 'old lava bomb'. Presumably, these were sedimentary deposits called 'concretions' – accreted balls of sand, silt and mud cemented with calcium carbonate that often contain beautifully preserved organic remains. These highly rounded rocks are found in abundance on certain islands along the Antarctic Peninsula.

Although contention still surrounds who actually collected the first fossils from Antarctica, credit for the historic discovery of once abundant ancient life in the highest southern latitudes usually goes to the Norwegian whaling Captain Carl Anton Larsen, who collected fossils on Seymour Island during the austral summer of 1892–93. Larsen, sailing in the whaler *Jason*, was mainly interested in catching right whales.

The *Jason*, after having been at sea for several months and short of supplies, met up with another whaling ship, the *Balaena*, on 28 December 1892 near Joinville Island. On board the *Balaena* were William Gordon Burn Murdoch, an artist and explorer, and a young Scot, William S Bruce, who was undertaking scientific work on the island. Bruce would later lead the 1901–1902 Scottish National Antarctic Expedition. In his book *Edinburgh to the Antarctic* ... (1894) Burn Murdoch made some interesting comments on the rocks and the burning desire to get closer to them:

> We have been close enough to the land to enable us to distinguish the colouring and form of the rocks. The low, reddish-coloured islands seem to be crossed by fissures running into each other obliquely, with a little snow in the crevices; but there is not the least sign of lichen or moss ... It is a marvel that no scientific expedition has been send down here since the days of Ross ... It is almost unbearable to see the land so close and yet have no means to land on it. We feel tempted to jump overboard and swim. All our boats hang idle on the davits, yet we are not allowed one to land with. Snowy petrels and penguins of all kinds evidently breed there. Captain Larsen of the *Jason* has landed, and he tells us he found beds of fossils on the beach, shells, and tree trunks. Some of the fossil shells he showed us resembled very large cockles ... [3]

Dundee whaling Captain Carl Anton Larsen (1860–1924) collected the first Antarctic fossils during the 1892–93 austral summer.

Burn Murdoch's famous book *From Edinburgh to the Antarctic ...* was published in 1894.

Icebergs, ships and penguins illustrated by Burn Murdoch, during the Scottish expedition of 1892–93.

Captain Larsen receiving his guests from the *Balaena* with a welcoming tune 'Way down the Swanee River' from an organ on deck during the voyage from Edinburgh to the Antarctic.
Reproduced from Burn Murdoch (1894).

Captain Larsen teasing a seal.
Reproduced from Andersson (1944).

Some of Larsen's crewmen traded their fossils for tobacco, but Larsen was shrewd enough to exchange some of his brandy for pipe tobacco while keeping hold of his fossils.[4] The scientist William Bruce relayed Captain Larsen's discoveries of the fossils and their inferred scientific significance:

> Captain Larsen landed on the South Orkneys and Seymour Island, and in the latter he found some fossils which had fallen from a decomposing cliff. These are the first fossils ever brought from Antarctica. There are specimens among them of *Cucullaea, Cytherea,* and *Natæa* [sic], and pieces of a coniferous tree. They are probably of the Tertiary age, and indicate a warmer climate than now prevails in these high southern latitudes ... [5]

These prophetic words, surmising that Antarctica has not always been in a perpetual deep freeze, are often overlooked. Bruce proudly relished in the discoveries made on Seymour Island, stating further that:

> ... We must investigate the nature and distribution of the rocks, which contain for the palaeontologist an entirely new fossil fauna and flora.[6]

Larsen's fossils ended up at the University of Oslo and, soon after their discovery, were formally described by English scientists G Sharman and E Newton in 1894 in *Transactions of the Royal Society of Edinburgh* as the very first fossils from the continent of Antarctica. The only clues as to the provenance of the fossils described by Sharman and Newton are from the fossils themselves. The eroded and abraded nature of the specimens indicates that they were collected along the fossiliferous cliffs of the east coast of Seymour Island near Cross Valley. All but one specimen of the collection was from the Eocene Epoch and about 50 millions years old. However, an anomalous, extremely abraded clam, described a few years later by Sharman and Newton in a 1898 paper,[7] belongs to the Cretaceous and was probably found by chance on the beach and taken from older deposits just over 65 million years old a few kilometres south of Cross Valley. Unaware that there were Cretaceous deposits on Seymour Island at the time, Sharman and Newton inadvertently described this clam as one from the Tertiary. Despite the fact that Larsen was able only to make a few brief visits to Seymour Island, this was the first time that any ancient remains of animals and plants were collected from the Antarctic continent. The significance of the discovery of fossils on Seymour Island led Sharman and Newton to write:

> ... the number of forms known [from Antarctica] is at present few, yet these give the

promise of a large field of fossiliferous rocks as yet unexplored ... and it is of extreme interest to find that in this distant southern area Lower Tertiary deposits occur with forms of life so very similar to those with which we are familiar in our own northern country.[8]

Another early report of fossils from Antarctica is one that has been largely overlooked. In February 1902, Frank Wilbert Stokes, a talented artist on the Belgian Antarctic expedition, collected a mere nine fossils at Admiralty Inlet, Louis Philippe Land.[9] These fossils, were described the following year by US conchologist Stuart Weller as '... the first which have ever been brought back by any of the expeditions to the far south ...', making no mention of the collections by Larsen in the 1890s.[10] Weller wrote:

... [the] collection is not large, containing scarcely more than a dozen specimens, but a sufficient number of species have been recognised to determine somewhat definitely the age of the strata containing them, and to permit of important conclusions as to the faunal geography of Upper Cretaceous time.

Although history repeated itself by the absence yet again of specific locality information in Weller's paper, it is most likely that the assemblages of macroinvertebrates were collected on Snow Hill Island in Admiralty Inlet, especially since Stokes painted the coast of this small isle in great detail.[11] Weller wrote that the specimens were probably from the same stratigraphic position (i.e. the same age) in concretions of a very dense, fine-grained

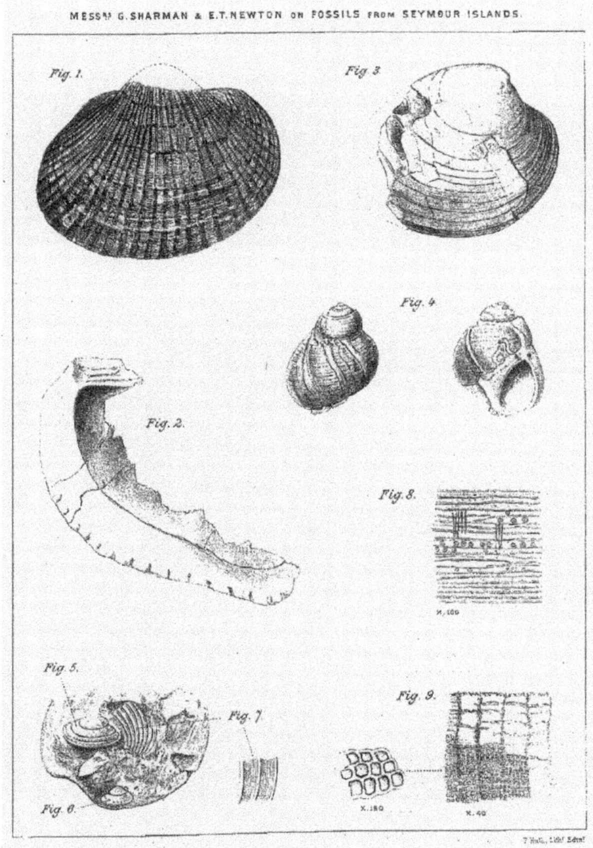

The first formal descriptions and illustration of fossils from Antarctica, collected during the voyage of the *Jason* in 1892 on Seymour Island. G Sharman and E Newton described these fossils in 1894, now known to be of Eocene age, in the *Transactions of the Royal Society of Edinburgh*.

A nearly forgotten but important contribution to the early palaeontology of Antarctic was published in 1903 by Stuart Weller. Weller was the first to describe Mesozoic fossils from Antarctica. These are ammonites from Snow Hill Island.

Source: Weller 1903.

brown sandstone and a coarser-grained glauconitic sandstone – all taken from a talus slope. Weller concluded that the assemblage was of a Cretaceous age because of the strong similarities of the fossils – at least three species of ammonites, one species each of clam and snail, a serpulid worm, and a lobster – to those forms in southern Patagonia and southern India that were dated to the Late Cretaceous. Weller also suggested the strong possibility of a shallow water connection between these regions at the close of the Cretaceous.

Meanwhile, the prophetic words of Sharman and Newton regarding the fossil discoveries on Seymour Island led Swedish geologist Dr Nils Otto Gustaf Nordenskjöld to target the James Ross Island Group, including Seymour Island, as a primary area of importance for the Swedish South Polar Expedition of 1901–03. This was one of the most far-reaching and significant expeditions in the history of Antarctic palaeontology, if not science in general.

In 1902 Nordenskjöld made the first major sled journey on Antarctica. He wintered over on Snow Hill Island and during his time there collected many fine fossils and geological specimens from the western islands off the Antarctic Peninsula. His ship, the *Antarctic*, dropped some men off at Hope Bay, and then was crushed in the pack ice. Three separate parties of this major expedition were isolated at this time and none of them had knowledge of the others' whereabouts or situations. Despite the tragedy, the men discovered Jurassic plant fossils at Hope Bay. The crew then spent a harsh winter stranded on Paulet Island, while Nordenskjöld and his men endured another savage winter on Snow Hill Island, waiting patiently until November 1903 when they were finally rescued by an Argentine ship *Uruguay*.

During his time spent in the Antarctic Peninsula, Nordenskjöld briefly visited Seymour Island where his discoveries included the bones of giant penguins nearly 1.5 metres tall, a great diversity of shells and other marine fauna, and well-preserved fossil leaves. This led him to conclude that, at one time in the distant past, Antarctica had been a much warmer place. Many of these early collected specimens are now housed in Stockholm.

Nordenskjöld was a ground-breaking scientist on all fronts. His keen insight of geology and palaeontology led to fundamental new ideas on tectonics, ancient climate, oceanic circulation, and evolution of Antarctic fauna and flora. Although he

Swedish geologist Otto Nordenskjöld (1869–1928) made ground-breaking palaeontological discoveries along the Antarctic Peninsula during the Swedish South Polar Expedition of 1901–03.

Team members of the Swedish South Polar Expedition of 1901–03 on their departure from Gothenburg, Sweden. Back row (L to R): Bodman, Skottsberg, KA Andersson. Front row (L to R): Ohlin, Nordenskjöld, Larsen, Ekelof.

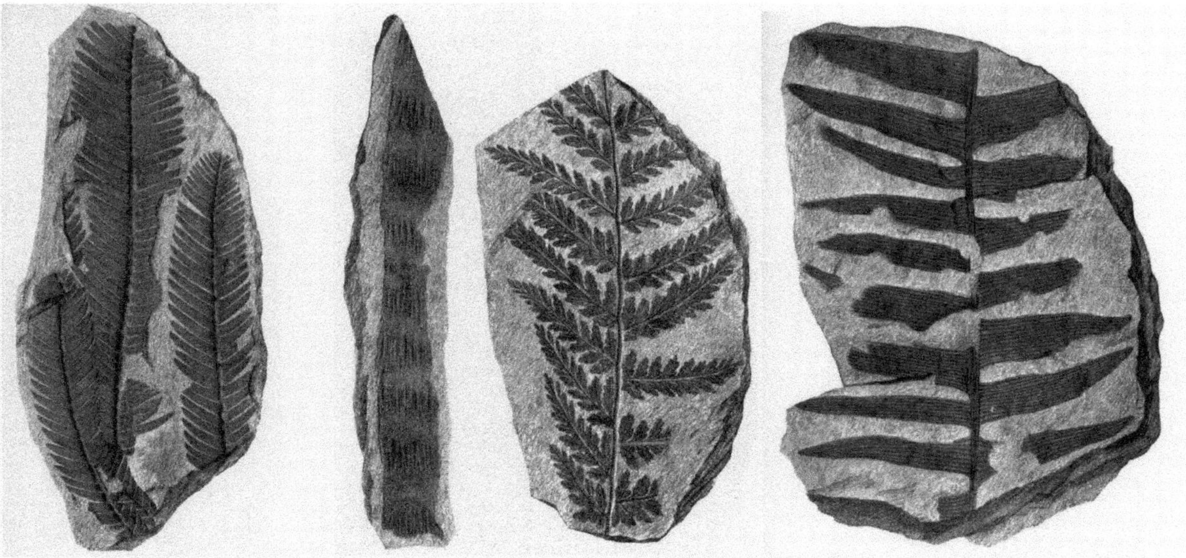

Jurassic and Eocene fossil flora from Hope Bay and Seymour Island were collected by Gunnar Andersson during the Swedish South Polar Expedition of 1901–03.

Otto Nordenskjöld's famous book, *Antarctic*, was published in Stockholm in two volumes in 1904. Contributions to the book were also made by Andersson, Larsen and Skottsberg.

Otto Nordenskjöld's ship *Antarctic* was crushed by pack ice during the Swedish South Polar Expedition of 1901–03. The crew survived on penguins and seals, and Nordenskjöld remains a hero to this day for not losing a single member of his crew during the ordeal.

spent nearly three years in the region, he and his men were able to visit Seymour Island only a few times.

Today we know that Seymour Island is a rich treasure trove of fossils. Nordenskjöld's expedition was able to make only brief and fleeting visits to Seymour Island, so the men did not have much time to make extensive collections. Also, the first time they visited there they missed the section of beach where fossils are abundant. This may have misled them to think that the island had little to offer in the way of fossils.[12]

In 1928 Nordenskjöld met an untimely death – he was killed by a bus. The centenary of his fossil discoveries was celebrated in May 2001 at an international symposium held in his honour at the University of Gothenburg, sponsored by the Royal

Otto Nordenskjöld's hut still stands on the northern end of Snow Hill Island where his party overwintered in 1902–1903. The large snowfield on one side and the mountain on the other resulted in this being a very windy place to spend the winter.
Photo: Rodney Feldmann.

A wooden sign on Seymour Island commemorated the rescue of Otto Nordenskjöld and his party by the *Uruguay* in 1903. Guarded by Adélie penguins, the sign was seen during a 1974 expedition, but disappeared shortly thereafter.
Photo: Bill Zinsmeister.

Society of Arts and Sciences. It is a tragedy that his life was cut short at the age of 58, especially after such an exciting and sensational career of Antarctic discovery.

§

The Scottish Antarctic Expedition of 1901–1902 led by William S Bruce set out to achieve a diversity of scientific research. Bruce had been offered a place as a naturalist on Robert Falcon Scott's *Discovery* expedition but had declined because he didn't want to be part of an expedition in which the primary goal was to reach the South Pole. So Bruce established his own Antarctic expedition, assembling a team of scientists who would focus largely on undertaking hydrographic work in the Weddell Sea region. During the bottom trawling of the region the first truly ancient fossils were collected in Antarctica – small archaeocyathids, ancient coral-like animals probably more closely allied to the sponges, and known only from the early part of the Cambrian Period, 540–520 million years ago. This find established the oldest age then known for the continent of Antarctica.

Ernest Shackleton marched alongside Scott on the 1901–1904 *Discovery* expedition which was the first British attempt to get to the South Pole. Shackleton suffered from scurvy and had to be invalided out of the expedition. With a burning desire to prove himself, he went on to lead his own assault on the Pole with the 1907–1909 *Nimrod* expedition, reaching 88°23'S on 9 January 1909. This was the furthest south that anyone had hitherto been.

In December 1908, Shackleton's team scaled the immense Beardmore Glacier on the edge of the polar plateau and made an unexpected discovery. They found coal, and lots of it. Not only coal, but impressions of early leaves in sandstone boulders and wood from a long-perished, ancient tree. From the plant fossils they had discovered on Seymour Island, Shackleton and his men came to the astounding conclusion made by William Bruce some 16 years earlier that Antarctica must have been forested long ago. Not only that, the ancient trees and leaves had to mean that Antarctica was warm in the geologic past. We know now that these plants existed some 250 million years ago, when the continent was somewhat closer to the equator than it is at present.

Shackleton's party was the first to collect geological samples from mainland Antarctica. They collected samples from erratic boulders found in moraine in the upper reaches of the Beardmore Glacier which also contained archaeocyathids. These fossils are highly characteristic of Cambrian age rocks, from the base of the Phanerozoic Eon when life first took off in a big way during the so-called 'Cambrian Explosion', about 540 million years ago.

In 1914, two of Shackleton's team – Australian professor of geology, Sir Edgeworth David and British geologist Raymond Priestley – published an excellent geological study of expedition specimens. Their book included a special section on the palaeontology written by Griffith Taylor and E J Goddard.

§

The first fossils of vertebrate animals to be collected on the mainland of Antarctica were fragmentary bones of 380-million-year-old fish found in erratic Devonian rocks near Granite Harbour. They were collected in the summer of 1911–1912 by Frank Debenham, an Australian scientist who was part of Scott's *Terra Nova* expedition. Debenham's samples were pieces of the fossiliferous Aztec Siltstone, probably originating further inland at Mt Suess, which had been carried along as glacial moraine – pebbles transported by the

A wooden cross erected in the memory of George T Vince of Scott's *Discovery* expedition, who drowned near Hut Point on 11 March 1902. Photo: Jeffrey Stilwell.

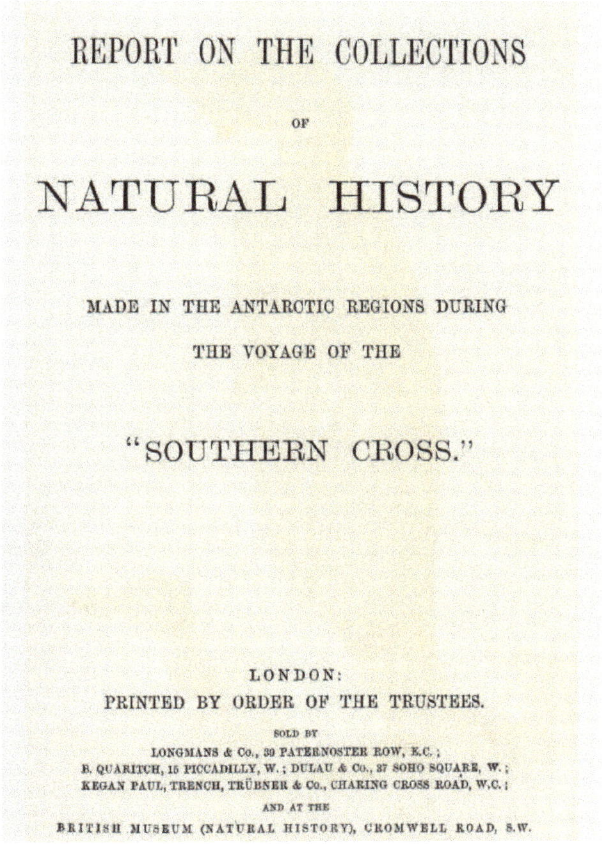

REPORT ON THE COLLECTIONS

OF

NATURAL HISTORY

MADE IN THE ANTARCTIC REGIONS DURING

THE VOYAGE OF THE

"SOUTHERN CROSS."

LONDON:
PRINTED BY ORDER OF THE TRUSTEES.

SOLD BY
LONGMANS & Co., 39 PATERNOSTER ROW, E.C.;
B. QUARITCH, 15 PICCADILLY, W.; DULAU & Co., 37 SOHO SQUARE, W.;
KEGAN PAUL, TRENCH, TRÜBNER & Co., CHARING CROSS ROAD, W.C.;
AND AT THE
BRITISH MUSEUM (NATURAL HISTORY), CROMWELL ROAD, S.W.

A 1902 report on the scientific results from Captain Scott's first expedition.

Ernest Shackleton's hut at Cape Royds on Ross Island was built during the 1907–08 expedition. Mt Erebus is in the background. Photo: Jeffrey Stilwell.

The interior of Ernest Shackleton's hut at Cape Royds, Ross Island, as it was in 1995. Photo: Jeffrey Stilwell.

Century-old provisions lost to time, but preserved here in Antarctica's icy conditions. Photo: Jeffrey Stilwell.

moving ice of glaciers.

The incredible significance of the study of Antarctic geology and the question of the age of the frozen continent is emphasised in the dramatic entries in Scott's last diary. In early 1912 Scott's party of five struggled back towards their base camp after reaching the Pole on 17 January and finding Amundsen's victory message, which revealed that he had beaten them there by a month. Struggling along, low on food, scarred and weather-beaten, they still managed to stop at the Beardmore Glacier and collect some 16 kg of rock and fossil samples. Scott and his fellow expedition members are applauded for recognising the significance of the plant fossils in 'this awful place'. It was the only day on Scott's final march homewards from the Pole that they were distracted from their arduous man-hauling regime and spent a day 'geologising'. Scott's entry for his diary on Thursday, 8 February 1912, exudes excitement as he recalls the discoveries they made:

> The moraine was obviously so interesting that when we had advanced some miles and got out of the wind, I decided to camp and spend the rest of the day geologising. It has been extremely interesting. We found ourselves under perpendicular cliffs of Beacon Sandstone, weathering rapidly and carrying veritable coal seams. From the last [Edward] Wilson, with his sharp eyes, has picked several plant impressions, the last a piece of coal with beautifully traced leaves in layers, also some excellently preserved impressions of thick stems, showing cellular structure. In one place we saw a cast of small waves in the sand. Tonight Bill has got a specimen of limestone with *archaeocyathus* – the trouble is one cannot imagine where the stone comes from; it is evidently rare, as few specimens occur in the moraine. There is a good deal of pure white quartz. Altogether we had a most interesting afternoon, but the sun has just reached us, a little obscured by night haze. A lot could be written on the delight of setting foot on rock after fourteen weeks of snow and ice and nearly seven out of sight of all else. It is like going ashore on a sea voyage.

Any knowledge of the inland, remote geology of the continent was considered of great importance to science in those times. Only nine days later, things were getting harsher, and Scott's conscience was to turn to the subject of lugging the rocks back with them. That day Petty Officer Edgar Evans had wandered off into the snow and was found frostbitten with 'a wild look in his eyes'. He was practically unconscious by the time they got him into the tent and he died quietly at 12.30 am the next morning.

On 17 or 18 March (Scott was unsure of the exact date), they felt despondent about making it back alive. They offloaded a lot of their equipment

A view up Beardmore Glacier, with Cloudmaker Bluff on right. Photo; Bert Rowell.

but, under Edward Wilson's urging, kept the geological specimens. Scott wrote again about the burden of the rocks:

> We are cold on the march now, and at all times except meals. Yesterday we had to lay up for a blizzard and today we move dreadfully slowly. We are at No.14 pony camp, only two pony marches from One Ton Depot. We leave here our theodolite, a camera, and Oates' sleeping bags. Diaries, &c., and geological specimens carried at Wilson's special request, will be found with us on our sledge.

Scott knew that they would not make it back alive. They carried the rocks onwards and the specimens were later discovered the following season with their frozen bodies at their last camp. Recent studies on the past climate trends of Antarctica have shown that the year in which Scott's party perished was an exceptionally cold year, and that if climatic conditions were average they quite probably would have survived.

Edward Wilson, with his acute observational skills, made reference in his diary to fossilised leaves. He wrote 'most of the bigger leaves are distinctly beech-like in shape and variation, like British beech but smaller, and the venation much more abundant and finer in character'. This locality was at the base of Mt Buckley, in debris that had fallen from above, and was undoubtedly close to the *Glossopteris* site on the Beardmore Glacier. In 1986, Dave Harwood and his team, from the University of Nebraska-Lincoln, found the Antarctic fossil beech *Nothofagus beardmorensis* locality, close to where Scott and his colleagues must have passed more than 70 years previously.

The rocks collected by Scott's party are today housed in the collections of the British Museum (Natural History) in London and at the Scott Polar Research Institute at Cambridge. Yet the specimens were not collected in vain. In 1914, British palaeontologist AC Seward described the fossil plant remains collected by Scott's party from the Beardmore Glacier region.[13] Seward was able to identify two

How fossils are formed

Fossils are the remains of prehistoric life on Earth. Fossils can be skeletons of animals such as fish, shells or other invertebrate hard parts, impressions or trackways. In some instances the whole organism can be completely preserved with little to no alteration through the millions of years, such as fossil insects entrapped in amber.

The burial of organic material within layers of sediment that becomes rock generally results in alteration or mineralisation of the original organic material. Bones, shells, wood and other porous hard parts of animals and plants can become replaced by different minerals. In the majority of cases bones are not fossilised. Instead they are eaten away by scavengers, or decay into dust through regular processes of weathering, abrasion and bacterial action. It is only in those rare cases when an organism is rapidly buried after death, such as in a flood, or by floating out to sea and sinking in quiet, muddy conditions, that its remains may become entrapped within layers of sediment, to later become impregnated with additional minerals and thus be preserved as fossil bones. Calcite and aragonite, both forms of calcium carbonate ($CaCO_3$), commonly replace shell material. In some cases, bone can be replaced by opal (cryptocrystalline silica), although in most fossil fishes the bones retain their original mineral composition but are added to by secondary minerals seeping into the pore spaces in the bones, changing the colour of fossil bones and teeth from white (or opaque) to brown or black.

Bone is made up of the mineral hydroxylapatite (calcium phosphate hydroxide). In favourable conditions, with an abundant source of calcite, the calcium can replace the phosphorous as water is removed, thus strengthening and preserving the bone; this is called calcification. In cases where groundwaters replace the bone with silica or enrich the bone with silica minerals, it is termed silicification. There are more than 20 known minerals that can replace original fossil material.

Captain Robert Falcon Scott's hut, built in 1902, at Hut Point, McMurdo Station, Ross Island. Photo: Jeffrey Stilwell..

genera of plants in Scott's specimens, *Glossopteris* and *Vertebraria*, both well known forms from countries now not close to Antarctica. *Glossopteris*, an extinct seed fern plant with distinctive leaf patterns, is an iconic Gondwanan plant fossil. In the early twentieth century, it was known only from India, Australia and South Africa. The first seeds of the concept of 'Gondwana' had been sown.

Today, we know of the similarities between the *Glossopteris*-bearing Permian rocks of the Transantarctic Mountains and Tasmania. These rocks in both regions sit on ancient granitic rocks, but there is one difference – the Antarctic Permian is exclusively non-marine and the Tasmanian Permian mostly a shallow marine deposit.

Edward Wilson's laboratory in Scott's Hut at Cape Evans, as it remained in October 1991. Photo: John Long.

In 1921, Sir Arthur Smith Woodward, the eminent British Museum palaeontologist, published a landmark paper describing the fossils from the *Terra Nova* expedition. He identified some eight different kinds of fossil fishes, most from isolated bones, scales or spines. Smith Woodward recognised a fauna that was essentially similar to that found in the Old Red Sandstone, which was well known throughout Scotland and much of Europe. This was the first recognition of such a fauna in the southern hemisphere, and suggested a Late Devonian age for the rocks. The cosmopolitan nature of the fish fossils was not taken to imply anything about the past positions of the continents in question, just that freshwater fishes were assumed to have once been widespread around the globe at that time.

§

US pilot Admiral Richard Byrd was the first person to fly over the South Pole in 1930.[14] Byrd also became the first person to overwinter alone on the continent in 1934. The lack of adequate ventilation almost caused his death, as did wandering about near crevasses. The psychological effects on his mind took several years to wear off, yet every day Byrd managed to make important scientific measurements of the weather conditions, and to record unusual atmospheric phenomena.

Byrd's privately financed expeditions of 1928–30 and 1933–35 put the USA in the forefront of Antarctic exploration and adventure, and were expounded in his immensely popular books. He went back to Antarctica in 1946 as part of Operation Highjump, which was then (and remains) by far the largest Antarctic expedition, with more than 4700 naval and marine personnel, 44 observers, 13 ships, and a number of aircraft. The expedition sighted more than 1.5 million square miles of Antarctica, half of it previously unexplored, and took 15 000 aerial mapping photographs. The following season the US Navy Second Antarctic Developments Project (Operation Windmill) used ship-based helicopters to get geodetic ground control for the Highjump photographs. The expedition contributed to production of the first medium-scale maps of the region and influenced decisions regarding locations of stations for the International Geophysical Year the following decade.

In the mid to late 1960s, Bert Rowell of the University of Kansas participated on some of the earliest expeditions to collect Paleozoic invertebrates from Antarctica, and later leading many expeditions that would focus on the earliest Cambrian faunas of the Transantarctic Mountains. His team revealed that Antarctica was once home to a giant archaeocyathid reef that spanned most of the continent, almost 540 million years ago.

The first drilling of ancient geologic strata on the Antarctic continent commenced in 1972 with Leg 28 of the Deep Sea Drilling Project (DSDP).

Collecting fossils in Antarctica

One of the most common questions a palaeontologist is asked about collecting fossils in Antarctica is: Where do you find them? A thick cover of ice has buried much of Antarctica's geological features, except for the high mountains that protrude above the 2.7-kilometre-thick sheet. Mountain exposures are the best places to find Antarctic fossils, but these are often hard-to-reach places, which restricts the amount of equipment one can carry.

Collecting fossils in normal weather conditions is often a routine job involving hammers, chisels, plaster for covering the larger specimens, and sometimes jackhammers or electric drills. In Antarctica, the palaeontologist usually goes into the field with just a few tools, as to reach the best sites one usually has to do a lot of climbing far from the base camp. On our expeditions collecting macroinvertebrates and fish fossils we used an Estwing geological hammer and series of chisels of varying sizes. The combination of extreme cold and working with hard sedimentary rocks can blunt the tools very easily, so sharpening them regularly was called for every few days. Some fossils look very easy to collect on the surface, but looks can be deceiving! Many times, the fossil may be partially buried in the permafrost layer and very difficult to collect. The collector must chip away at the frozen ground while trying not to break the fossil, which can take many hours of hard work to free the treasured specimen.

If the palaeontologist is working at a site that a generator can be carried to, a jackhammer or drill can be employed. When US palaeontologist Bill Hammer was called in to dig up a dinosaur skeleton near the top of Mt Kirkpatrick such techniques were called for, and used successfully.

Jeffrey Stilwell as a young Purdue University student in 1985 with a small colony of Adélie penguins on Seymour Island.
Photo: Luis Paredes-Mejia.

The famous ship *Glomar Challenger* drilled four holes into continuous marine glacial sediment at depths of 200 and 440 metres below the sea floor on the Ross continental shelf. These results were published in 1975 in the *Initial Report*, and indicated that glaciers were calving ice at sea level some 25 million years ago. This challenged the commonly held notion of the time that the ice sheet was a mere two million years old. The research was extended into 1975 by studying the Dry Valleys, which involved researchers in the USA, New Zealand and Japan. This investigation cored over 300 metres of fjord sediments (dating back six million years) in the lower Taylor Valley.

In 1974 William Zinsmeister of Purdue University and David Elliot of the Ohio Polar Institute organised (together with the Argentinian Antarctic Institute) the first systematic palaeontological exploration of the fossil sites of Seymour Island. Fortunately, Argentina established a permanent research station ('Isla Vicecomodoro Marambio') on this special island. This work has since revealed much about the crucial role that high southern latitudes have played in understanding the evolution of Earth's entire biosphere.

Seymour Island has an extraordinary preservation of sediments ranging from 70 million years ago through to about 40 million years ago, including the great extinction event at the end of the Cretaceous that wiped out the dinosaurs. Many US expeditions to Seymour Island have collected diverse assemblages of Cretaceous to Eocene marine invertebrate faunas and also brought back discoveries of marine reptiles and some isolated teeth and bones of dinosaurs and early mammals. Seymour Island is bursting with fossils for some special reasons – one being that it is free of permanent ice year-round (it is in a precipitation shadow, and thus sheltered from high mountains to the west along the Peninsula) and another because it has near complete rock

Mike Woodburne and Bill Daily examine the remains of a large tidal channel deposit in the Eocene La Meseta Formation of Seymour Island. This site is famous for the first discovery of fossil mammals on the Antarctic continent in early 1982. Photo: Bill Zinsmeister.

exposure of bad-land type topography of rolling, yet dissected and treeless hills, with sediments of the right age and palaeoenvironment for an abundance of ancient life. Seymour Island ranks as one of the top regions on the planet for studies of the events of the end of the Cretaceous Period, culminating in the mass extinction event 65 million years ago, and after. It is, in fact, the best on the planet for detailed research on this important event in the history of life, especially with rock exposures of some 70 km^2.

As a participant on two expeditions to Seymour and surrounding islands in the mid-1980s, Jeffrey Stilwell can attest to the excitement of visiting this isolated part of Antarctica. Despite severe storms with winds clocked up to some 200 km/hour and heavy dumps of snow at times, Seymour Island still remains relatively snow free because of 'warm', sunny days.

Over the past few decades many geological expeditions sponsored by many countries around the globe have explored parts of the Transantarctic Mountains and the Antarctic Peninsula, both containing rich exposures of fossil bearing strata. Their discoveries have filled out many of the missing gaps in the fossil record of the continent. Antarctica is now revealing itself as being the home to an almost continuous pageant of life, represented by fossils ranging from simple invertebrates and algae at the beginning of the Cambrian Period, 540 million years ago, through to recently discovered fossil whales and dolphins about three million years old.

OPPOSITE: Outcrops and geology in Beardmore Glacier area are select areas for geologists and palaeontologists to research the ancient Antarctic past. Photo: John Long.

4
ANTARCTICA'S OLDEST FOSSILS

Our way lies to the south. How one wishes for time and unlimited provisions! Then indeed we could penetrate the secrets of this great lonely continent ... The closer observation of these mountains ought to give geological results of importance. We may even have the good fortune to discover fossils, or at any rate to bring back specimens that will determine the geological history of the country and provide a connection between the granite boulders lying on the slopes of Erebus and Terror and the land lying far to the south.

Ernest Shackleton, *South*, entry for 3 December 1908.

When Ernest Shackleton's team reached the rocky outcrops of the Beardmore Glacier on 7 December 1908, they felt the exhilaration of new and untold scientific treasures to be found. Despite losing their pony Socks down a deep crevasse that day, they were able to spend the next day collecting fossil specimens that would later prove to be crucial in dating the continent back to the dawn of life, to the Cambrian Period.

Like the other continents, Antarctica has a thick crustal succession of sedimentary, volcanic and metamorphic rocks. However, its rocks are not covered by plants or soil – wherever they are exposed they show their pure geological, unweathered form. And although only a small percentage of the Antarctic continent is exposed above the ice sheet, its geology is now reasonably well known.

The oldest rocks found on Antarctica are of Precambrian age – more than 540 million years old. These form basement rocks that once were areas of ancient continental crust called 'cratons'. There are thick sedimentary sequences of both Archaean and

Eon	Era		Period / Epoch	mya	Antarctic record
Phanerozoic Eon	Cenozoic Era Age of Mammals	Neogene	Holocene	0.01	shell assemblages – McMurdo Sound
			Pleistocene	1.8	fossil whales – Marine Plain, Vestfold Hills
			Pliocene	5.0	*Nothofagus* – Beardmore Glacier
			Miocene	23	
		Paleogene	Oligocene	34	marine invertebrates, giant penguins, early mammals from Seymour Island and nearby
			Eocene	56	
			Paleocene	65.5	
					extinction of dinosaurs, etc.
	Mesozoic Era Age of Dinosaurs		Cretaceous	145	marine invertebrates and reptiles from Seymour Island and nearby
			Jurassic	200	plants – Alexander Island dinosaurs, pterosaurs, etc. from Mt Kirkpatrick (Hanson Formation)
			Triassic	251	amphibians and reptiles – Fremouw Formation
					many extinctions
	Paleozoic Era Age of Fish and Invertebrates		Permian	299	fossil plants – Prince Charles Mountains
			Carboniferous	359	many extinctions
			Devonian	416	fossil fishes – Aztec Siltstone, Victoria Land invertebrates and fossil fishes – Ellsworth Mountains, etc.
			Silurian	444	many extinctions
			Ordovician	488	
			Cambrian	542	archaeocyaths, trilobites, etc. – early reefs, Transantarctic Mountains
Precambrian Eon			Proterozoic	2500	no record of life in Antarctica yet found
			Archaean	4600	

This fossiliferous limestone from the Queen Maud Mountains contains the eroded, ovate tops of the world's first sponge-like invertebrates called 'archaeocyathids' – some of the oldest preserved life from the Antarctic continent. These early sponges comprise some 30 per cent of Cambrian deposits in the Queen Maud and Bender mountains of East Antarctica. Photo: Bert Rowell.

Proterozoic ages, but no definite fossils from these strata have yet been found.[1]

Many good fossils are to be found in exposures of Cambrian sedimentary rocks, such as the five-kilometre-thick succession of the Shackleton Limestone exposed in the Beardmore Glacier region. In Northern Victoria Land there are small outcrops of Cambrian limestone as megaclasts in younger conglomeratic rocks, some of which are richly fossiliferous. The Ordovician Period is represented by a lot of coarsely crystalline igneous rocks, mostly intrusive volcanic rocks in Southern Victoria Land around the Dry Valleys, as in the Granite Harbour Intrusives.

Antarctica in the Cambrian Period

The Cambrian Period marks a worldwide explosion of life as fossils of many kinds appear widely in the fossil record. At the start of the Cambrian the atmosphere was warm, and richer than today in both and oxygen and carbon dioxide – with perhaps as much as 18 per cent atmospheric carbon dioxide.[2] During the Precambrian, the level of oxygen in the atmosphere

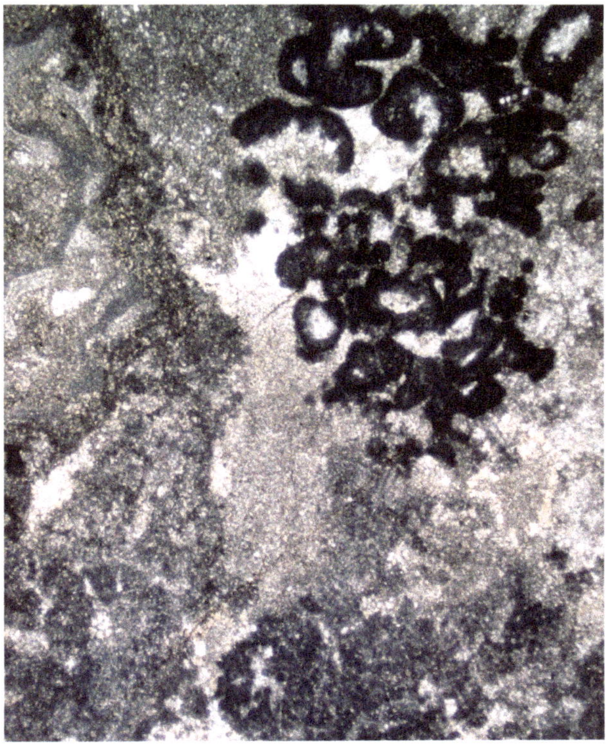

A sample of Lower Cambrian Shackleton Limestone revealing cross-sections of fossils. Photo: Bert Rowell.

was low. Around 750 million years ago it was a mere 0.5 per cent, but increased to 5 per cent at the start of the Cambrian. Many organisms living in the seas took advantage of the excess of carbon dioxide to make calcium carbonate shells or hard parts of calcium phosphate. This meant that animals evolving muscles had a major breakthrough as muscles require adenosine triphosphate (ATP) to work effectively, and ATP requires an oxygen-rich environment. There were no plants on land at this time, so the oxygen was there for any organism in the seas that could make good use of it. Life thrived at this time, and we see the appearance of nearly all modern phyla of animals (except the Bryozoa or 'moss animals') from joint-legged arthropods, echinoderms and molluscs, to the first fish-like vertebrates.

During the Cambrian Period there was a counterclockwise rotation of Gondwana around an axis near Northern Victoria Land. This rotation stopped by the end of the Cambrian and Antarctica was situated with its northern margin, at what is today called Wilkes Land, in contact with southern Australia at equatorial latitudes. Kemp Land was conjoined to India and the southern margin, Queen Maud Land was joined to the south-eastern margin of Africa at around 30 degrees south. Antarctica therefore had a climatic regime that ranged from tropical to temperate. The region where the Transantarctic Mountains exist today once formed the coastal margin to the Palaeo-Pacific Ocean. This eastern seaboard of Antarctica was home to the first great reefs Earth had ever seen.

Some 510 million years ago there was a global extinction event in which many of the Early Cambrian groups were wiped out, possibly due to anoxia, or oxygen starvation, hitting hard the bottom-dwelling faunas.[3] However, the groups that survived flourished and diversified during the rest of the Cambrian, except for one notable exception, the sponge-like archaeocyaths, which survived only until the end of that period.

Nearly all of the Cambrian Antarctic localities fall along the length of the Transantarctic Mountains and within the Ellsworth Mountains, but also from erratics in moraine near Mt Provender in the western part of the Shackleton Range and King George Island, Antarctic Peninsula.

OPPOSITE: Warm, shallow seas of Antarctica during the Early Cambrian more than 500 million years ago with sun shining on glistening reefs composed of archaeocyaths, the great reef builders of the Cambrian, which had all but disappeared by the close of the Cambrian. Their demise, after flourishing for so many millions of years, is believed to have been related to the evolution of 'bulldozing' organisms such as the trilobite (depicted here on the sea floor), disturbing the archaeocyaths' habitats, among other factors. Illustration: Tony Windberg.

Early Cambrian life

Archaeocyaths were cup-shaped organisms, usually around 10 cm or so in length, which were similar to modern day sponges, but the exact nature of their living component is not really known. They were sedentary creatures that filtered water for particles of food. The structure of a typical archaeocyath is a cone-shaped layer of porous calcium carbonate with another similar cone inside it, the two walls joined by thin septae. In cross-section they resembled wheels with spokes joining the outer and inner layers.

The first archaeocyath appeared in the Early Cambrian, roughly 530 million years ago. During the subsequent 10 million years they were incredibly successful, diversifying into an amazing array of forms and playing a dominant role in the construction of Earth's first reefs. We know of approximately 120 families of archaeocyaths from this time. Their rapid demise was just as dramatic as their diversification, and probably relates in part to the evolutionary expansion of 'bulldozing' organisms such as trilobites churning up the sea floor. By the end of that 10 million year period, these reef-builders had all but disappeared. The last recorded archaeocyath on the planet is a single species from the Late Cambrian of Antarctica. Competition with the diversification of Cambrian sponges may have played a part in the decline and eventual demise of the archaeocyaths.

Primitive corals appeared in the Early Cambrian along with other groups of cnidarians (or coelenterates), worms (annelids), primitive sea lilies (pelmatozoans), although they are not a major part of the Antarctic Cambrian faunas. Bryozoa ('moss animals') is the only phylum which has no recorded fossil history in Cambrian or Upper Precambrian rocks. Bryozoans have their first documented occurrence in

The invertebrate family

Invertebrates are animals lacking a backbone. They include the vast majority of all animal life on earth and in our seas, such as worms, shells, insects, crabs, corals, and jellyfish. The major groups of invertebrates are classified according to their symmetry, their degree of complex anatomy and the nature of their hard parts. The following are the main groups known today and some of the major extinct groups which are dealt with in this book.

Protozoa are microscopic single-celled animals. Some secrete shells or make 'tests' out of sand grains, such as the foraminifers, a diverse group commonly found as microfossils in marine sediments.

Porifera or sponges are simple, filter-feeding organisms that secrete hard spicules to support their framework (either as silica or calcareous units). These may well be related to the extinct cup-shaped organisms called archaeocyathids.

Cnidaria (Coelenterata) are organisms with five-fold symmetry, and simple anatomical systems. This group includes the jellyfishes as well as coral polyps that secrete hard calcareous homes for themselves.

Graptolites, extinct free-floating colonial organisms, were thought to have been related to cnidarians, too, but are now believed to be more allied with hemichordates (so more related to us!).

Annelida – this group includes marine worms such as polychaetes and terrestrial earthworms. The common fossils of these creatures are their burrows and traces, and sometimes minute hard parts that form the jaws (called scolecodonts).

Echinodermata are creatures with five-fold symmetry and secrete calcareous plates to make up their hard outer shell. The group includes starfish (asteroids and ophuroids), sea cucumbers (holothuroideans) sea urchins (echinoids), sea lilies (crinoid) and many extinct groups, especially in the Paleozoic, such as mitrates, blastoids and cystoids.

Mollusca are soft-bodied animals with well-developed internal anatomy. These include snails, clams, nudibranchs, octopus and other squid-like creatures such as the nautilus, argonaut, squid, and the extinct ammonoids and belemnoids.

Arthropoda are animals with jointed legs, such as crustaceans (crabs, shrimp and lobsters), insects, centipedes, millipedes, arachnids (spiders, mites, ticks and horseshoe crabs) and many extinct groups, including the trilobites, bradoriids, conchostracans, etc.

This is only a basic guide to the major groups of invertebrate animals. Many extinct groups found in the fossil record have no clear bearing to living creatures.

Postage stamps commemorating the British Antarctic Survey and depicting a range of Cambrian invertebrates: (from left to right) the archaeocyathid *Monocyathus*, the inarticulate lamp shell *Lingulella* (one of the oldest brachiopods on the planet), and the trilobite *Lyriaspis*.

The Nelson Limestone found in the Neptune Range of the Pensacola Mountains contains a diversity of Middle Cambrian trilobites and other organisms, which have been recovered over the many decades of dedicated research by several scientists.

the Early Ordovician, but it is quite plausible they were present in the Cambrian – perhaps they were either rare, soft-bodied or not preserved for some uncertain reason, or we just have not discovered them yet.

Apart from the archaeocyaths, the most abundant creatures found as fossils at this time are the brachiopods, molluscs and trilobites. One of the only major invertebrate groups not to have a fossil record in Antarctica (as far as we know) is the colonial invertebrate graptolites (Phylum Hemichordata). These invertebrates are often found in graptolithic black shales, where they are usually the only recorded fossil, nothing else.

Brachiopods or 'lamp shells' are bivalved creatures that resemble a clam with a different plane of symmetry. In brachiopods the plane is through the midline of the shell, in clams it is through the horizontal hinge plane. The animal inside the shells has tentacles that filter the water for food – these are called 'lophophores'. The brachiopod attaches itself to the substrate through a foot or pedicle that comes out of a circular hole in one of the shells. The shells resemble the ancient Roman oil lamps, hence the name. Brachiopods evolved into two main lines: the inarticulates, which have two shells of equal size that do not hinge together but are held by ligaments, and the most diverse group, the articulates, which have properly hinged valves.

Molluscs first appeared at the end of the Precambrian and underwent a great radiation during the Cambrian Period. The ancestral mollusc may have been without a shell and could have resembled something like the Late Precambrian *Kimberella*. The most primitive group of molluscs could be the monoplacophorans which bore a simple, low, cone-like shell on their backs and paired muscles. These would become the dominant molluscs of the Cambrian Period.

Another primitive fossil group are the rostroconchs, with slightly twisted simple shells and fused dorsal margin, reminiscent of clams but very different all the same. True snails or gastropods also appear at this time, and forms with high conical shells, such as *Knightoconus*, were common.

Bivalves, the group containing the clams and oysters, also appeared in the Cambrian as did the first cephalopods – among the most advanced of all invertebrates – represented by forms such as *Plectroceras*. This group would diversify to include squid, cuttlefish, octopuses, and a variety of shelled forms such as the nautiloids and ammonoids.

The chitons (Amphineura or Polyplacophora) with their many plates first appeared in the Late Cambrian and typify another primitive molluscan group that has not changed much in appearance since then.

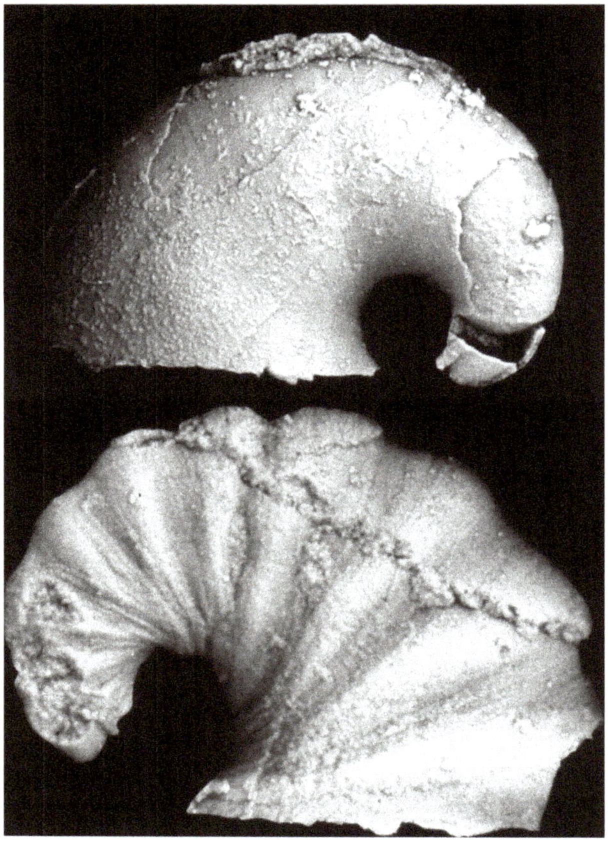

This early monoplacophoran, *Latouchella*, from the Shackleton Limestone of the Beardmore Glacier region has marked growth ribs and a slightly coiled shell. Images: Bert Rowell.

Trilobites were a very successful group of arthropods that appeared, worldwide, at the beginning of the Cambrian and disappeared at the end of the Permian Period. Early on in their evolution, some trilobites attained gargantuan sizes, such as the king of the Cambrian sea bottom, *Acadoparadoxides*, which commonly reached more than 30 cm in length. Trilobites are so named because of the three-part carapace that covered their bodies comprising a head (cephalon), middle section (thorax) and tail (pygidium). Some had large, compound eyes, while others that burrowed lacked eyes – they were blind. They all had a backwards facing mouth and fed largely by ingesting organic-rich material in the sediment. Like all arthropods they regularly moulted, shedding their old carapace, which is often the part of them found as a fossil.

Other primitive arthropods found in the Cambrian include the small, bivalved ostracods. These are tiny shrimp-like creatures which lived inside two hinged valves made of phosphatic material. These valves are readily fossilised and easily identified by their external features and internal muscle scars. Conchonstracans are another early group of bivalved arthropods whose shells are simpler in structure than the ostracods and, like the ostracods, are quite useful palaeoenvironmental indicators, sensitive to their surrounds.

Life on the great Cambrian reefs

Fossils spanning most parts of the Cambrian Period are present in patchy areas in the Transantarctic Mountains and Ellsworth Mountains in limestones that comprise a high diversity of marine organisms, mostly trilobites (arthropods), conodonts, early molluscs and brachiopods. The Shackleton Limestone is the modern expression of what was once an extensive platform of carbonate sediments that fringed the greater Antarctic Craton. It has produced an assemblage containing algae, cyanobacterial mounds and archeocyathids, which in places represent the first reef communities.

The reefs in the Shackleton Limestone of Antarctica are arguably the first early true reefs as they formed topographic mounds with geological evidence showing that waves broke over the top of

Oncolites are lithified sedimentary deposits found in the Shackleton Limestone that contain many spherical structures (oncoids) formed by cyanobacterial growth. Photo: Margaret Rees.

them. Individual reefs were rarely more than two metres high, although successions up to 50 metres have been identified which represent many phases of cyclical reef growth. These reefs developed mainly on low-energy, sub-tidal shelves, occasionally growing in higher energy environments, as evidenced by their coexistence with shoals of spherically shaped pellets called 'ooids'. Such depositional settings were stable for long periods of time.

Three types of Cambrian reefs have been identified in the Shackleton Limestone by their microbial compositions.[4] Cup-shaped archaeocyathan sponges comprise about 30 per cent of the reefs, although commonly they represent less than 10 per cent or may be entirely absent from some reefs. The second type of reef is formed from *Stromatactis*-bearing microbial growths. The third and most common of these early reefs is defined by its high content of *Epiphyton*, but also includes archaeocyathans and the microbial growths, *Renalcis* and *Girvanella*, as well as cryptomicrobial clotted muddy limestone. In such reefs the framework is quite open and porous, and the algae *Epiphyton* and *Renalcis* encrusted much of the core framework material. Fossil evidence shows that these great Cambrian reefs once spanned the entire length of the continent of Antarctica, from Northern Victoria Land through to the Weddell Sea region, a distance of several thousand kilometres, comparable to the Great Barrier Reef of Australia today.

Early Cambrian erratic boulders found in Miocene glacial deposits of the Cape Melville Formation on King George Island, at the tip of the Antarctic Peninsula, provide a fascinating window into the diversity of primitive life on the continent at this time. The fauna was first collected in the late 1980s by members of the Polish Antarctic Survey. The rocks were dissolved in weak acetic acid to leave phosphatic microfossils, preserved in three-dimensional form in the residues.

The Ruthven Bluff, Argentina Range, at the end of the Transantarctic Mountains near the Weddell Sea, contains fossil-rich deposits of crustaceans called bradoriids which were common in the Cambrian and became extinct in the Early Ordovician, when similar bivalved crustaceans called ostracods apparently took over their niche. Photo: Margaret Rees.

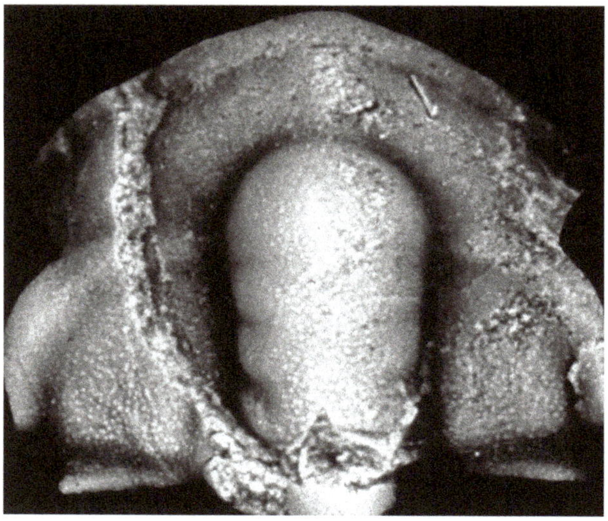

LEFT: The head shield (cephalon) of one of the world's oldest arthropod trilobite animals is preserved in the Shackleton Limestone. It is named *Hsuaspis* cf. *bilobata* after scientist Professor Hsu. RIGHT: A well-preserved fossil of *Dailyatia,* belonging to the family Kennardiidae. *Dailyatia* is characterised by having a bilaterally symmetrical, bipolar body with tightly fitting 'sclerites', a term commonly used for the hardened parts of arthropod exoskeletons. Fossils such as these have been recorded in Antarctica and several places in Australia, reflecting ties between these two regions in the Cambrian. Images: Bert Rowell.

The transition from Beardmore Group Precambrian greywackes to Shackleton and Byrd Group archaeocyathine Limestones takes place at the prominent bluff on the left and also at the dark Cambrian Bluff on the right in the Holyoak Range. Photo: John Long.

The King George Island fossils include archaeocyathids, inarticulate brachiopods similar to the living form *Lingula,* and brachiopods such as *Eoobolus* sp. (aff. *E. elatus*), *Karathele napuru* and *Vandalotreta djagoran.* The erratics have yielded small shelly fossils comprising over two dozen species of molluscs, bivalved arthropods and enigmatic forms known from sclerites and tubes appropriately called 'Problematica', which is basically a waste paper basket term for an organism that defies proper placement in any known taxonomic group.

The mollusc fauna of this age includes an extinct group called the helcionellids, which had simple cap-shaped shells. These include five forms: *Paraconus* sp. (cf. *staitorum*), *Yochelcionella* sp., *Anabarella* sp. (cf. *argus*), *Pelagiella madianensis* and *Beshtashella tortilis.*

Hyoliths were another primitive molluscan group that had conical shells with a small lid to close the opening. Hyoliths from the fauna include *Conotheca australiensis, Microcornus petalus, Hyptiotheca carraculum* and '*Hyolithes*' *conularioides,* plus one hyolith-like form called *Cupitheca holocyclata.*

The bradoriids were a group of small, bivalved crustaceans (less than 18 mm) which inhabited a broad range of habitats in the Cambrian and early part of the Ordovician Period, and which subsequently became extinct. Bradoriids from West Antarctica include *Albrunnicola bengstoni,* a form thought to be very close to *Hipponicharion* sp. from the Yorke Pensinsula of South Australia.

The tommotiids are recognised by isolated plates or sclerites that probably fitted over the back of the animals to form a shield. This enigmatic group of unknown affinity is represented by *Dailyatia ajax, Lapworthella fasculata* and *Shetlandia multiplicata.*[5] Other sclerite covered forms belonging to obscure extinct groups include: *Chancelloria racemifundis, Allonnia* cf. *tripodophora, A.* cf. *tetrathallis, Archiasterella*(?) sp. and the halkieriid *Halkieria parva.* There are also tube-like enigmatic fossils belonging

Kennardia (and the closely allied *Dailyatia*) are both members of the family Kennardiidae (Order Tommotiida). *Kennardia* differs from *Dailyatia* in having coarse, concentric ribs that in profile appear more rectangular. A few species of *Kennardia* have been recorded from the Shackleton Limestone.

Photo: Bert Rowell.

to *Hyolithellus filiformis* and *H. micans*, *Torellella* sp. and *Byronia*(?) *bifida*.

The Early Cambrian of the Argentina Range in the north-eastern Pensacola Mountains has also produced some marine fossils. *Bicarinella evansi* is a bradoriid found in the upper horizons of the Schneider Hills. Its shell is hemispherical with sharp anterior and posterior lobes on the carapace. In this respect it closely resembles other bradoriids from the family Hipponicharionidae, a group not known from any sites in Gondwana. This suggests that the Antarctic bradoriid arthropods may have entered Gondwana via the palaeo-Pacific route from Laurentia, bypassing Australia.[6]

Late Cambrian life

In the west of Antarctica, in the Ellsworth Mountains, the Heritage Group contains lots of limestone with a rich fauna representing a warm tropical shallow marine setting, probably in the low latitudes. The fauna has affinities with Australia and the Australo-Sinian Province, and includes a variety of early molluscs, including the early snails *Knightoconus* and *Kingerella*, early monoplacophorans such as *Marocella* and rostroconchs such as *Kennardia*, as well as trilobites.[7] The last occurrence on the planet of archaeocyathids, which had their heyday in the Early and Middle Cambrian, is *Antarcticocyathus webberi* in the Ellsworth Mountains.

The richly fossiliferous Robertson Bay Group in Northern Victoria Land has yielded a fauna of trilobites that show a close affinity to Australian and Asian fossil assemblages of this age. The fauna also contains brachiopods, molluscs, trace fossils and microfossils including small bivalved arthropods called ostracods.[8] This fauna contains inarcticulate brachiopods such as *Schizambon* and *Prototreta*, and an articulate brachiopod *Billingsella*. The latter genus has its closest resemblance with the species *Billengsella perfecta and B. coloradensis* from northern America, reinforcing the 'SWEAT' hypothesis that North America was in contact with or proximity to Gondwana during the Cambrian.

Molluscs from the Bowers Group include an ancient snail *Scaevogyra*, a genus commonly found also in North America and Asia, and an enigmatic mollusc called *Pelagiella*. This creature had a simple coiled shell in which the whorl contacts itself, suggesting it was significantly more advanced than the Early Cambrian forms.

The trilobite fauna from the Bowers Group is represented by about a dozen species from some seven different families. The simple blind trilobites called agnostids are the commonest group. *Pseudagnostus* is a widespread genus known from nearly 90 species worldwide, and the Antarctic species most closely resembles *Pseudagnostus communis* from China. The genus *Stigmatoa* is known from several species throughout Australia but one species, as yet unnamed also occurs here in Northern Victoria Land. Other genera recorded from here include *Olentella, Irvingella, Pedincephalus, Talbotinella, Prochuangia* and *Proceratopyge. Talbotinella* is otherwise only found in Argentina and Australia; and both *Proceratopyge* and *Prochuangia* have closest affinities with Chinese trilobite faunas.

The enigmatic, Early Cambrian mollusc named *Marocella* from the Beardmore Glacier region in the Shackleton Limestone is interpreted as an ancient limpet-like mollusc that was mobile, living in the limey mud at the bottom of the sea.

Images: Bert Rowell.

Links between Australia, China and Antarctica are also seen in other events taking place at that time. The oldest known chordates are found in upper Lower Cambrian sediments near Chengjiang, Yunnan, China. Forms such as *Myllokungmingia* and *Haikouichthys* are simple eel-like creatures with well-developed gills, V-shaped muscle blocks called myodomes and even possible vertebral structures around the notochord. In essence they are very fish-like and are classified as the first vertebrates. There is another interesting connection here if we consider where these first vertebrates lived. In Early Cambrian time we see that the South China Terrane where these 'protofishes' lived was in close proximity to Australia, the continent where the oldest bony tissues appear.[9]

It seems that East Gondwana may have been the likely centre for the origin of the biggest step in vertebrate evolution, the appearance of the first vertebrate with bone. Whether you want to pin it down to the first fishes without bone from the nearby Chinese terrane, or place the origin at the point where bony tissues first appear (central Australia), either way it is within or adjacent to East Gondwana. This suggests that if we keep searching the Cambrian deposit of Antarctica for microfossils, perhaps vertebrate remains will one day be found there, too.

Ordovician life

During the Ordovician Period, warm water deposits, such as limestones and salt, were formed in the equatorial regions of Gondwana (Australia, India, China and Antarctica), while glacial deposits and ice-rafted debris built up in the south polar areas of Gondwana (Africa and South America). A new seaway, the Iapetus Ocean, widened between the ancient continents of Laurentia (North America), Baltica (Northern Europe) and Siberia around 480 million years ago. Very few fossils of this age are known from Antarctica and those we have found are present within a shale-sandstone-conglomerate sequence in Robertson Bay (Northern Victoria Land) that contain arthropods and conodonts, and may have

An impressive one-metre-wide giant arthropod trackway in the Hatherton Sandstone at Escalade Peak, Skelton Névé region. Photo: John Long.

been reworked. Trace fossils representing burrows and arthropod trackways of phyllocarid crustaceans are recorded from the Blaiklock Glacier Group in the west part of the Shackleton Range, in sandstones inferred to have been deposited in shallow waters.

Sedimentary and volcanic rocks, with rare fossils, are recorded from the Shackleton Range.[10] The Upper Cambrian–Ordovician sequence in Marie Byrd Land, Swanson Formation, contains *Palaeodictyon*;[11] and conodonts from the start of the Ordovician are also known from Northern Victoria Land.[12] However, there is a great gap in our knowledge of prehistoric Antarctica spanning some 60 million years, from the end of the Early Ordovician through to the start of the Devonian Period, without any known fossil record. During this time Gondwana rotated and moved further northwards to be embraced by warmer climes, setting the scene for its great radiation of life in the middle part of the Paleozoic Era.

OPPOSITE: The spine of the acanthodian fish, *Gyracanthides warreni*, from the 'Fish Hotel' site, Cook Mountains, mid-Devonian Aztec Siltstone. Photo: John Long.

5
THE AGE OF FISH AND FERNS

Every fossil collector is delighted when he comes across a fossil fish; one gets rather tired of everlasting mollusca, and a fish cropping up now and then is a great encouragement to a geologist.

Rev. H N Hutchinson, *Creatures of Other Days*, 1894.

Two major events occurred in the Silurian and Devonian periods that would alter the face of Earth for all time. The first is that plants left the seas by the Late Silurian and 60 million years later had become well and truly established on land. By the Late Devonian, large lycophytes, today known as 'horsetails', formed primeval forests up to 20 metres high, while ferns dominated the undergrowth and coiled down from the treetops.

The other event that would one day enable animal life to go forth from the seas and fill those primeval forests was that fish with jaws underwent a great evolutionary radiation during the Silurian Period. By the Middle Devonian, advanced lobe-finned fish capable of breathing air and scampering onto land had appeared. The first tetrapods – fish-like creatures with digits on their limbs – evolved in the Late Devonian, but did not begin their wholesale invasion of land until 30 million years later.

The Devonian world was mostly hot and steamy. The amount of oxygen in the atmosphere varied, being around 20 per cent at the start of the Devonian and possibly dipping in the Middle Devonian. At the same time, carbon dioxide rose to around 10 per cent, creating a massive greenhouse event that came to a peak during the main coal-forming events of the Early Carboniferous Period. By the end of the Devonian, oxygen was sitting at around 19 per cent but rose in the Early Carboniferous to a peak of nearly 29 per cent. As the atmosphere warmed

A very large Devonian trace fossil, *Beaconites,* from the Hatherton Sandstone, near Gorgon's Head, East Antarctica. The large, segmented burrow, which is reminiscent of a skeleton of a snake, was probably created by a large, crustacean 'shovelling' the substrate out of its way. The producer of the trace is thought to have burrowed to a depth of no more than a few tens of centimetres. Photo: John Long .

and thickened, creatures ranging from large land invertebrates and air-breathing fish took advantage of the fact that it was now easier to breathe outside the water than ever before. But life was not so sunny in the Devonian, as two mass extinction events around 360 million years ago devastated life and brought about a radical realignment for its survivors.

Antarctica remained as the central pivotal region of Gondwana throughout the Paleozoic Era. During the Silurian Period, Antarctica was positioned in temperate latitudes with the present-day South Pole being at approximately 30°S. This did not change much during the Devonian Period, apart from the slight rotation of Gondwana causing Antarctica to dip further south, with average latitudes being around 45°S. In the Devonian seas the major invertebrate groups also kept evolving into a multitude of new families. Some of these marine communities formed gigantic algal-stromatoporoid reefs, one of which still stands exposed today in the Kimberley region of north Western Australia.

Antarctica has a questionable record of life from the Silurian Period, as only trace fossils of this age are known, and the dating on these fossils is problematic. The Devonian, however, is richly fossiliferous both with marine faunas of Early Devonian age and fish in the Middle Devonian Aztec Siltstone. These deposits also contain Antarctica's first record of land plants and the world's oldest known soils.

During the Silurian Period the marine invertebrates continued diversifying into more discrete niches. Simple mites and arachnid-like creatures began invading the landscape colonised by the first land plants, and primitive fish had acquired jaws and teeth. Although many of the neighbouring lands around Antarctica contain key fossil sites documenting many of these great evolutionary steps, to date no definite Silurian rocks have been confirmed from Antarctica.[1]

Silurian–Early Devonian trace fossils

Since the early 1980s Margaret Bradshaw of Canterbury, New Zealand, has led many Antarctic expeditions to collect invertebrate and fish fossils and to study *in situ* trace fossils in the Ellsworth Mountains and Ohio Range as well as in the central Transantarctic Mountains. Her discoveries include many diverse trace fossil types, including trackways of giant arthropods such as sea scorpions (eurypterids), and many kinds of burrows and feeding traces. Comparisons with the Tumblagooda Sandstone trace fossil assemblages of Western Australia, now known to be of latest Ordovician or basal Silurian age, suggest that these sequences in Antarctica could be a lot older than first estimated.

The Hatherton Sandstone contains abundant large trace fossils most probably made by a digging crustacean that lined each section of its burrow, making it appear chambered in cross-section. One large form, *Beaconites barretti*, made wide burrows comparable in size to a human arm, while another form, *Beaconites antarcticus*, made narrower burrows. Although there has been much speculation over which creature made the burrows, the authors have found a resting trace of the *Beaconites* animal on top of a burrow at Mt Gudmundson. The trace is still being studied by Margaret Bradshaw, but we can

Skolithos-type trace fossils near Mt Odin. *Skolithos* is a simple, tube-like, vertically oriented burrow that usually reveals a much greater length versus width. It is interpreted as a dwelling burrow made by a suspension-feeding animal in marine environments. Photo: John Long.

John Long searching for microvertebrates in 1991 at Gorgon's Head in the Cook Mountains, East Antarctica. Photo: Brian Staite.

A large bedding plane of trace fossils, possibly Late Silurian or Early Devonian, near Mt Odin, Dry Valleys, Transantarctic Mountains. Photo: John Long.

at least confirm that it appears to be from a large crustacean.

Other common trace fossils in the Hatherton Sandstone include very large trackways made by a giant sea scorpion (eurypterid), possibly up to two metres in length. Excellent examples of these are found on top of Mt Perseus in the southern Cook Mountains, and at Escalade Peak in the Skelton Névé region. The latter trackway measured almost 90 cm in width with individual appendage 'footprints' up to 20 cm across. Other commonly found traces include 'spaghetti rock', *Heimdallia*, made by intensive burrowing of the sediment, an ancient form of bioturbation. *Heimdallia* was first named for its discovery in the Heimdall Glacier area of the Dry Valleys region.

The Devonian Period

There is a rich record of Devonian life known from Antarctica, from marine invertebrates living in shallows to primitive fish of many kinds from large-scale freshwater river and lakes systems. At this time Antarctica was blanketed in primeval forests of huge horsetails, the lycophytes, and low shrubbery made up of ferns and primitive stem-like plants called psilophytes. The East Gondwana Province, comprising Australia and eastern Antarctica, may have been the centre of a major evolutionary radiation which took place in the Devonian Period. Sharks first began to diversify in number and size in this region.

The Middle Devonian Aztec Siltstone of Southern Victoria Land holds the richest diversity

Trace fossils

Sometimes animals leave marks in the sediment that indicate they were once present in an environment. These can be trackways (footprints), burrows, feeding trails or resting places. If conditions are just right some of these traces can be preserved as fossils, but in general this is rare. For footprints to be preserved the trace must be carefully and gradually buried by sediments. Wind-blown silts and sands infilling a hard, sun-baked, clay-rich substrate act to preserve the trackway.

Burrows of organisms can be infilled by other sediment after the burrow is abandoned. Because they already extend well into the sediment and often have a harder compacted lining made by the burrowing creature, these are readily preserved as fossils. Sometimes the burrower may be preserved at the bottom of its burrow – crayfish are sometimes found as fossils preserved in this way. Trace fossils are known from sedimentary rocks of all geological ages. They are particularly common from marine environments where there was marked biological activity. In some shallow marine deposits the intensity of burrowing or feeding trails can be very dense – this is called bioturbation, or biological churning up of the sediments. Earthworms do this in soil.

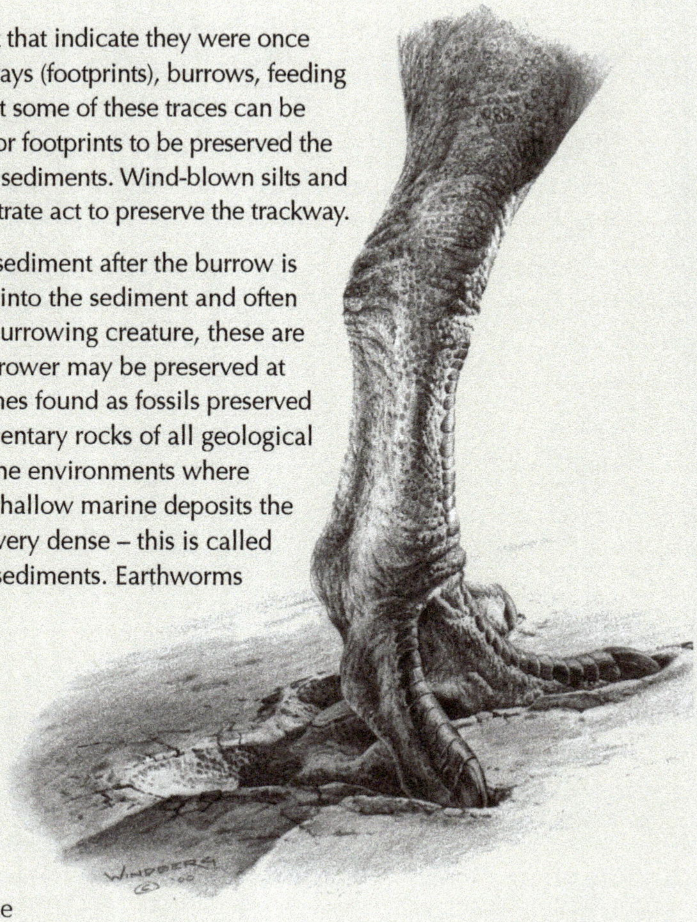

The study of tracks, trails and traces in general is called ichnology, and fossilised traces are called ichnofossils. Ichnofossils can be named according to their type, but their name has no relationship to other animal fossils. For example a trackway of a dinosaur with foot shape similar to *Tyrannosaurus* is named as "*Tyrannosauropus*" meaning *Tyrannosaurus*-like foot trace. Whether or not *Tyrannosaurus* or another similar dinosaur really made these tracks cannot be verified, so the name only applies to the trace. Illustration: Tony Windberg.

Palaeontologist Margaret Bradshaw working at the base of the mid-Devonian Aztec Siltstone on Mt Gudmundson East Antarctica.
Photo: John Long.

of fish of this age from anywhere in the southern hemisphere. It includes representatives of the earliest occurrence of several endemic lineages of primitive sarcopterygian (lobe-finned) fish. These evolutionary events were followed by global dispersion and faunal interchange with Laurasia. Such groups of fish were soon to give rise to more advanced northern hemisphere sarcopterygians, like the elpistostegalids, now regarded as the immediate precursors to the first four-legged land animals, the tetrapods.

Many fish fossils of the Devonian Period are preserved in red-coloured sandstones, mudstones and shales that represent river, lake and estuarine deposits (often termed 'red beds'). Many famous sites in Scotland occur in such rocks which were first named the 'Old Red Sandstone'. This term now applies to a widespread belt of Devonian river and shallow marine rocks occurring throughout the ancient supercontinent of Laurasia (sometimes called 'The Old Red Continent'). Red beds get their colour from the oxidation of iron-rich sediments in warm to tropical dry climates. In most cases they may include green shales and mudstones as well as ancient soil deposits. Fish fossils may accumulate in ancient channel and lake deposits within these rock sequences, as occurs in the Aztec Silstone. The first fish fossils described from Antarctica were referred to by Sir Arthur Smith Woodward as coming from the Upper Old Red Sandstone of Antarctica, making direct comparison to the well known fish fossil sites in Scotland.

There are many exposures of Devonian sedimentary sequences throughout the Transantarctic Mountains representing a diverse range of environments, from shallow marine seaways through to large meandering river plains, delta and lake systems. The terrestrial habitats are seen in the Aztec Siltstone, which has several ancient soil horizons, some with plant rootlets in life position. The first abundant vertebrate fossils occur in Lower Devonian marine sediments of the Ohio Range, with well preserved and diverse fish faunas occurring in the Middle–Upper Devonian Aztec Siltstone, represented by over 30 species of fish and a few land plants.

Early Devonian marine life

During the 1960s there were some remarkable early expeditions to the far western extent of the Transantarctic Mountains, including the Horlick and Ohio ranges. They were found to be rich in brachiopods, bivalves, and rare fish remains, characterised by the close association with similar assemblages from the eastern United States, the Falkland Islands and southern Africa.[2]

The Horlick Formation is a sequence of horizontally bedded sandstones, siltstones and shales which crops out in the Ohio Ranges in central Antarctica. It contains Antarctica's most diverse assemblage of Early Devonian shelly marine fossils and remains of early fish.[3]

The marine invertebrate fossils of the Horlick Formation are mostly preserved in sandstone layers. There are molluscs (16 species) and brachiopods (7 species), with rarer remains of trilobites, tentaculitids, bryozoans, crinoids and hyolithids. The most abundant group of invertebrate fossils is the brachiopods, and one species, the terebratulid *Pleurothyrella antarctica*, dominates the fauna.

In the Mt Wyatt Earp Formation of the Crashsite Group in the Ellsworth Mountains in West Antarctica there are brachiopods and bivalves, particularly the abundant brachiopod *Orbiculoidea* cf. *falklandensis*,[4] the bivalves *Nuculites* cf. *cuneiformis* and *Grammysoidea*? sp., both of which have close affinities with species from New York State. Cephalopod nautiloids (orthoceratids) are also present, but indeterminable to genus. The rostroconch mollusc *Hippocardia* sp. is poorly preserved from this site and so cannot be compared with other known species.

Early Devonian fish from the Ohio Range and Ellsworth Mountains are not well known – they are represented by a few acanthodian spines of the genus *Machaeracanthus*, and some placoderm plates which have been identified as belonging to arctolepid arthrodires.[5] *Machaeracanthus* is well known from the Early Devonian of Europe and North America, so was apparently widespread at this time, as were the arctolepid arthrodires. Thus, the limited Early Devonian fish fauna from Antarctica provides no real clues as to the affinities of the vertebrates living at this time.

Teeth of the Devonian shark *Aztecodus*, from the Portal Mountains in the Skelton Névé region. Photo: John Long.

A beautifully preserved jaw of the onychodont fish *Luckius* with a shark's tooth, *Mcmurdodus,* in the centre of the specimen, from the Cravers Peak Limestone in Queensland (c. 385–390 mya). The first record of this shark was found at Mt Feather, Skelton Névé in the 1950s.
Photo: John Long.

Middle Devonian fish

We have a good picture of what the Middle Devonian fish faunas of Antarctica were like through a diverse range of species found in the Aztec Siltstone of Southern Victoria Land. Here there are representatives of all the major groups of fish that had appeared by this time, from the jawless agnathans, the spiny acanthodians, the armoured placoderms, and the first of the true bony fish, the osteichthyans.

The first fossil fish were collected from erratic boulders found in the Mackay Glacier region by Frank Debenham during Robert Falcon Scott's 1912 expedition.[6] Since then various expeditions have collected more of the fossil fish from many localities as far south as the Cook Mountains, and as far north as the Allan Hills.

The 1970–71 Victoria University Wellington Antarctic Expedition (VUWAE 13) led by New Zealander Peter Barrett saw Australian palaeontologists Gavin Young and Alex Ritchie make the first systematic collections of the Aztec Siltstone fish sites throughout the Skeleton Névé region. The expedition visited many sites from the southern Warren Range through to Mt Crean, and discovered the first complete skulls of sarcopterygian fish, articulated whole palaeoniscoid fish, many well-preserved placoderms, articulated acanthodians, and a partially articulated shark.

More than 30 different species of fish have been identified from various horizons within the Aztec Siltstone, now considered to be of Middle–Late Devonian age. The climate during these times remained temperate with regular seasonal rainfall and occasional summer droughts indicated by the sedimentary structures preserved in the Aztec Siltstone, such as desiccation cracks and varved lake deposits.[7]

Jawless fish (*Agnatha*)

The jawless fish are today represented by the parasitic lampreys and scavenging hagfish. During the Silurian and Devonian periods there was a great radiation of the jawless fish into many different families. Many of them developed bony plates covering the head and trunk, such as the heterostracans and osteostracans, groups well-known in the northern hemisphere but absent from Gondwana.

The lowermost section of the Aztec Siltstone contains the scales of one group of agnathans, the thelodonts. Unlike other agnathans, the thelodonts lacked large bony plates, instead having their bodies covered by many thousands of small scales. Each scale

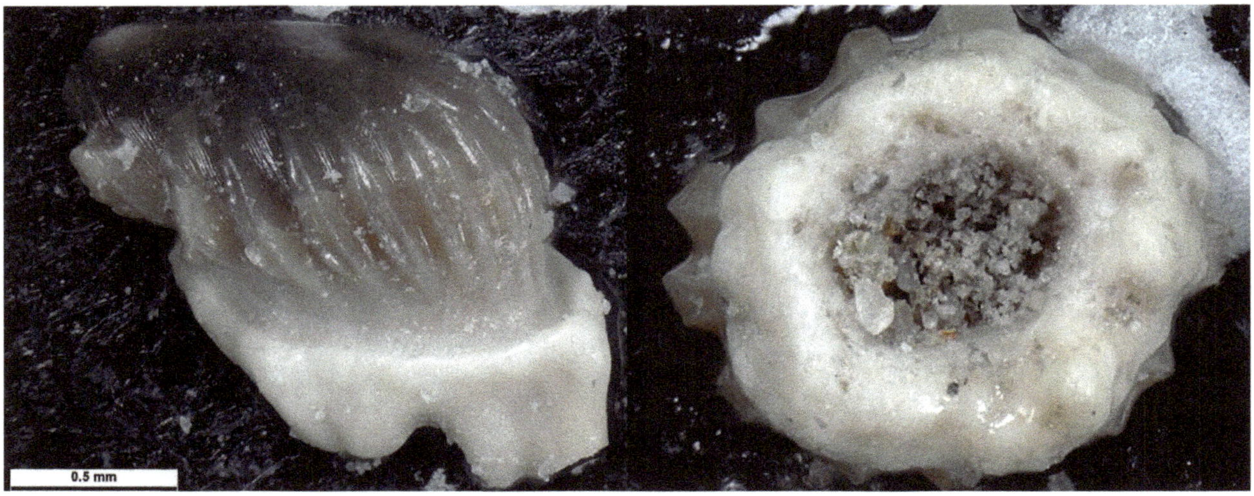

A single scale of the thelodont *Turinia antarctica* from the Devonian Period of Antarctica. Photo: Ken Walker.

had an elaborate crown of ornamented dentine, and an acellular bony base with a central pulp cavity, almost resembling a basic vertebrate tooth. The thelodont *Turinia antarctica* has been recorded from various levels at the base of the Aztec Siltstone succession.[8] The scales of *Turinia antarctica* most closely resemble the species *T. gondwana* from Bolivia and *T. hutkensis* from Iran in having a complex three-tiered crown, but other species recently recorded from Australia also share this characteristic.[9] The genus *Turinia* is also known from a complete specimen of *T. pagei* from the United Kingdom. It was most likely a scavenging fish that scooped up organic detritus from the bottom of the rivers and lakes in which it lived.

Sharks (*Chondrichthyes*)

Sharks have an enigmatic beginning in their fossil record. We have what resembles shark placoid scales dating back to the Late Ordovician Period, 460 million years ago, but no true sharks' teeth occur until the start of the Devonian Period. Some scientists have suggested that the earliest sharks may well have been toothless.

Because sharks are cartilaginous fish that lack true bone, their fossil record is known largely from their teeth, scales and spines. The oldest articulated fossil shark, *Doliodus problematicus*, from the Early Devonian of Canada was described only in 2004.[10] It has paired fin spines, a feature otherwise thought to typify the acanthodian fish. Complicating things further is that a fish from the Early Devonian of Britain, *Ptomacanthus*, previously regarded as a typical early 'acanthodian', has now been shown to be an early shark.[11] Both *Doliodus* and *Ptomacanthus* have distinctive shark-like teeth, each with a well-developed root and set in rows on the jaws. Such enigmatic forms are now regarded as 'stem' gnathostomes (jawed vertebrates). They reveal that the sequence of character acquisition leading to the modern sharks and bony fish was more complicated then previously thought.

Antarctica has a fascinating record of sharks because all fossilised sharks' teeth older than the Aztec Siltstone specimens tend to be much smaller in size and less diverse for any one particular fauna. Some of the Early Devonian sharks' teeth include the 1–2 mm bifurcated teeth of *Leonodus* from Spain, and the tiny, twin cuspid teeth of *Doliodus* (5–6 mm). The Aztec Siltstone sharks include at least four genera and include the first large sharks of the Devonian Period. *Portalodus*, for example, had teeth almost 20 mm in size, suggesting it had an overall length around two to three metres.

Fossil shark remains from the Devonian of Antarctica were first recognised from the samples collected in the mid-1950s by New Zealander Bernie Gunn and Guyon Warren. The samples were sent to Errol White of the British Museum (Natural History), London. White erected a new genus *Mcmurdodus* and gave the name *Mcmurdodus featherensis* for one peculiar little tooth, only 5 mm wide, which has a sideways directed series of triangular cusps. White suggested the tooth might be related to the extant notidanoid sharks, such as the Cow Shark *Notidanus* or Six-gilled Sharks *Hexanchus*. In the mid-1980s another tooth of *Mcmurdodus* turned up from Lower–Middle Devonian limestones in central Australia, reinforcing the Gondwanan affinity of this genus.[12] *Mcmurdodus* was probably only about 50–75 cm in length. Its teeth show adaptation for gripping, so it was most likely a predator that fed on other smaller fish.

Not all sharks from the Aztec Siltstone are known from just their teeth. The most complete specimen found so far is *Antarctilamna prisca*, known from a partially articulated specimen discovered by Gavin Young in the Lashly Mountains.[13] *Antarctilamna* is also known from the Bunga Beds in south-eastern Australia and is represented by almost complete braincases and jaw cartilages, fin spines and teeth identical with the Antarctic specimens.[14] Since its initial description in 1982, isolated teeth and fin spines of *Antarctilamna* have been identified from South Africa, Arabia, Iran, and from both North and South America.

A tooth of *Portalodus bradshawae*, a fossil shark from the Portal Mountains, Skelton Névé region, Antarctica. Photo: John Long.

Antarctilamna was a small shark that lived in the Middle Devonian rivers and lakes of Antarctica, Australia, South Africa and North America. Illustration: William Stout.

The origins of fish

The oldest nearly complete austral fossil fish found so far is *Arandaspis* from central Australia, dated from around 480 million years ago. It bore a thick, armour-plated body with headlight-like eyes pointing to the front of its face, and a pair of light sensitive 'pineal' eyes on top of its head. Its comb-like scales covered the tail which bore no median or paired fins, just a slightly expanded tail fin. *Arandaspis* and other similar armoured fish from this site show no evidence of an internally ossified skeleton.

Once bone had become an established essential of the vertebrate wardrobe, many different kinds of fishy fashion elaborately decorated the bony plates of early fish. The jawless fish (termed 'agnathans') evolved and flourished in the Silurian Period (435–410 million years ago). These included bizarre forms with weird bony head armours, such as the bizarre galeaspids from China and the pituriaspids from Australia. However, the most abundant groups, widespread in Europe, Russia and North America, were the heterostracans and the osteostracans. Heterostracans had an armour made up of many plates that fitted together to cover their head and trunk regions, but the plates did not have extensive overlap surfaces, as did the plates of the jawed placoderms. Another group of jawless fish found throughout the world were the thelodonts. They lacked elaborate bony plates but instead had a dermal skeleton made up of thousands of small scales, each having a simple to complex dentine crown and a bony base pierced by a conspicuous pulp cavity. Superficially these scales have resemblances to the basic vertebrate tooth structure, but they are derived from the ectoderm and are growing structures. Teeth are derived from the endoderm and do not grow after their formation in the dental lamina.

Today the only living representatives of a once widespread and diverse array of jawless fish are the lampreys and hagfish, both of which lack bone. Anaspids, another extinct group that lacked a thick bony skeleton, may be the closest fossil forms to lampreys.

Three other fossil sharks, known only from their teeth, occur in the Aztec Siltstone. These are *Portalodus*, *Aztecodus* and *Anareodus*.[15] *Portalodus*, the largest, has characteristic two-pronged teeth up to 20 mm in size, with one cusp only marginally larger than the other side. It was initially identified as a *Xenacanthus* species by Gavin Young in 1982 but subsequently named *Portalodus bradshawae* in honour of Margaret Bradshaw. *Aztecodus harmsenae*, named after Fraka Harmsen, has widely splayed cusps with a serrated area in the middle of them. Its root is broad and rectangular. The third fossil shark, *Anareodus statei*, named in honour of the Australian National Antarctic Research Expeditions (ANARE) and New Zealand field survival expert Brian Staite, has symmetrical teeth with one large cusp and one smaller cusp. All of these sharks occur together in the same levels of the Aztec Siltstone. Each of them was well adapted for gripping struggling prey. Recent research casts doubt on the validity of *Anareodus* as a separate genus as it could, alternatively, be a *Portalodus* tooth from a different jaw position.[16]

Margaret Bradshaw and Fraha Harmsen at work near Mt Portal, Skelton Névé region, in 1992. Photo: John Long.

Spiny sharks

The 'acanthodians' were a mixed group of early jawed fish which are loosely characterised by having paired spines in front of all their fins and were covered with tiny scales with strongly convex bases. Today they are widely regarded by most specialists in early vertebrate groups to be a paraphyletic collection of stem gnathostomes, more primitive than sharks and bony fish but more derived than the armoured placoderms. Three main groups of acanthodians are traditionally recognised: the ischnacanthiforms, which have strong jaw bones fused to the jaw cartilages, the filter-feeding acanthodiforms and the heavily armoured climatiforms. However, many other 'acanthodian' taxa show combinations of features that make them difficult to place in any of these groupings.

Acanthodians were first recognised in 1921 by Sir Arthur Smith Woodward who identified isolated spines collected in erratics from the Aztec Siltstone in the Mackay Glacier region. Smith Woodward erected

A mid-Devonian acanthodian fish spine from the Aztec Siltstone, Mt Crean, East Antarctica. Photo: John Long.

A set of jaws and a gill of an acanthodian fish from the Aztec Siltstone, Mt Ritchie, Skelton Névé region. Photo: John Long.

the new genus *Byssacanthoides* and named the find *Byssacanthoides debenhami* in honour of geologist Frank Debenham. The spines were robust with distinct thick ridges running down them.

In 1962 Errol White described further acanthodians from the Aztec Silstone. He recognised the larger more elaborately ornamented spines of *Gyracanthides*, a genus previously known only from Victoria, Australia. This was a large acanthodian, perhaps up to 60 cm in length, which has since been identified also from Africa and Iran. White also described a third acanthodian in the fauna, which he named *Antarctonchus*, and is defined by having up to 19 ribs along the spines.

In 1989 Gavin Young described an isolated cheek plate of a new acanthodian from the Aztec Siltstone as *Culmacanthus antarctica. Culmacanthus* is a genus described by John Long in 1983, based on complete specimens of a peculiar deep-bodied acanthodian found at Mt Howitt in Victoria, Australia. It was then recognised as a member of the diplacanthoid superfamily (a group typified by *Diplacanthus* from Europe), and placed in the new family Culmacanthidae. The discovery of both *Gyracanthides* and *Culmacanthus* from Antarctica was strong support for Gavin Young's earlier hypothesis of a distinct East Gondwanan faunal province.

A review of these identifications reinforced the taxonomic validity of the various species based on isolated spines, and also described a new genus based on the most complete fossil yet found of any Antarctic acanthodian.[17] The type specimen, a relatively complete fish, was collected in 1971 from the Portal Mountains by Gavin Young on the Victoria University Wellington Antarctic Expedition (VUWAE 13). He also found further isolated remains of it from Mt Crean. This new acanthodian,

A spine of the acanthodian *Gyracanthides* from the so-called 'Fish Hotel' site in the Cook Mountains, East Antarctica. 'Fish Hotel' was given its name as the 'beds were full'! Photo: John Long.

Part of the cheek plate of the acanthodian *Culmacanthus antarcticus* from the mid-Devonian Aztec Siltstone, Mt Ritchie, Skelton Névé region. Photo: John Long.

The original fossil of *Milesacanthus* from the Antarctic Devonian. Image courtesy John Long.

Milesacanthus antarctica, is a deep-bodied fish about 22 cm in length, with long, curved dorsal fin spines. It has four pairs of ornamented branchiostegal rays covering the gill chamber. Its spines, ornamented with up to six smooth ridges on each side, all have large pulp cavities inside them. It has approximately 450 diagonal rows of scales covering its body with about 150 scales in each row. The scales are flat crowned with 14–24 short ridges on top. Its small mouth is supported by a bony mandibular splint lacking teeth. It seems that *Milesacanthus* may have been a specialised browser of lake vegetation.

And the Aztec Siltstone still keeps giving. Well-preserved and diverse acanthodian and chondrichthyan remains in residues of calcareous grey siltstone from the upper Middle Devonian cropping out at Mt Crean, Lashly Mountains, include scales of an acritolepid *Pechoralepis juozasi,* a climatiid *Nostolepis* sp., the diplacanthid *Milesacanthus antarctica,* and an indeterminate acanthodiform.[18] Interestingly this assemblage is comparable to those acanthodians from central Iran. The findings also include rare chondrichthyan teeth of *Aztecodus harmsenae, 'Anareodus' statei* and *Antarctilamna* sp. There are also ctenacanthoid-type scales and branchial denticles which may belong to *Antarctilamna,* and further scales of a yet-to-be identified chondrichthyan.[19]

Armoured fish

Armoured placoderms were the dominant group of fish throughout the Devonian Period and constitute the most diverse group of fish in the Aztec Siltstone fauna. Placoderms are characterised by having a head shield and a trunk shield made up of overlapping plates of bone. The shields may articulate with a ball-and-socket joint, as in the dominant group of placoderms, the arthrodires (meaning 'jointed neck'). We now know that some of these fish reproduced by internal fertilisation, as modern sharks do.

The Portal Mountains looking north. The Aztec Siltstone rocks are topped by Permian Weller Coal Measures. Photo: John Long.

Arthrodires reached an acme of evolutionary diversification in the Devonian and some forms grew to enormous sizes, such as the eight-metre-long *Titanichthys* of Morocco. There are seven main orders of placoderms, each with its own peculiar array of plate patterns. The most common in Antarctica are the antiarchs, a group having a long trunk shield, a head shield with a single ovoid opening for eyes and nostrils, and peculiar bone-enclosed pectoral fins. The most advanced antiarchs are the bothriolepidids, and the most common genus in that group, *Bothriolepis*, is well represented in the Antarctic fauna by at least 10 species.

Bothriolepis was first recognised from Antarctica in 1921 by Arthur Smith-Woodward, who erected the new species *Bothriolepis antarctica* based on some fragments of armour collected on Scott's 1912 expedition from the Mackay Glacier region. Much later, extensive collecting by VUWAE 13 scientists enabled Gavin Young to revise and describe new species of *Bothriolepis* from several localities throughout the Skelton Névé region. Young erected nine new species of *Bothriolepis* from Antarctica, including one large form, *Bothriolepis mawsoni*, which may have reached nearly 80 cm in length.[20] Identification of various species of *Bothriolepis* enabled Young to form a biostratigraphic framework for subdividing parts of the Aztec Siltstone. Species such as *Bothriolepis askinae* occurred near the base of the Aztec Siltstone, followed by a fauna dominated by *B. kohni*, *B. portalensis* and *B. karawaka* near the top of the succession. Above these *Bothriolepis* fauna, he identified another antiarch which had otherwise only been found in south-eastern Australia, a genus he had earlier described as '*Pambulaspis*'. In a revision of the strange antiarch faunas occurring in Australia, Antarctica and South America, Gavin Young referred the Antarctic species of '*Pambulaspis*' to the new genus, *Venezuelepis*. This reflects the close faunal affinities of the placoderms of Antarctica with those of other distant Gondwanan regions, like the northern margin of South America.

A detailed analysis of the anatomical characters of these *Bothriolepis* species revealed that they shared close affinities mostly to the *Bothriolepis* species from south-eastern Australia (*B. cullodenensis*, *B. gippslandiensis*, *B. fergusoni*) and to one species in southern China (*B. shaokuanensis*). *Bothriolepis* appears much earlier in China, Australia and Antarctica (Middle Devonian) than for Laurasia or Siberian terranes where it is exclusively Late Devonian. The positioning of the south China Terrane close to East Gondwana in the Late Devonian was suggested from various lines of fossil evidence, so it is highly possible that some species interchange may have taken place by the late Middle Devonian via shallow seaways.[21] Although *Bothriolepis* fossils are largely found in freshwater deposits, a few species have definite marine habitats, as in the *Bothriolepis* species from the Gogo Formation of Western Australia and from marine limestones at Kokenhusen in Latvia.

It is highly likely that fish like *Bothriolepis* had marine larval stages that enabled efficient dispersal of the species, as otherwise a rapid global distribution of the genus would be difficult to explain. One needs to remember that fossils tell us only where species died, not where or how it lived most of its life. For example, the salmon today lives much of its life in the open sea but goes upstream to breed and die, ultimately leaving its skeletal remains in a freshwater environment.

A reconstruction of the armoured placoderm *Bothriolepis askinae*. Illustration: William Stout.

A well-preserved head shield from a skull of *Bothriolepis* from the Aztec Siltstone of Mt Metschel, Skelton Névé region, East Antarctica. Photo: John Long.

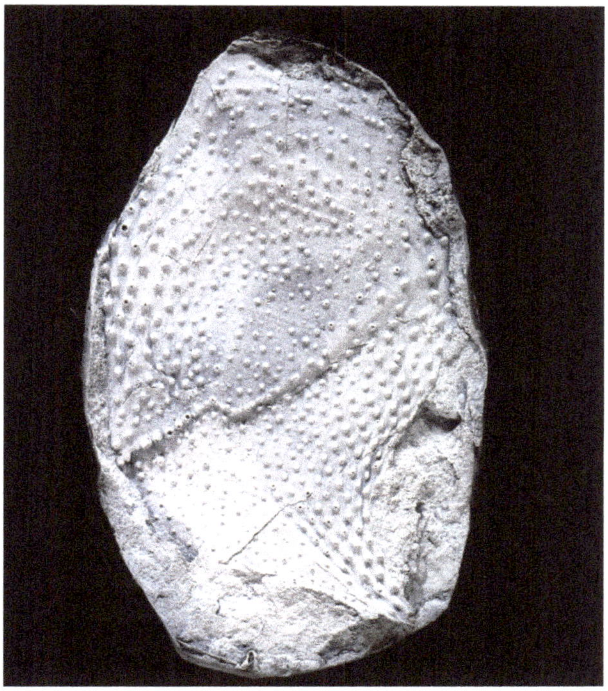

The placoderm plate of an undescribed arthrodire from the 'Fish Hotel' site, Cook Mountains. Photo: John Long.

The median dorsal plate impression of the mid-Devonian fish *Groenlandaspis*, from the Aztec Siltstone of Alligator Ridge, Skelton Névé region. Photo: John Long.

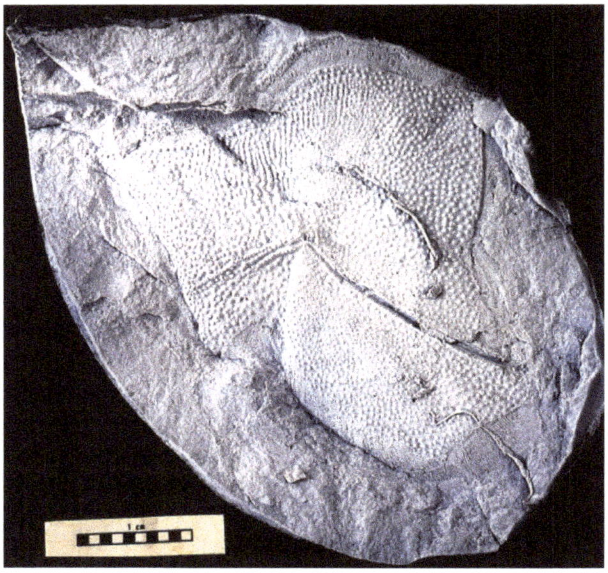

A trunk shield plate of the arthrodire placoderm *Boomeraspis goujeti* from the Boomerang Range, Skelton Névé region. Photo: John Long.

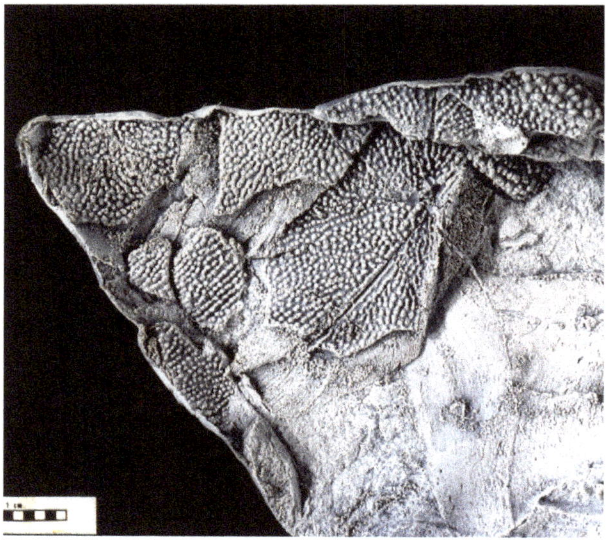

Head shield of groenlandaspid placoderm from the mid-Devonian Aztec Siltstone, Mt Crean, East Antarctica. Photo: John Long.

The median dorsal plate of the placoderm *Groenlandaspis* from the 'Fish Hotel site', Cook Mountains. Photo: John Long.

Another common placoderm from the Aztec Siltstone is the arthrodire *Groenlandaspis*, known from one named species *G. antarcticus*, plus several other new species which are still being studied.[22] Like *Bothriolepis*, *Groenlandaspis* had a widespread distribution in the latest Devonian of the northern hemisphere lands from where it derived its name – meaning 'shield from Greenland'. In 1975 Alex Ritchie first recognised this form from Antarctica and also noted its first occurrence in Australia. We now know that *Groenlandaspis* occurs in older Middle Devonian strata in many Gondwanan regions.

In 1995 John Long described a new primitive member of the groelandaspid family, *Boomeraspis goujeti*, from isolated plates he had collected near the top of Alligator Peak in the Boomerang Range in late 1991. *Boomeraspis*, like the primitive genus *Mulgaspis*, shows that the groenlandaspid family is represented by several primitive forms in East Gondwana.[23]

In 1962 Errol White from the Natural History Museum of London described an unusual arthrodire which he named *Antarctaspis mcmurdoensis*, based on a partial head shield which had been found in the upper section of Aztec Siltstone in the Lashley Mountains. For many years this genus has been thought by placoderm specialist to be related to an enigmatic group of flattended placoderms called phyllolepids (meaning 'leaf scale'). Phyllolepids were known largely from isolated plates having distinctive concentric patterns of ornament and one articulated species from Scotland. Like *Groenlandaspis* they characterised the uppermost part of the Late Devonian wherever they were found – Scotland, Belgium, Russia and North America. The discovery of this group in Australia in 1931 suggested to Australian geologist Edwin Sherbon Hills that the strata in which the fossil occurred must therefore be Late Devonian age. Since then, new genera of phyllolepids have been described from Australia, including *Austrophyllolepis* and *Placolepis*, both occurring in much older rocks than does the northern hemisphere *Phyllolepis*.[24] Recently new forms of arthrodires similar to *Antarctaspis* have been described from much older rocks in central Australia, and the affinity to this strange genus is no longer resting with the phyllolepids, but as a distinct group of endemic arthrodires known only from east Gondwana.[25]

The phyllolepids *Austrophyllolepis* and *Placolepis* have both been identified from isolated plates in the Aztec Siltstone fauna. At first it was thought that phyllolepids occurred only in the uppermost layers of the Aztec Siltstone but more recent finds in some of the older layers of the Aztec Siltstone have revealed two new species of these Australian genera.[26]

The head shield plate of the phyllolepid *Placolepis quiltyi*, from the 'Fish Hotel' site, Cook Mountains. Photo: John Long.

The diversification of jawed fish

The first jawed fish were probably the sharks, whose fossil record can be traced back to the Late Ordovician, around 450 million years ago. Because sharks have a largely cartilaginous skeleton, they are rarely found as complete fossils, and most of their fossil record is based upon the study of their teeth, spines and scales.

As in modern sharks, fossil sharks had multiple rows of teeth which were shed throughout their life, so their teeth are abundant as fossils in the sedimentary record. Recognisable sharks' teeth do not appear in the fossil record until much later in the beginning of the Devonian Period, so it is possible that the first sharks may have been toothless.[27]

By the Late Devonian many kinds of sharks had evolved but most were small (under a metre) and there are very few species found together at any single site. The best-preserved, most complete fossil sharks of this age are those from the Upper Devonian Cleveland Shale of Ohio, such as *Cladoselache*, whose overall shape and fin spines show a general similarity to the modern Mandarin dogfish, *Cirrhigaleus barbifer*. Throughout the entire evolution of sharks their major body plan only diverged twice from this standard pattern – once to become the holocephalans (e.g. modern chimaera or rabbit fish) and the second time to develop into the flattened rays.

Another group of primitive jawed fish that first appears in the early Silurian is the placoderm group. Placoderms are shark-like fish in their overall body shape, with broad pectoral fins and a mostly cartilaginous internal skeleton, but differ noticeably from sharks in having evolved cellular bony plates that encircled the head and trunk region. Placoderms also have a toothed palate bone (the parasphenoid) which is otherwise only found in the osteichthyans – the bony fish.

The most diverse group of placoderms were the arthrodires, which mostly had a well-developed neck joint with a rounded knobby process on the trunk, and a concave socket on the back of the head. The arthrodires had two sets of upper jaw tooth plates (called superognathal bones) and were the largest of all placoderms, some reaching an estimated eight metres in length.

The acanthodians are another group of Paleozoic fish that appeared in the beginning of the Silurian Period. The group is still relatively poorly known – their fossil remains are largely represented by spines and scales. Acanthodians were once characterised by having paired fin spines and a fine micrometric scale cover. The best-known member of the group is *Acanthodes*, the last and perhaps one of the most specialised members of the entire group. The group reached a peak of diversity in the Middle to Late Devonian and was extinct by the close of the Paleozoic Era, some 250 million years ago.

The first bony fish, the osteichthyans, have an obscure early fossil record, known only from isolated scales dating back to the Late Silurian, around 415 million years ago. These simple rhombic scales have a peg-and-socket type of articulation and belong to lophosteiform fish. Shortly after the appearance of these scales, we find remains of other primitive bony fish whose bones are covered by a unique shiny tissue called cosmine. This tissue is usually linked with the so-called 'sarcopterygian' or fleshy finned fish. Today the Sarcopterygii is represented by the three living lungfish and one coelacanth.

A reconstruction from recovered skeletal remains of the mid-Devonian Antarctic fish *Donnrosenia*. Illustration: Brian Choo.

Bony fish

The osteichthyans are the major group of fish alive today. They are characterised by having jaws and palates which are lined with many additional tooth-bearing bones, a skull with numerous bones forming a set basic pattern and, in later forms, a well ossified internal skeleton. They comprise two major groups, the ray-finned fish (Actinopterygii) and the lobe-finned fish (Sarcopterygii). The Actinopterygii today comprise over 29 000 species, including commonly known species like the trout, salmon, swordfish and goldfish.

Only one Devonian actinopterygian, *Donnrosenia schaefferi*, has been described from Antarctica, from the Aztec Siltstone fauna.[28] Basically trout-like in appearance, *Donnrosenia* grew to about 25 cm in length and swam in the large rivers and lakes of Middle Devonian Antarctica, hunting smaller fish and invertebrates, probably near the top of the water column. With its torpedo-like body shape, it was most likely the fastest of its time. *Donnrosenia* was an early form, and has a close affinity with *Howqualepis*, of similar age, from Mt Howitt in Victoria, Australia.

The original specimen skull of *Donnrosenia*. Photo: John Long.

The Sarcopterygii have few living representatives but include the lungfish (Dipnoi) and the coelacanth *Latimeria* (Coelacanthiformes). These are the living remnants of the once great diversity of extinct groups once collectively known as 'crossopterygians' and including the 'rhipidistians'.

The lungfish today live in freshwater habitats and have the ability to breathe air. Some forms, like the South African and South American genera *Proptopterus* and *Lepidosiren* can aestivate; that is, they bury themselves in a mucous-lined burrow and wait out the dry season for the next year's rains to

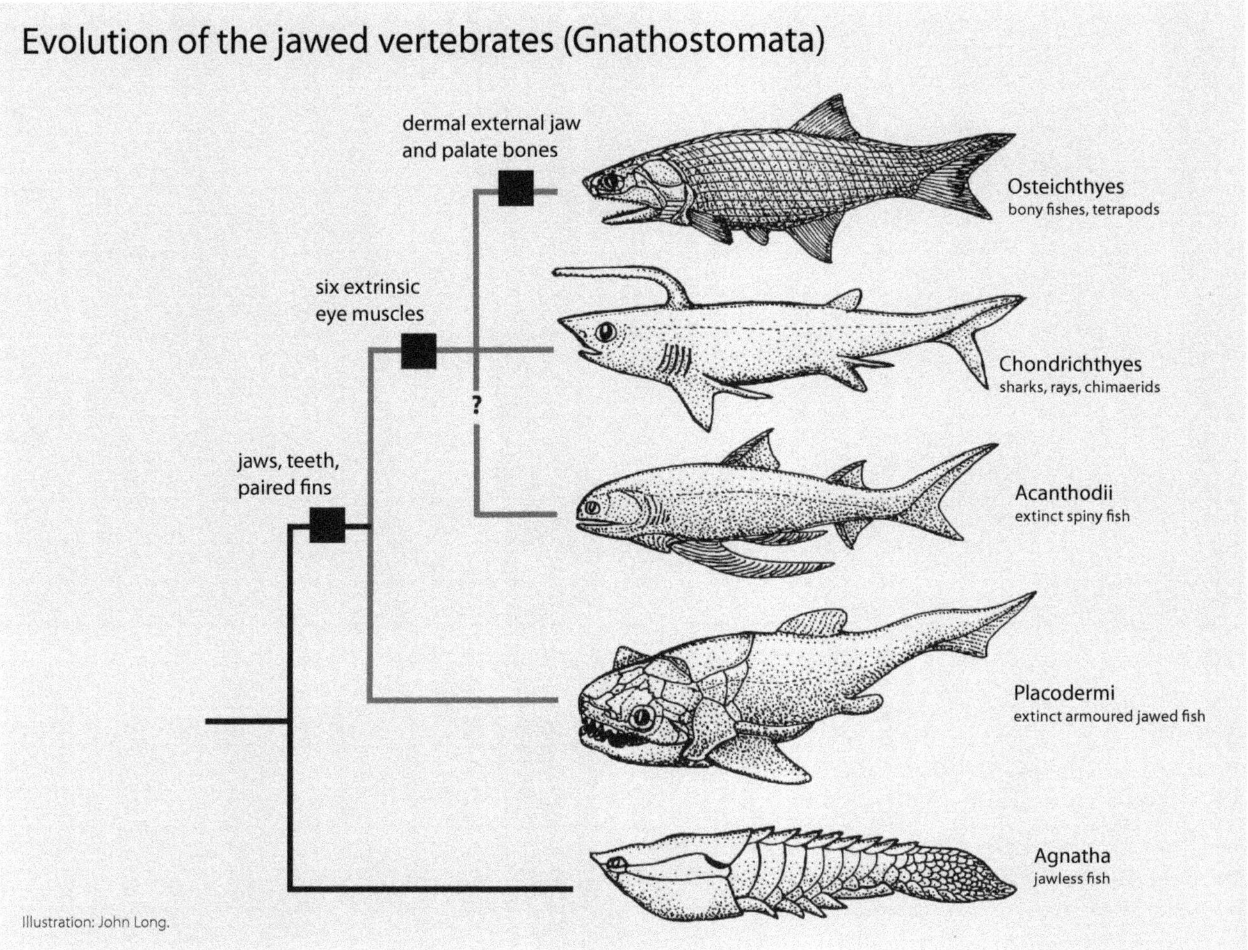

Illustration: John Long.

come. All living lungfish have hard tooth plates for grinding up or crushing food, but older fossil species show a great range of dentitions adapted for rasping, triturating or predation.

Lungfish are poorly represented in the Aztec Siltstone fauna; their only fossil remains are isolated tooth plates, which demonstrate that at least three forms were present. A peculiar tooth plate with lots of sharp tooth ridges having many pointed cusps has been found at Mt Crean. It resembles the tooth plates commonly seen in Early Carboniferous members of the Family Ctenodontidae, and could possibly represent an early member of that lineage. Isolated tooth plates resembling those of the well known European and American genus *Dipterus* have been found from Mt Ritchie and Mt Gudmundson. A few isolated tooth plates from the latter site appear to be the same as the genus *Howidipterus*, also known from the Middle Devonian Mt Howitt site in Victoria, Australia.

The sarcopterygians also include a paraphyletic mixed group called the 'Osteolepiformes'. These had a shining surface tissue called cosmine covering the bones (as did most sarcopterygians) and thus their bones and scales are easily recognisable in the field. One species, now named *Koharolepis jarviki* is known from a complete skull, and the front half of the shoulder girdle and pectoral fin, collected on Mt Crean.[29] *Koharolepis* was a flat-headed predator with very small eyes that would have grown to at least a metre in length. Its shiny, cosmine-covered bones and broad skull indicate its primitive position within the osteolepiform family tree.

In 1973 a similar broad-headed fish, *Canowindra grossi*, was described from Canowindra in New South Wales. It was an enigmatic fish which didn't fit easily into any of the known early sarcopterygian groups and was considered to be an aberrant but primitive osteolepiform. In 1987, *Beelarongia,* another of these fish, also with cosmine scales and bones, and similar in skull form to *Canowindra,* was described. Clearly a pattern was emerging of a distinctly endemic family of east Gondwanan osteolepiform fish. In 1992, all three of these enigmatic fish, *Koharolepis, Canowindra* and *Beelarongia* were placed in a new family Canowindridae. Most analyses of the osteolepiform fish place this group at the base of the osteolepiform radiation, indicating that East Gondwana could have been home to one of the first major evolutionary events for the lineage leading to the first tetrapods, the amphibians. *Koharolepis* is still the most complete skull of the group, and is the only specimen known from actual, well-preserved bone, the others being studied from natural moulds in rock.

Other members of the 'osteolepiform' group described from the Aztec Siltstone are represented only by their snouts: *Vorobjevia, Mahalalepis* and

A lungfish tooth plate discovered by Alex Ritchie in 1971 from the mid-Devonian Aztec Siltstone of Mt Ritchie, Skelton Névé region. Photo: John Long.

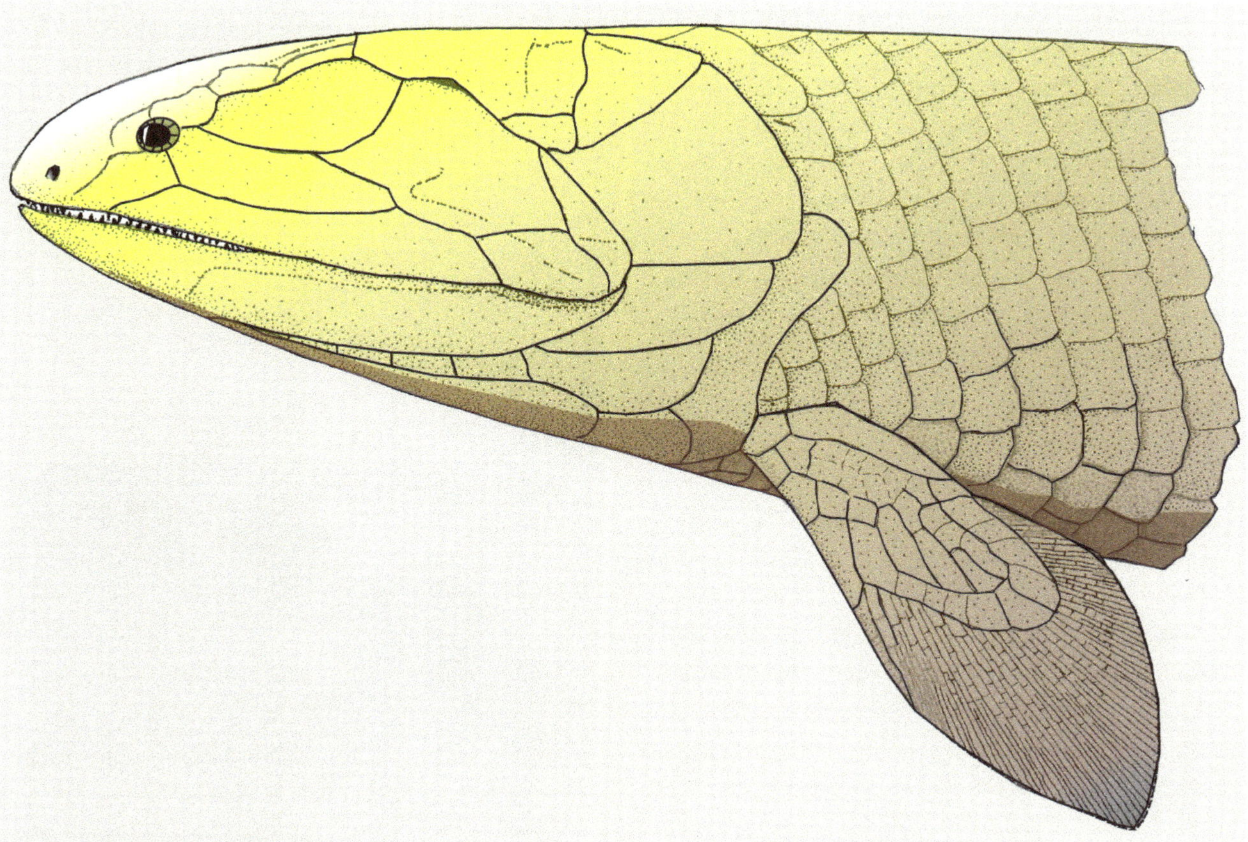

The osteolepiform fish *Koharolepis.* Illustration: John Long.

Platyethmoidea. One of these, *Mahalalepis* can be placed within the family Megalichthyidae, a group having northern hemisphere affinities. Megalichthyids are well known from the Carboniferous deposits of Scotland, Europe and North America but their occurrence in older Devonian rocks is known only from Gondwana. This suggests the group most likely arose in Gondwana before migrating later via shallow seaways to the northern hemisphere continents, further reinforcing the significance of the Antarctic fish fossils to understanding the bigger picture of global migration and evolution of fish through time.

In 1971 a particularly large sarcopterygian predator, collected by Alex Ritchie and Gavin Young from near the top of Mt Ritchie, was named *Notorhizodon mackelveyi* and assigned to a group of fish

Devonian mass extinctions

Towards the end of the Devonian Period, Earth suffered a series of massive meteorite impacts which led to a series of extinctions known as the Kellwasser Events. Around 368 million years ago, a monster rock from space struck at Siljan, Sweden, forming a crater 53 km in diameter. Further impacts resulted in the 15-km-wide Kaluga Crater in Russia, and the Charlevoix Crater, some 54 km wide, in Canada. Whatever triggered the extinctions, they are thought to have been caused by either global oceanic anoxia or rapid global cooling. Anoxia is the loss of oxygen from the bottom of the sea, which combined with receding sea levels, causes the upwelling of toxic bottom waters into shallow sea basins. This kills off all of the creatures that are burrowers or are sedentary feeders, and thus affects the overall food chain for the whole marine environment. A similar devastating effect can also be caused by rapid cooling of the seas, affecting the livelihood of all species adapted to warm water.

Organisms living in the warm Devonian seas were hit the hardest, and the ancient reefs were the biggest victims.[37] Up to 90 per cent of the known phytoplankton in the world oceans was suddenly wiped out at this time. One of the major groups of animals that succumbed to this event was the homoctenids – small cone-shaped organisms whose planktonic life was similar to modern pteropod snails. Tabulate corals and layered stromatoporoids also died out, along with 75 per cent of the known brachiopod species. Over half the known species of placoderm fish died out, and the rest would linger only until the close of the Devonian Period, some eight million years later. Those fish that survived this killer event did so mainly by moving into freshwater habitats, getting away from the deadly seas.

The end of the Devonian Period saw another mass extinction of smaller proportion when many higher groups of vertebrates died out. The worldwide extinction of all placoderms was the most dramatic of these events, but this time also marks the end of other higher orders of fish such as the heterostracans, the porolepiform sarcopterygians, and the ischnacanthid acanthodians. There is no clear evidence for cosmic impact or climatic change as trends of increasing oxygen and the greenhouse effect building in the Late Devonian continued into the Early Carboniferous. The extinction of these groups of fish could be due to ecological replacement by the rapid diversification of ray-finned fish and sharks at this time.

A journey to the Portal Mountains in search of Devonian fishes.
Photo: John Long.

The lower jaw of the large tristichopterid fish *Notorhizodon mackelveyi* from mid-Devonian Aztec Siltstone, Mt Ritchie, Skelton Névé.
Photo: John Long.

Part of a skeleton of a rhizodont fish called *Aztecia australis,* from East Antarctica, collected by Alex Ritchie and Gavin Young in 1971.
Photo: John Long.

called the Rhizodontiformes. The members of this group are well known as gigantic predators which inhabited the Carboniferous coal swamps of Europe, Some forms like *Rhizodus* reached possibly six metres in length. *Notorhizodon* was only incompletely known, but its huge lower jaws and palate lined with teeth nearly 4 cm high clearly identified this fish as the biggest predator in the Aztec Siltstone ecosystem, possibly reaching close to three metres in length.

The original description of *Notorhizodon* was based on two separate sets of material: one comprising a large partial skull from the top of Mt Ritchie and some associated lower jaw and shoulder girdle material collected by Alex Ritchie and Gavin Young from Mt Warren. In 2001, a revision of the material by Zerina Johanson and Per Ahlberg revealed that the large skull from Mt Ritchie is probably not a rhizodont but belongs with another more specialised group of osteolepiform fish called the tristichopterids. Tristichopterids were narrow-headed predators with elongate bodies, much like the pike of the Devonian. As the oldest members of this family also occur in the northern continent of Laurasia, it is not possible to determine where the group may have first evolved. The fact that the initial radiation of osteolepiforms appeared to have taken place in East Gondwana would favour the scenario that tristichopterids also arose here, then rapidly dispersed to Laurasia by the end of the Middle Devonian. This hypothesis is in accord with the fact that the oldest member of the group from Laurasia is *Tristichopterus* from the late Middle Devonian of Scotland.

The scant material orginally assigned to *Notorhizodon* from Mt Warren was found to belong to a true rhizodontid which has been named *Aztecia mahalaa.*[30] *Aztecia* is regarded as a very primitive member of the rhizodontids, which appears to be another major group of sarcopterygians that evolved in East Gondwana and later radiated into northern hemisphere environments.

The Aztec Siltstone fish fauna is thus seen to be intrinsic to understanding the newly emerging patterns of evolutionary radiations in Devonian fish. It contains some highly endemic genera which are otherwise only known from south-eastern Australia. These include *Culmacanthus, Venezuelepis, Austrophyllolepis* and possibly the dipnoan *Howidipterus.* Intercontinental correlations of these taxa between Australia and Antarctica, and between other regions of Gondwana, are useful when testing conflicting models of palaeogeographic reconstructions.

Devonian plants

Devonian plants were first collected from Antarctica from the Beacon Heights Orthoquartzite unit which underlies the Aztec Siltstone. The hard stems and wood of lycopods are preserved as natural impressions in the clean white sandstone.[31] These

A reconstruction of the acanthodian fish *Culmacanthus.* Illustration: William Stout.

Intrusions of the Jurassic Ferrar Dolerite (the dark rock) into Beacon Supergroup rocks in the Transantarctic Mountains.

Photo: Jeffrey Stilwell.

The impression of an early plant called a lycopod from the Beacon Heights Orthoquartzite, Antarctic Devonian.

Photo: John Long.

Psilophyte plants from the Beacon Heights Orthoquartzite, Middle Devonian, Gorgon's Head, Cook Mountains, East Antarctica.

Photo: John Long.

The plant stems of a mid-Devonian psilophyte. Photo: John Long.

include *Haplostigma irregulare* and a new species of *Drepanophycus* from erratic blocks of dark argillite near Mt Hartkopf (Milan Rock), near the head of the Laird Glacier, Marie Byrd Land.[32]

Palynomorphs, or microfossils of pollen and spores from plants, can be extracted from sedimentary rocks by dissolving them in hydrofluoric acid and then sifting through the insoluble residues. Palynomorphs have been identified from the Beacon Supergroup and the Lower Devonian Horlick Formation of the Ohio Ranges.[33] The only macroscopic plant fossils of this age from Antarctica are some stems of the psilophyte *Hostinella*. This indicates that the Early Devonian flora of Antarctica appeared to be similar to other simple land plant communities elsewhere at this time.

Devonian plants have also been collected from layers of siltstones on the northern slope of Mt Black Prince, Northern Victoria Land, at an elevation of approximately 3050 metres. The specimens were first thought to be the stems of *Lepidodendropsis pulchra*. Although the preservation of the specimens is poor, the identification is consistent with a Middle Devonian age assignment.[34]

Further fossil plants have been found in dark carbonaceous shaley rocks near Mt Hartkopf in Marie Byrd Land. The flora included the psilophtye *Drepanophycus schopfi*, the lycopsid *Haplostigma irregularis*, a possible *Protolepidodendron* stem and indeterminate psilophyte axes.[35]

Most recently a small collection of plants from the Beacon Heights Orthoquartzite and the Aztec Siltstone was found to include the lycopsids *Haplostigma lineare*, *Malanzania* sp. and *Archaeosigillaria* sp. cf. *A. caespitosa*. The slightly younger Aztec Siltstone flora also contains *Haplostigma lineare* and the stem-like psilophyte *Praeramunculus alternatiramus*.[36]

The picture we get of the late Middle Devonian forest developing on Antarctica at this time is thus characterised by low, sprawling herbaceous psilophytes and low shrubby lycopsids, with occasional medium-sized arborescent forms developing. There are no large stems found from either the Aztec Siltstone or Beacon Heights Orthoquartzite to indicate that larger trees, such as those that would typify the Late Devonian forests, had yet existed.

OPPOSITE: The floor of the Dry Valleys in East Antarctica is choked with large boulders from the majestic cliffs composed of Beacon Sandstone and Ferrar Dolerite.
Photo: Jeffrey Stilwell.

6
A TIME OF SOUTHERN COAL SWAMPS

With all the luxuriance of the foliage, and the denseness and stature of the trees which overspread the great lagoons of the Carboniferous period, the general effect must have been sad and sombre in the extreme. But it persisted through long ages in unspeakable loneliness and silence, echoing neither voice nor sound, except when some giant of the forest snapped in twain and fell heavily into the arms of its companions.

E Hull, *The Coalfields of Great Britain*, 1881.

Towards the end of the Carboniferous Period, around 300 million years ago, a global event occurred that would have devastating effect on the world's climate. Gondwana had collided with Laurasia in the north to form a single supercontinent, Pangaea (meaning 'all of Earth'). Antarctica was positioned on the south-eastern margin, its centre at dead south over the pole. Many of the continents of today were joined to form Pangaea which stretched from the northern to the southern pole. Most of the rest of the surface area of Earth was occupied by a corresponding single ocean, known as Panthalassa ('Goddess of the Sea'), with a smaller sea to the east of Pangaea known as Tethys.

Upper Paleozoic and Lower Mesozoic rocks in the eastern Transantarctic Mounatins were deposited during Gondwana's drift across the South Pole, and during a transition from 'icehouse' to 'greenhouse' conditions following the cessation of Late Paleozoic glaciation. These regions were located higher than 75°S from 320 to 210 million years ago. The Southern Victoria Land and Darwin Mountains sedimentary successions record a change from Lower Permian glacial deposits through to Permian fluvial coal measures. Present climatic simulations suggest seasonal climatic extremes within Pangaea's polar interior. The assemblage and drift of Pangaea resulted in heightened mountain-building activity and associated development of numerous depositional basins. One of the largest basins, more than 10 000 km long, was known as the 'Gondwanide foredeep'. It extended across southern South America, South Africa, the Falkland Islands, Antarctica and Australia.

The ice age that had grown swiftly from the Late Carboniferous embraced most of Earth and reached its peak, causing a major lowering of sea levels. The northwards rotation of the supercontinent would have formed a climatic regime favourable to monsoon development. The interior of northern Pangaea at this time was relatively dry, typical of full monsoonal conditions. Gondwana would remain subsumed within Pangaea for another 140 million years before breaking away.

The frigid Carboniferous Period left no record of sedimentation in Antarctica, but there is a very good record of Permian rocks left by the retreating glaciers. These formed large accumulations of tillites, with interspersed fluvial and lacustrine deposits, many of which contain excellent fossils. Ancient soil deposits in both Antarctica and Australia show that in high latitude regions (70°–80° South) the soils that developed were typical of equatorial to temperate regions today (25–38°C). The geology of Antarctica plays a pivotal role in understanding this great shift in global climate.

Life in and around the lakes

There are very few good animal fossils known from the Carboniferous and Permian deposits of Antarctica, but there are many well-preserved fossil plant sites, especially from the Permian. Most of the Permian sediments at this time were deposited by melting glaciers or from rivers fuelled by glacial meltwater.

There is evidence of mid-Carboniferous to earliest Permian life in the Mt King Formation of Alexander Island, which has a diverse assemblage of bryozoans, brachiopods, crinoids and molluscs. However, in all likelihood, these fossils are from a so-called 'exotic' terrane, and their origin is uncertain.

Vertical animal burrows, known as *Skolithos*, have been found in the lower part of the Feather Conglomerate near Mt Bastion in Southern Victoria Land. These burrows range in size from a diameter of 5–20 mm and a length of 5–30 cm. Some of them show the presence of circular mounds at their tops. These Permian trace fossils are believed to have developed in a fluvial environment, for example, an ancient river flood plain.[1]

A nicely preserved *Glossopteris* seed fern leaf from the Permian of East Antarctica. Photo: David Cantrill.

The Mt Glossopteris Formation, which outcrops in the Ohio Range, has produced non-marine invertebrate fossils of Permian age, first collected in the early 1960s on Mt Schopf by Ohio University field parties. They found clams (bivalves), early crustaceans (conchostracans) and an insect, as well as trace fossils.[2] The fauna includes the conchostracans *Leaia gondwanella* and *Cyzicus (Liostheria) doumanii* from about 570 metres above the base of the Mercer Ridge Formation. They also recorded ostracods which have yet to be described. There are remains of the insect *Uralonympha schopfi* on the same bedding plane as the conchostracan *Leaia* along with fragments of the plant *Glossopteris*, indicating a depositional environment that was probably a flood plain swamp setting.[3]

Permian bivalves known as *Palaeanodonta fisheri* have also been recorded from the Ohio Range.[4] These are small clams up to 18 mm long and 9 mm wide, which lived in small ponds. Trace fossils called *Conchlichnus antarcticus* have also been identified from two horizons, one at the base of the formation, and also from a layer 458 metres above the base.[5] Early Permian body fossils of a clawed crustacean were discovered in the Pagoda Formation outcropping at Mt Butters in the Shackleton Glacier area. These represent the oldest decapod fossils found in Antarctica, and prove that crayfish had begun to radiate into freshwater habitats by the Early Permian.

A *Glossopteris* leaf and associated scale from the Mt Glossopteris Formation. Photo: David Cantrill.

Early studies of Permian wood

Fossilised wood collected by Scott's 1901–04 expedition was initially described by Arber in 1907 but later re-described by Edwards (1928), placing the wood into a form genus he named *Dadoxylon*. Halle (1911) described some fossil wood fragments from the Falkland Islands and named two new species, *Dadoxylon lafoniense* and *D.* cf. *angustum*. In 1914 British palaeontologist A C Seward described a poorly preserved specimen from an erratic boulder near the Priestly Glacier in Northern Victoria Land collected by men of Scott's Terra Nova Expedition in 1910. Seward called this wood '*Antarcticoxylon*' although, as in the case of *Dadoxylon*, the lack of diagnostic features also casts doubt on the validity of the genus.[6]

A comprehensive study of Antarctic Permian wood has identified three species of *Araucarioxylon*, including two indeterminate forms from Mt Schopf (Ohio Range), and one species from Southern Victoria Land, *A. allanii; Dadoxylon weaverense* from the Queen Maud Mountains; *Protophyllocladoxylon dolianitii, Damudoxylon, Megasporoxylon antarcticum, M. canalosum,* and *Polysolenoxylon kraeuselii* from the Ohio Range, Horlick Mountains; and a possible occurrence of *Antarcticoxylon* from the Mt Nansen Massiff, Northern Victoria Land.[7]

Further detailed studies of Antarctic fossil woods shifted emphasis onto what the wood could tell us about ancient climates. The growth rings, up to 8 mm wide, found in *Araucarioxylon* from the central Transantarctic Mountains, combined with other data

The end Permian mass extinction

Around 250 million years ago, the large scale glaciation that covered much of the southern continents began to retreat. The end of the Permian witnessed by far the greatest extinction event life on Earth has ever endured. More than half of all animal families in both marine and terrestrial habitats disappeared. Sponge reefs were completely wiped out and diverse marine assemblages were replaced with simple communities dominated by just a few species of bivalves and brachiopods. On land the insects took a devastating blow and almost never recovered. The widespread *Glossopteris* forests that dominated Gondwana were also doomed at this time. Several groups of mammal-like reptiles died out but the surviving species flourished and radiated after the event. One of these, the pig-like dicynodont *Lystrosaurus,* spread rapidly throughout Pangaea.

Upper Permian rocks contain a huge diversity of fossil shells and skeletons, showing that the seas teemed with life. Lowermost Triassic rocks, in contrast, mostly lack burrows, and marine benthic invertebrates are extremely rare – suggesting a dramatic change in oceanic conditions from well-oxygenated bottom waters to widespread anoxia. A key question of great concern for biologists and palaeontologists is: What possible combination of environmental changes could possibly have had such a devastating effect?

Sometime before 251 million years ago massive amounts of molten rock spewed forth in Siberia – a volcanic event that might have been related to weakening of Earth's crust from a massive impact by an asteroid or large meteorite. The eruptions continued over a period of about one million years and basalt lava covered 1.6 million km^2 of eastern Russia to a depth of 400–3000 metres. The effects of such volcanism of this scale can be two-fold. Huge volumes of greenhouse gases can bring about a warming of the lower atmosphere, but the injection of aerosols into the upper atmosphere can destroy ozone and bring about a cooling effect. Whatever caused the Permian extinction, global climates at this time drastically altered from humid, cool temperatures to warmer, drier conditions. Moreover, the initial global warming induced by the Siberian eruptions melted frozen gas hydrates and massive volumes of methane rose to the surface of the oceans in huge bubbles (called the 'Big Burp'). This, in turn, created a 'runaway greenhouse' phenomenon that catastrophically increased global temperatures by some 6°C.

How bad was the extinction? Really bad! Estimates indicate that only 10 per cent of all species living survived the end Permian mass extinction, compared with 23-40 per cent of species at the Cretaceous–Tertiary boundary 65 million years ago (see page 132). At the close of the Permian, some 75 per cent of all amphibian families, and 80 per cent of all reptile families were wiped out. With such a small survivorship, great branches of the tree of life were chopped off – many groups of animals, plants, and micro-organisms were gone forever and there were few founders available to repopulate the biosphere. It took another 100 million years for global diversity at a family level to return to pre-extinction levels. Ecological recovery appeared to be quicker within reefs, which bounced back by the Middle Triassic, some 10 million years after the crisis. The massive eruptions along with the supposed extra-terrestrial impact were, in all likelihood, a significant factor in accounting for most of these annihilations. However, it is probable that the loss of near-shore habitat caused by the joining of the continents was also responsible for some of the species loss.

What happened in Antarctica? During the post-apocalyptic greenhouse period, temperatures and carbon dioxide levels soared over a very short interval of geologic time, and Antarctica was colonised by conifers such as *Volziopsis* and seed ferns such as *Lepidopteris*, which immigrated from Australia or Madagascar to the north. How *Lystrosaurus* survived the calamity is a mystery, but it may have been preadapted to high carbon dioxide levels from a variety of anatomical features and a probable burrowing habit. We can see from reading the rock record of the frozen south that Antarctic sites were overrun with large-scale palaeo-channel sandstones of braided streams that are interpreted as having occurred from widespread lowland deforestation.

The Transantarctic Mountains. Photo: Jeffrey Stilwell.

from water sensitivity calculated from the plant fossils, suggests that the plants lived in a warm, ice-free summer climate.[8]

Late Permian *Glossopteris* floras

Glossopteris was the dominant southern hemisphere plant in the Permian Period – its distinctive leaves and woody stems have been found in Antarctica, Australia, India, South America and Africa, and for many years it was one of the main lines of fossil evidence for the existence of Gondwana. *Glossopteris* is a member of the seed fern group, or Glossopteridales. These plants were fern-like in appearance but bore true seeds which are sometimes found inside ovule-bearing organs resembling small pine cones. They were deciduous trees that shed their leaves each autumn and may have had an overall appearance not unlike the modern gingko tree, possibly growing to around 20 metres in height. The leaves vary in length from around 10 cm to one metre. The roots of *Glossopteris* have been named as *Vertebraria* because they have regularly spaced nodes that makes them resemble an animal's backbone. Many different species of *Glossopteris* are known – India alone has over 70 species.

Glossopteris occurs throughout Antarctica, almost at every known outcrop of Permian sediments.[9] Exceptionally well-preserved permineralised leaves and wood from *Glossopteris* and other plants are known from two silicified peat localities, one at Skaar Ridge in the central Transantarctic Mountains, and the other in the Beaver Lake area of the Prince Charles Mountains. Both of these localities contain three-dimensional remains of plants which reveal detailed anatomical features of their leaves and wood structures. The Skaar Ridge plants occur within a peaty layer of the Upper Permian Buckley Formation, but the fossils are known only from loose boulders. The flora found at Skaar Ridge also includes the remains of an ancient moss, *Merceria*.[10]

The Beaver Lake area of the Prince Charles Mountains contains Permian–Triassic non-marine strata of the Amery Group. The Amery Group comprises a basal conglomerate of late Early Permian age followed by coal-rich Upper Permian unit called the Bainmedart Coal Measures, which contain the

A beautiful slab of seed fern leaf specimens of *Glossopteris* from the Permian rocks of Beacon Supergroup, Transantarctic Mountains. Photo: Bill Zinsmeister.

well-preserved leaf locality. Above these coal measures is an Upper Permian–Lower Triassic red bed succession called the Flagstone Bench Formation. Earlier studies of the palynology from the Amery Group have suggested a Middle–Late Permian age for the layer containing the fossil leaves.[11] The flora preserved in this layer includes leaves of *Glossopteris* and its associated root *Vertebraria*.

Beautifully preserved leaves of the plant *Noeggerathiopsis* have been found in a swamp deposit in eastern Antarctica.[12] By immersing the samples in hydrofluoric acid for 3–5 days, the actual leaf material was freed from the rock and could be studied by thin section. These leaves are similar to northern hemisphere plants and suggest that there must have been Late Paleozoic floral interchange between the plants from opposing high-latitude floral provinces.

These discoveries suggest that, in Late Permian times, the climate was consistently wet for this part of Antarctica. The growth rings found in the fossilised wood show a strongly seasonal climate, with seasonality of wood production being regulated by availablity of light in these high latitude areas.[13] As the succession reaches the Permian–Triassic boundary, defined by the start of the Flagstone Bench Formation, the abundance of red bed deposition tells of predominantly drier conditions. This boundary is marked by the presence of the first appearance of the spores of *Aratrisporites* and *Lepidopteris*.

In December 2003, a discovery in the Beardmore Glacier region by Molly Miller from Vanderbilt University provided the first solid evidence of Permian Antarctic forests living *in situ*. Dr Miller and her team discovered three stands of very large *Glossopteris* trees (some poking up through the snowfield) reaching at least 25 metres in height, based on estimates from trunk diameters. The preservation is so astounding that the forests of tree trunks along with fossils of fallen leaves scattered around the tree trunks have survived the ravages of time. These very early deciduous trees, lived at a time when the Antarctic climate was much warmer, but they still had to endure extreme conditions of low sunlight regimes for about half of the year, and darkness for the rest.

From studies of their rings, palaeobotanist David Cantrill believes that the *Glossopteris* trees grew steadily each summer and stopped abruptly during the winter, akin to a light switch being turned off and on. The most likely reason for this is a reaction to light levels, rather than temperatures. There are no modern analogues to these polar forests. Superficially they resembled big Christmas trees, except they had broad, lance-shaped leaves and not needles. According to Dr Miller it is possible that the leaves may have even turned colour with the seasons.

The discovery of whole stands of trees (an inferred 1000 or so per acre) about 25°–20° from the South Pole is a first for Antarctic palaeobotany, and gives us an insight into the Permian climate.

OPPOSITE: A reconstruction of the Early Triassic cat-sized carnivorous reptile *Thrinaxodon*.
Courtesy: Journal of Vertebrate Paleontology and Christian Sidor; © Jude Swales.

7
GIANT AMPHIBIANS AND HAIRY REPTILES

On the first true mammals from the Triassic, our far distant ancestors whose evolutionary roots began from the few survivors of the biggest mass extinction event the Earth has ever seen – the Permian-Triassic 251 million years ago … All these [Triassic] mammals are most distinctly mammalian. They do not look like first attempts of nature. There is nothing transitional about them. They bring with them no reminiscence of reptiles, birds, or fishes. If they had descended from humbler forms, it must have been by many generations ; and many connecting links must be totally lost … Where we are sure of the disappearance of so many remains, how easy to believe the remains of different creatures – of lower mammals – have also disappeared …

Alexander Winchell, *Walks and Talks in the Geological Field*, 1886.

For most of the Triassic Period the continents were locked into the supercontinent of Pangaea and remained above sea level. The great ocean Panthalassa surrounded this massive continent but the area of the continental shelf was severely reduced, which meant that many of the richly diverse habitats of the continental shelves were not preserved in the geologic record. Thus, the Pangaean interval of the Triassic was an important period in which continental beds of sand, pigmented by red ferric oxide, were deposited by rivers and deserts, and terrestrial organisms had little to impede their dispersal across this great land. In the interior of Pangaea extensive evaporate deposits of salt and gypsum formed from desiccated salt flats and lakes.

The survivors of the Permian–Triassic mass extinction, the greatest loss of life of the entire Phanerozoic Eon, were biologically robust enough to diversify and increase in abundance during the Early Triassic. These opportunists lived in a time of relatively low diversity on land and in the sea, and despite the near annihilation of life on Earth, significant evolutionary events occurred throughout the Triassic Period. On land, dinosaurs arose during the Late Triassic, the first mammals emerged, and forests were dominated by conifers, cycads and gingkos. The therapsids, which had started to develop mammal-like features, diversified and became the dominant reptile group of this period. Mammal fossils are rare in the Triassic, undoubtedly reflecting their small, rodent-like nature and probable nocturnal habit. The reptiles stood triumphant in the hot and dry conditions of the Permian and Triassic and some were able to return to marine environments, especially since ecological opportunities presented themselves with vacant niches after the mass extinction.

The polar regions were not excluded from the widespread desert conditions. The geography of the continents profoundly affected conditions on Earth and arid conditions dominated during the Triassic. As the clockwise rotation of Pangaea continued through the Triassic, Antarctica, along with Australia and India, progressed into southerly, polar latitudes, and mountain chains developed along the entire south-western margin of the supercontinent. Living in the Pangaean Antarctic habitats were tetrapods such as *Cynognathus* and *Lystrosaurus*, which was arguably the most successful and widespread therapsids of all time. Seed ferns such as *Dicroidium* flourished on land, while in the seas there is little known from the continent, apart from rare bivalves and other marine macroinvertebrates.

The Triassic Period marks the dawn of the Mesozoic Era, often called the 'Age of Dinosaurs'. It was during this time that dinosaurs rose to dominate life on land. It was also the time when the flying reptiles, the pterosaurs appeared, as did the fish-like marine reptiles, the ichthyosaurs, in the seas. Global climates were warm and dry, and strongly seasonal in the Triassic with thick red bed successions developing on most continents, and these sediments, from ancient river systems, are the main source of terrestrial fossils. The end of the period, approximately 210 million years ago, saw the break-up of the supercontinent and the eruption of basaltic rock occurred along the margins between present-day North America and north-west Africa. What was the trigger to break-up such an enormous, long-lived supercontinent such as Pangaea? It is believed that a series of geologic domes, which were caused by upswelling from the upper mantle, caused the initial fracturing of the crust.

OPPOSITE: The post-apocalyptic terrestrial ecosystem of Antarctica during the Triassic Period included tough survivors such as the pig-like *Lystrosaurus* and the dog-like *Thrinaxodon*. Illustration: Tony Windberg.

Triassic animals of Antarctica

The date 28 December 1967 is an important one for Antarctic palaeontology. Peter Barrett, from the Institute of Polar Studies, Ohio, discovered a fossil bone fragment near Graphite Peak in the upper Beardmore Glacier area of East Antarctica, near latitude 85°3.3'S longitude 172°19'E at an elevation of 2780 metres. The fossil was entombed in pebbly quartz of fluvial (river) origin, in the lower part of the Fremouw Formation which contains the Triassic plant *Dicroidium*. The horizon containing the bones is only about 76 metres stratigraphically above the *Glossopteris*-bearing Buckley Formation, a Permian sequence containing coal. Barrett's team knew they had found something very important and thought that it resembled part of a jawbone. Identification of the specimen was confirmed by palaeontologist Donald Baird of Princeton University. The bone turned out to be the back portion of a left lower jaw of a temnospondyl amphibian. Subsequent expeditions led by Ed Colbert of the American Museum of Natural History and Bill Hammer of Augustana College, Illinois, have greatly extended our knowledge of Antarctic life in the Triassic Period. This includes the major discovery at Coalsack Bluff in 1969 of fossil bones of the pig-like *Lystrosaurus*, which then provided important evidence for continental drift.

> The occurrence of this form [*Lystrosaurus*] in now isolated Antarctica is not explainable except in terms of the existence during the Triassic of a

former supercontinent, Gondwanaland, which later split into the present southern continents and peninsula India.[1]

The bones of Antarctic *Lystrosaurus* made it possible to tie in the sequence from Coalsack Bluff to that of the Lower Triassic *Lystrosaurus* zone of the Beaufort Series in South Africa. Reconstruction of the fossils indicatied an environment in which giant semi-aquatic amphibians, up to four metres in length, swam in large river systems hunting lungfish, while on land lived a variety of strange hairy therapsids. The three Early Triassic temnospondyl amphibians recorded are relatively small: *Austrobrachyops jenseni*, *Crybatrachus kitchingi*, and an unidentified rhytidosteid.

A reconstruction of the Early Triassic *Lystrosaurus* from the Fremouw Formation, East Antarctica. Some illustrators prefer to cover its body with hair, but this particular painting is represented as hairless. Illustration: Tony Windberg.

Preserved sequences from the Early Mesozoic in Antarctica relate to the intermittent subduction of Pacific Ocean sea floor beneath the Antarctic craton which commenced in the Late Paleozoic. An extensive line of subduction, marked by an island arc, spanned what is now Marie Byrd Land, through to Thurston Island, to the Antarctic Peninsula and Scotia Arc. Thick sequences of deep-sea sediments accumulated as they were scraped off the subducting plate, forming a complex rock unit called an accretionary prism. Differentially metamorphosed from intense to low-grade deformation, the accretionary prism lithologies are dominated by feldspathic sedimentary rocks of turbidite origin, together with minor amounts of basaltic lavas, tuffs, phyllites and cherts.[3] They are, on the whole, largely unfossiliferous, apart from some Late Triassic radiolarians from the Greywacke Shale Formation of the South Orkneys and shallow water sandstones and mudstones of the Legoupil Formation of the Trinity Peninsula Group, north-west Antarctic Peninsula, which yield an Early to Middle Triassic bivalve fauna comprising *Bakevelloides* and *Neoschizodus* of New Zealand and Japanese affinities – together with other austral regions indicating that a distinct faunal province ('Maorian') can be traced within the southern high latitudes.

A hand-painted plaster reconstruction of *Lystrosaurus*, reputedly made by the wife of Edwin Colbert, then Curator Emeritus of the American Museum of Natural History. Image: Bill Zinsmeister.

The primitive amphibian *Austrobrachyops jenseni* was the first Triassic vertebrate fossils to be discovered in Antarctica in 1967 at Graphite Peak, Beardmore Glacier region.[2] Illustration: Tony Windberg.

Part of the cranial and post-cranial skeleton of *Lystrosaurus* now housed in Beijing, China. Photo: John Long.

Passing to the craton-ward side of the Triassic orogenic belt, sedimentary rocks of the Beacon Supergroup continued to accumulate in an immense foreland basin running the length of the Transantarctic Mountains. A thick sequence of terrestrial sediments was deposited by a major fluvial system draining both the orogen and the Antarctic craton in the Beardmore Glacier region. One of the most important vertebrate-bearing units, the Lower–Middle Triassic Fremouw Formation, comprises interbedded braided stream channel sandstones and finer grained flood plain deposits, entombing a significant diversity of reptiles including surviving groups from the Permian–Triassic mass extinction event and the new diversification of groups that followed well into the Triassic. Near the base of the formation, the fauna yielded include a number of therapsids including *Lystrosaurus*, *Thrinaxodon* and *Ericiolacerta*, together with other reptiles and temnospondyl amphibians. These therapsids and temnospondyls are generally much larger than the Early Triassic taxa, and a large kannemeyerid dicynodont occurs alongside the large *Cynognathus*. There is also a gomphodont – an herbivorous cynodont. The relatively advanced gomphodonts were the group that reputedly led to true mammals.

Lystrosaurus was one of the most widespread and successful reptiles of the Early Triassic. It can be likened to a large, plant-eating pig that was apparently semi-aquatic and probably foraged for a variety of food types. Its impressive tusks would have been ideal for rooting in tough vegetation and would

Molly Miller in the Allan Hills region holding part of a trace fossil, probably a burrow from around 250 million years ago. There are similar burrows in South Africa containing the fossilised skeleton of *Thrinaxodon*. Photo: Christian Sidor.

have aided greatly in a foraging type of lifestyle. Given the cosmopolitan distribution of *Lystrosaurus*, this well-known dicynodont can perhaps be considered the most successful therapsids of the Late Paleozoic–earliest Mesozoic. A younger *Cynognathus* fauna from the upper Fremouw Formation comprises this large, meat-eating cynodont reptile among others with at least three types of temnospondyl amphibians. There are also two very large capitosaurs recorded in the upper part of the formation, both of which have skulls exceeding one metre length! In 1999, during a short reconnaissance from McMurdo Station, some vertebrate material was recovered from Southern Victoria Land, which comprises mostly bone fragments dated as Late Triassic from the Lashly Formation. A single dicynodont tusk and indeterminate small limb elements were collected.

An unusual occurrence of tetrapod traces from the Triassic came to light in 2008 following research conducted by Chistian Sidor, Molly Miller and John Isbell.[4] They found two sets of very distinct traces, one described as a giant terminal chamber collected from the lower Fremouw Formation dated as Early Triassic at the Wahl Glacier in the Beardmore Glacier region, and the other provides the first important evidence of Early Triassic tetrapods from Antarctica, which until now has

A reconstruction of the Permian lizard-shaped parareptile *Procolophon trigoniceps*. A group of procolophonids survived the Permian–Triassic mass extinction event, but died out not long afterwards. Illustration: Tony Windberg.

Cranial reconstructions of the cat-sized, carnivorous reptile *Thrinaxodon liorhinus*, depicting this Early Triassic animal with and without hair. In total, 14 specimens of this species were collected during the 1970–71 austral summer season. Illustration: Tony Windberg.

Christian Sidor searching for Early Triassic animals and traces in the Allan Hills region. Photo: Cara Fritz.

been devoid of body or trace tetrapod remains. The giant chamber is inferred to have been made by an animal of comparable size to the South African cynodont *Thrinaxodon liorhinus*, which from fossil evidence is known to have burrowed. The identity of the maker of the second set can only be guessed due to sparse data, but may have been dug by a procolophonid. These researchers proposed that the underground dens and burrows were important shelters for Antarctic terrestrial animals, even though the climate for the time is thought to have been rather mild.

Triassic plants of Antarctica

Among the most important palaeontological discoveries made in Antarctica, is the realisation that fossil forests must have existed on the Antarctic mainland. Since their initial discovery in the late 19th century, a large number of fossil forests have been documented, and some are preserved *in situ*, that is with their petrified logs and stumps in living position. Forests have been recognised in the Permian and Early Triassic of the Transantarctic Mountains, the Early Cretaceous of Alexander Island, the Late Cretaceous–Early Paleogene of the

Triassic Antarctic foliage dominated by the conifer *Rissikia* and the corystosperm *Dicroidium* from Mt Bumstead, East Antarctica. Photo: David Cantrill.

northern Antarctic Peninsula, and the Eocene of McMurdo Sound, East Antarctica. Some preserved stumps bear clearly defined annual growth rings.[5]

The combination of macropalaeobotany and an abundance of palynological information has confirmed that many of the key elements of today's southern hemisphere forests could be traced back into the Mesozoic Era. Lineages of some lycopods, ferns, araucarians and podocarps may originate as far back as the Late Triassic, and the Late Jurassic evergreen, coniferous rainforests that covered much of the southern Gondwana margins have persisted to the present day. Thus, a garden of Antarctic delights is not only possible, but actually abounds in several special gardens that contain plants which are known to be descended from Antarctic ancestors.

Fossil plants from the Triassic, ranging in size from tiny rootlet structures to giant silicified logs, occur throughout the Fremouw Formation and its stratigraphic equivalents. The uppermost levels of the formation have also produced some exquisitely preserved silicified peats. Through detailed research on the Late Paleozoic–Early Mesozoic plant record, the glossopterid flora had been reduced to a few isolated relicts resulting from the Permian–Triassic mass extinction. These relicts were replaced by a flora of transitional affinities, and evidence from fine-grained flood plain deposits and coal seams provides insight that a uniform *Dicroidium* flora was present in Antarctica by the end of the Early Triassic.

OPPOSITE: A cast of the skull of the Early Jurassic theropod, *Cryolophosaurus ellioti*, with Mary Walters from the Monash University Science Centre for scale. Photo: Jeffrey Stilwell.

8
VOLCANIC LAKES AND EARLY DINOSAURS

[Dinosaurs] have lived their day, and played their part; let them rest in peace. Doubtless some enterprising showman, catering for the amusement of an idle public, might have been tempted to advertise in big posters 'A jumping Dinosaur,' to follow after 'The Boxing Kangaroo' or the 'Fighting Lion'; but we are glad to think that the Dinosaur has been spared a fate so ignoble, and, instead of this, affords to the biologist a splendid subject on which to exercise his imagination.

Rev. H N Hutchinson, *Creatures of Other Days*, 1894.

The fragmentation of Pangaea in the Triassic escalated during the Jurassic. The rising sea level during this time has resulted in marine sediments of the Jurassic being better preserved in the rock record than those of the Triassic. Among the most important animals for dating Jurassic rocks are the ammonites – cephalopod molluscs with coiled shells – which evolved and dispersed rapidly in the seas. Climates of the Jurassic were mild and seasonal, and many groups of terrestrial and marine organisms flourished. Dominating the land were the largest animals that ever walked the planet – the dinosaurs – with some behemoth, plant-eating forms reaching more than 30 metres in length. During the Jurassic, the gymnosperms (seed-bearing plants) blanketed mild, seasonal forests around the globe, including Antarctica. The angiosperms (flowering plants) were to evolve millions of years later in the Cretaceous.

Dinosaurs reigned the planet for more than 160 million years and took advantage of a huge spectrum

of ecological opportunities available to them, some of which are quite surprising.

Reptiles dominated both land and sea, and even the air, as pterosaurs, the great flying reptiles, emerged to take to the skies for many millions of years into the Cretaceous. Birds evolved also in the Late Jurassic, but did not become widespread and specialised until the Cretaceous. Until the early 1990s, dinosaurs had been recovered from every continent except mainland Antarctica, but that changed with a spectacular discovery in the remote, high altitudes of Mt Kirkpatrick along the Transantarctic Mountains. The find of an assemblage of Early Jurassic dinosaurs in a south polar Antarctic environment at 65–70°S filled in a major gap in dinosaur palaeobiogeography and delighted scientists with its bizarre forms, such as the head-crested *Cryolophosaurus*. This theropod pre-dates the next oldest known allosauroid by 50 million years – yet clearly belongs to this successful carnivorous group with few differences. Some 150 million years ago at the close of the Jurassic, the southern continent of Gondwana was on the move again and Africa and South America (conjoined for at least half a billion years) split off from North America, starting a cascading series of tectonic fragmentation events, but Antarctica remained connected to Gondwana in high southern latitudes for millions of years to come.

Antarctica's first Jurassic dinosaur find

It was late in the austral field season of 1990–91 when veteran Antarctic geologist David Elliot of the Polar Research Institute, Ohio University, climbed up to the higher levels of Mt Kirkpatrick to look at the rocks. Elliot was part of a team of geologists working on basalts of one of the continent's highest peaks. In a scree slope near the top of the 4000-metre mountain, Elliot found the fossil bones of a dinosaur approximately 190 million years old. Word of the find was passed by radio to another field team in East Antarctica and the news eventually reached around the globe of the discovery of one of the most bizarre meat-eating dinosaurs found anywhere on the planet.

The sequence of events leading to the find was that vertebrate palaeontologists working in the Gordon Valley area of the Transantarctic Mountains were conducting fieldwork on late Early to early Middle Triassic terrestrial assemblages, and had just completed a month of field work. A few days later the team received a radio message that the Ohio State geologists working on the basalts had found what they thought were bones eroding out of a relatively small outcrop of the Hanson Formation below the lava flows – a palaeontologist's lucky card as this means that the rock capping the deposit could be dated radiometrically, and thus, would provide a minimum age for the fossils. Elliot noted the site and collected some fragment of bone, and handed the job on to his colleague Bill Hammer, a vertebrate palaeontologist from Augustana College in Illinois, who was well aware that extracting the large bones from the rock would prove to be quite a chore. Given the importance of the find, he relished

OPPOSITE: A reconstruction based on fossil evidence of Early Jurassic Antarctic temperate climate environment, dominated by the king of the dinosaurs at the top of the food chain, *Cryolophosaurus ellioti*. Illustration: Tony Windberg

A reconstructed skeleton and skull of *Cryolophosaurus ellioti* at the Monash University Science Centre. Photo: Jeffrey Stilwell.

A Triassic *Dicroidium* fossil from the Lashly Formation, Shapeless Mountain. Photo: David Cantrill.

the opportunity to delve into Antarctica's secrets of the world of Jurassic dinosaurs and took on the task of removing the bones from one of the most remote, mountainous regions on the planet. To say that the removal task was logistically challenging is definitely an understatement! The 100 or so bones that were collected came from a single section of the cliff face that was no more than four metres wide and only one metre thick. More than two tonnes of fossil-bearing matrix were removed from the locality using a gasoline-powered generator and a jackhammer. Because of the high altitude, the site was too dangerous for the large US helicopters to fly to, and a smaller helicopter had to be chartered from the neighbouring Italian Terra Nova base to collect the material.

The age of the fossils from the Hanson Formation was deduced from radiometric dating and biostratigraphy. A date of 177 million years ago for the basalt flows, provides an upper limit of early Middle Jurassic for the extrusive volcanic event, and as such the underlying sediments of the Hanson Formation must be older, but by how much? Studies of the palaeobiogeographic distributions and geologic time ranges of similar groups of terrestrial animals around the globe have revealed that the fauna is Early Jurassic in age, consistent with world data on the various groups. Plants belonging to *Dicroidium* occurring 300 metres lower in the section, constrains the age to Late Triassic or younger.[1] The affinities of the Antarctic terrestrial assemblage lies mainly with northern groups, which suggests that there were no major geographic barriers to dispersal between southern and northern continents into the early part of the Jurassic. During this time Antarctic was positioned at about 60–65°S, at a latitude comparable to the Antarctic Circle today. Based on palaeogeographic and climatic constraints, the Antarctic environment was mild enough to allow dinosaurs and other terrestrial organisms to flourish in high latitude realms. The presence of preserved fossil tree trunks four metres above the bone bed indicates a well forested high latitude forest and seasonally mild temperatures.

This major discovery in the history of Antarctic palaeontology represents the only known Jurassic dinosaurs from the continent and provides new clues on the origin of southern high-latitude terrestrial reptiles – with surprising new conclusions of global significance.

Early dinosaurs, synapsids and flying reptiles

The partial skeleton of the large, crested theropod *Cryolophosaurus ellioti* represents the only Jurassic allosauroid from any southern continent.[2] Astoundingly, it is some 50 million years older than other known members of this group of meat-eaters and scavengers. Moreover, *Cryolophosaurus* probably represents and documents an early to intermediate stage of gigantism among the great meat-eating dinosaurs.

Early fossil carnivores are poorly known from the fossil record, so finding such a rare, early beast from Antarctica means not only that *Cryolophosaurus* will be 'showcased' for many years to come, but there will be more concerted efforts to obtain more material of this dinosaur and search for other groups that lived before, during and after the Early Jurassic. Being such an early theropod, it is not surprising from a vertebrate palaeontologist's point of view that *Cryolophosaurus* (literally meaning 'frozen crested reptile') retains in its osteology some quite primitive features compared with other allosauroids, particularly with regard to its femur and foot morphology. The majority of bones collected from the unit includes cranial and post-cranial remains of *Cryolophosaurus ellioti* and represent a partial skull, missing bottom of jaw and snout; femur, ilium, ischium, pubis, fibula, two articulated metatarsals and 30 vertebrae. The most prominent feature of the skull – and hence the name – is the unique bony display crest that sits over the eyes and extends across the skull, perpendicular to its length.[3] A very detailed 2007 study of *Cryolophosaurus* provided much needed insight into the relationships of this Antarctic beast, compared with other described taxa, with allied groups such as *Dilophosaurus* and *Dracovenator*.[4] The bony crest is so ornamental that it was possibly used in some sort of visual display, perhaps even during a mating ritual, which despite seeming rather surprising, has been suggested for other related theropod dinosaurs. Another unusual aspect of the skull is in the morphology of the region behind the eye, which is distinct from the usual diapsid-type recognised in most dinosaurs. Apart from these different aspects, *Cryolophosaurus* looks like most other members of the group, despite being the 'grandfather' of allosauroids. Its body is relatively large, but not huge, at about 7–8 metres in length, with a skull length of 65 cm. It is definitely related to much younger Late Jurassic theropods such as *Allosaurus* from North America and *Monolophosaurus* from China.

The original skull of Antarctica's oldest and biggest meat-eating dinosaur, *Cryolophosaurus ellioti*, named after its discoverer, David Elliot of the Institute of Polar Studies at Ohio State University. Photo: John Long.

Found with *Cryolophosaurus* were at least two other scavenging theropods that are known only by their teeth. These theropods were probably scavenging the skeleton of *Cryolophosaurus*. Studies of the site on the position of the skeleton suggest that the bones of the skeleton were disarticulated and scattered over the small area by these unknown theropods. In addition, there is evidence of gnaw marks on the bones. The teeth cannot be identified, but the serration patterns and density of the teeth suggest that they belong to two different groups. Other remains of dinosaurs with the large skeleton of *Cryolophosaurus ellioti* indicate that the beast was feeding on an herbivorous prosauropod when it died and then was scavenged or killed by other theropods. Given the position of the prosauropod ribs extending between the articulated lower jaws all the way back to the throat regions, we can deduce that the carnivore choked to death on its meal and consequently was eaten itself when dying or dead. Around eight metres in length, this relatively large herbivore (based on limited evidence of a partially articulated foot and the distal end of a femur) belongs to a family called the Plateosauridae, the distant early relatives of the sauropod giants of the later Jurassic.

Palaeontologists have also encountered the bones from other tetrapod groups, mixed in with the bones of *Cryolophosaurus*, which were possibly part of a former meal. The finds include a beautifully

A partial foot cast of the sauropod *Glacialisaurus* (left) and a fragment of wood (right) recovered in association with the Jurassic dinosaur remains from the Hanson Formation of Mt Kirkpatrick. Photos: Jeffrey Stilwell.

A small tritylodont tooth (left) and a limb bone of a pterosaur (right) found in association with *Cryolophosaurus* remains from the Hanson Formation at Mt Kirkpatrick. Photos: Jeffrey Stilwell.

preserved tooth of a synapsid called a tritylodont. Synapsids belong to a large groups of animals that include the therapsids, which were the ancestors of modern mammals. The tritylodonts represent the last of the therapsids, and were akin to a Mesozoic beaver in both life habits and appearance.[6] Smaller than the dinosaurs, tritylodonts were diverse and abundant. They arose during the latest Triassic and survived until the Late Jurassic. The Antarctic tritylodont was relatively large for its group and was the only type of therapsid that survived into the Jurassic, and may have been related to *Bienotheroides* from China.[7]

Also found in association with the remains of *Cryolophosaurus* was a single humerus of a pterosaur which belongs to a primitive group of pterosaurs called the rhamphorhynchoids. The Antarctic pterosaur was small-to-medium-sized and probably had a wingspan of about one to two metres. In any event, *Cryolophosaurus* was the king carnivore at the top of the food chain in this environment, and this exciting find in the highest elevation of Antarctica marks the beginning of big meat-eaters as a dominant predator of the Mesozoic Era.

However, Jurassic fossils had been recognised in Antarctic sequences much earlier spanning the spectrum of ancient life – from plants found at Hope Bay to four beetle wing cases collected by WN Croft of the British Museum (Natural History) in 1946.[8] A lacustrine ray-finned fish collected during the 1960s by David Elliot from the interbedded sediments in the Mt Kirkpatrick basalt was named *Oreochina ellioti*. Many different sorts of macroinvertebrates from the Early to Late Jurassic are also known from marine deposits.

In late 2003, Bill Hammer and his team collected the remains of a sauropod, consisting of a three-metre-wide pelvis and an ilium, indicating an animal that would have been nine metres in length. In December 2007, Hammer's team, identified a new sauropod, *Glacialisaurus hammeri*, from a partial ankle, leg and foot bone which they had unearthed at Mt Kirkpatrick, using chisels, jackhammers and rock saws. Nathan Smith from Chicago's Field Museum estimated that the new sauropod would have weighed as much as six tonnes and may have reached some 7.5 metres in length.[5]

A reconstruction of the lacustrine ray-finned fish *Oreochina ellioti*.

Illustration: William Stout.

Glacialisaurus was a 'middle-of-the-road' sauropodomorph belonging to the family Massospondylidae, which may represent a secondary radiation of basal sauropodomorphs during the early part of the Jurassic Period. *Glacialisaurus* is most closely allied with *Lufengosaurus* from the Lower Jurassic Lufeng Formation of China, indicating that the earliest sauropods co-existed with basal sauropodomorph herbivores (often referred to as 'prosauropods') for a protracted period of time. The development and evolutionary relationships of the sauropodomorph dinosaurs are hotly debated by vertebrate palaeontologists.

Sauropods were generally huge and are mostly represented by incomplete skeletons. They were widely distributed in the Early Jurassic – not only in China, South Africa, South America and North America, but also in remote parts of Antarctica.

The primary reason for this widespread geographic distribution of the big herbivores is the fact, based on current configurations of land and sea, that major land connections between the continents still existed at that time, and because climates were more equable across latitudes than they are today. The discovery of *Glacialisaurus hammeri* seems to indicate that primitive sauropodomorphs coexisted with true sauropods.

A more recent discovery of a possible sauropod at roughly the same location in Antarctica lends additional credence to the hypothesis that the earliest sauropods coexisted with their basal sauropodomorph cousins. Interestingly, the long-held notion that sauropods were exclusively plant-eaters may not necessarily hold true for all early members of the group, as some may have been scavengers that ate whatever they could find, along with plant matter. With some luck, more bones from the important Mt Kirkpatrick site will yield more clues about prosauropod-sauropod evolutionary relationships.

Early Jurassic invertebrates and plants

In the 1980s a discovery of fossils by Mike Thomson and his colleagues from the British Antarctic Survey shed new light on Early Jurassic Antarctic life, as previously the Lully Foothills localities of the LeMay Group of Alexander Island was considered to be Triassic in age. These fossils are recognised as the first known Early Jurassic life from the Antarctic Peninsula region, and along with sedimentological evidence, indicate that the fossils were deposited in a shallow water environment with reworked volcaniclastic material. The present of ammonites belonging tentatively to *Epophioceras* indicated an Early Jurassic age, contrary to an older Triassic age assignment.[9]

Apart from a moderate diversity of other molluscs such as gastropods (at least 10 species) and a few bivalve species, the only other noteworthy fossils comprise a newly described genus of a starfish *Protremaster*, and some arm fragments and part of a disc of a species of brittle star (ophiuroid).[10] Few Jurassic asteroids have been collected in Antarctic

A Middle Jurassic dipertiaceous fern *Goepertella* from the Botany Bay Group, Antarctic Peninsula. Photo: David Cantrill.

A species of *Zamites* from the Early to Middle Jurassic, Botany Bay Group. Photo: David Cantrill.

Sagenopteris nilssoniana, Early–Middle Jurassic, Botany Bay Group. Photo: David Cantrill.

rocks of this age, so any specimens, even such as these which are superbly preserved external moulds, are important as they provide information on the early starfish. The folded nature of the preserved starfish suggests that it was entombed into the shallow marine environment and overwhelmed by a slurry of moving sediment.[11] Other fossils too poorly preserved, but worthy of mention, include a probable solitary coral, a sea lily (crinoid) ossicle, belemnite phragmocone, and some poorly preserved fern-like leaf fragments. In the same rocks, a gymnospermous shoot was collected, and it, too, represents an external mould with rhomboid-shaped leaves, and more than likely belongs to the characteristic Jurassic group of scale-leaved conifers, such as *Brachyphyllum*.

Late Jurassic tectonics and life

The Antarctic Peninsula was an active volcanic arc during the Late Jurassic to Early Paleogene and was formed by the south-eastward subduction of the proto-Pacific plate. It was a large mountain range comparable to the Andes with immense rivers draining into a massive basin which is now called the James Ross Basin. Resulting from this subduction, a very thick stratigraphic sequence estimated to be 5–6 km thick was deposited in a retro-arc basin to the east of the arc.[12] The base of the James Ross succession (also known as the Larsen Basin) is characterised by a distinctive sequence of air-fall ashes and black mudstones of Late Jurassic to Early Cretaceous age. We can collect fossils from this sequence because the rocks were geologically uplifted through millions of years, and are now being weathered. The Upper Jurassic Nordenskjöld Formation of the Antarctic Peninsula contains diverse marine life that signifies a more widespread distribution of many macroinvertebrate groups than previously recognised. The presence of rare forms of cephalopod such as teuthids (or sometimes referred to as 'vampyromorphs') indicate that they

Middle Jurassic invertebrate fossils from the Behrendt Mountains, eastern Ellsworth Land. From the top left, clockwise: trigoniid bivalve, *Vaugonia kawhiana;* the bivalve, *Pholadomya anterumbonis; the* ammonite *Perisphinctes* (*Discosphinctes*) cf. *antillarum;* the ammonite, *Teloceras* n. sp. Photos: Pat Quilty.

were preserved in fine-grained sediments deposited under anoxic or otherwise restricted conditions.[13] These rare cephalopods are the only known Jurassic fossils from any of the Gondwana continents, hence the distribution patterns for this groups is rather anomalous due to the paucity of preserved specimens.

Along with cephalopods such as ammonites and belemnites and other mollusc groups such as clams and snails, this diverse macroinvertebrate group dominates the known Late Jurassic shallow sea fauna of the Antarctic region.[14] Work conducted from the 1970s through to the 1990s has resulted in many new records of shells that were hitherto known only from the northern hemisphere fossil record. An example is the Late Jurassic 'paper pecten' bivalve *Aulacomyella* from the Antarctic Peninsula, which is preserved in black shale and mudstone facies, similar to records from the northern hemisphere.[15] *Aulacomyella* is a Middle to Late Jurassic homeomorph of Triassic *Halobia* and *Daonella*, all three groups having similar equivalve, thin shells, and all occupying niches within black shale environments. Other common groups include the buchiid, oxytomid, pectinid and inoceramid bivalves.[16] These discoveries add further weight to the concept of a Late Jurassic bi-polar bivalve fauna, especially since the presence of the distinctive Siberian genus *Arctotis* from the family Oxytomidae, was found to occur in the Late Jurassic Antarctic fauna.[17] Other invertebrate fossils from the Late Jurassic have been reported from Adelaide Island and from the South Shetland Islands.[18]

OPPOSITE: A Cretaceous ornithopod feeds on a ginko tree.
Illustration: Tony Windberg.

9
WHEN DINOSAURS RULED

Professor Huxley, in his late memoir on the order of reptiles called Dinosaurians, which are largely represented in all the formations, from the Neocomian to the Triassic inclusive, has shown that they present in their structure many remarkable affinities to birds …

Sir Charles Lyell, *The Student's Elements of Geology*, 1878.

The Cretaceous Period, 145–65 million years ago, represents the last major greenhouse episode of the Phanerozoic Eon, when Earth was warm and humid from pole to pole – a veritable hothouse – with no glaciers. At the beginning of the period the sea level was some 25 metres higher than at present, and rose to such an extent that almost every part of the planet was submerged at some stage during the Late Cretaceous. Ninety million years ago the level of the sea was abount 200 metres higher than it is today. The break-up of the continents continued with Africa, India and parts of Asia breaking away from Antarctica, then South America and Africa parted, leading to the isolation of major groups of animals and plants as reflected in their increasing diversity in the fossil record.

The remains of Cretaceous dinosaurs have now been found on every continent. Dinosaurs were still dominant when the angiosperms – flowering plants such as magnolias, figs, willows, poplars and plane trees – emerged and spread over the mostly separated continents. Angiosperms were the most diverse plant group, having evolved during the middle of the period, and became the principal food sources for many dinosaurs and mammals.

On a small part of Alexander Island, on the west coast of the Antarctic Peninsula, fossil forests with logs up to seven metres high can be found

Subdivision of the Cretaceous Period

Epoch	Age at base of stage (top 65.5)	Stage
Late Cretaceous	65.5–70.6	Maastrichtian
Late Cretaceous	70.6–83.5	Campanian
Late Cretaceous	83.5–85.8	Santonian
Late Cretaceous	85.8–89.3	Coniacian
Late Cretaceous	89.3–93.5	Turonian
Late Cretaceous	93.5–99.6	Cenomanian
Early Cretaceous	99.6–112.0	Albian
Early Cretaceous	112.0–125.0	Aptian
Early Cretaceous	125.0–130.0	Barremian
Early Cretaceous	130.0–136.4	Hauterivian
Early Cretaceous	136.4–140.2	Valanginian
Early Cretaceous	140.2–145.5	Berriasian

still standing upright, frozen in time from the day they died, 100 million years ago. In 2008–09 Jeffrey Stilwell and his team discovered slightly younger forests of some 90–80 million years old on Pitt Island (Chatham Islands), and these very high latitude forests were at the time of their growth attached to Marie Byrd Land, at the highest recorded south polar latitudes of some 75°–80°S. These evergreen forests of polar Antarctica were the habitat of dinosaurs, and the nearby seas harboured giant marine reptiles. At this most significant time, before the break-up of Gondwana, Antarctica was a thoroughfare for animals and plants, enabling them to disperse to and from the northern hemisphere into Australasia.

Palaeontologists are interested in reconstructing habitats and interpreting climate in the high latitudes during the Cretaceous, the world's last great greenhouse phase, especially in the deep south. Mean annual temperatures were substantially different at a given polar location during the Mesozoic than at comparable latitudes today. Although oxygen isotope records of oceanic temperatures are unfortunately patchy (especially in the southern regions), both palaeontologic and isotopic studies have suggested that the overall warm Early Cretaceous was punctuated by cold spells, with a major warming trend from the Early Aptian (c. 120 mya) to a peak in the Late Cenomanian–Early Turonian (c. 93 mya), followed by a gradual cooling trend through to the end of the Maastrichtian (c. 66 mya) and into the Early Paleogene (c. 60 mya). Organisms living near the two poles have not been well known or studied until the last two decades. Knowing that the polar regions were once warmer than they are now with magnificent forests growing at latitudes that now support only tundra, and also cognisant that the lack of light was still a limiting factor, palaeontologists are intrigued by the biotic composition of these ancient polar areas. Some animals and plants were excluded, while others prospered and diversified. Polar regions were the most extreme for the time, inhabited by dinosaurs and other land animals, and these served as both refuges and nurseries for some groups. The highest southern latitudes offer the best opportunity for deducing both terrestrial and marine environments during the Cretaceous hothouse.

In the early 1900s an important piece of evidence that the poles were once ice-free in the distant past was determined when fossil breadfruit trees, similar to those found today in Hawaii, were discovered in the Cretaceous deposits of Greenland. At the same time interval in Antarctica, the waters surrounding the continent were a mild 15°C or so, compared with the frigid −2°C to 0°C temperature of now. Work by Jane Francis in the United Kingdom and her colleagues has shown that, in the latter part of the Early Cretaceous, some 100 million years ago, West Antarctica was a really lush paradise with ferns and cycads, along with conifers that probably looked much like today's *Araucaria* species. Similar

Ula Point on the north-east coast of James Ross Island is a rich Cretaceous fossil site. Photo: Jeffrey Stilwell.

plant fossils on Pitt Island (Chatham Islands) of comparable age means that the Antarctic Peninsula and other previously joined continental pieces such as the Chathams enjoyed warm temperatures that reached near sub-tropical levels (maybe even up to 19°C, similar to South Africa). Ongoing research by UK, USA, Argentine and Australian teams is sure to refine our palaeobotanical knowledge of the highest southern latitudes.

Cretaceous dinosaurs from Antarctica

Dinosaurs reached a peak of global diversity in the later half of the Cretaceous Period. Large landmasses in the northern hemisphere, such as North America and central Asia, were home to a diverse range of habitats and many species of animals and plants. Antarctica and Australasia at this time had extreme climates, a time of cold polar forests with long periods of darkness, a time when only specialised creatures could survive in harsh conditions (but still much warmer than today). Although few dinosaurs have been found from Antarctica in the Cretaceous Period, and all of these come from the latest Cretaceous of James Ross, Vega and Seymour islands in Western Antarctica, a diverse Early Cretaceous dinosaur fauna is well documented from southern Australia and Late Cretaceous sub-Antarctic forms are now known from New Zealand and the Chatham Islands, among other parts of the southern hemisphere.[1]

The marine record of vertebrates and invertebrates is so much more diverse and rich compared to the meagre terrestrial record. Of the dinosaurian remains recovered from the latest Cretaceous in Antarctica, all have been collected from marine deposits in the James Ross Island group. There were undoubtedly pterosaurs in the Antarctic skies but, unfortunately, none has been recovered thus far. No Mesozoic crocodiles or turtles have been discovered either.

Palaeontologist Eduardo Olivero of Ushuaia, Patagonia, with the remains of the armoured dinosaur, *Antarctopelta oliveroi*, which he discovered on James Ross Island. All Antarctic Cretaceous dinosaur fossils, as scarce as they are, are found in marine sediments. Inset: The lingual tooth of the ankylosaur *Antarctopelta*, the first dinosaur remains found in Antarctica. Photos: Roberto Scasso & John Long.

Despite Antarctica's place as a centrepiece in the history of dinosaur dispersal around the Gondwana supercontinent, no fossils of this dominant group were discovered until 1986, when Eduardo Olivero found the first dinosaur in upper Campanian sediments.

Argentinian expeditions led by Zulma Gasparini resulted in the first discoveries of Antarctic dinosaurs from James Ross Island. The Argentinian scientists made world headlines when they recovered undisputed cranial and post-cranial skeletal pieces of an ankylosaur (armoured dinosaur) of the family Nodosauridae.[2] These ankylosaur elements, recently named as *Antarctopelta*, were derived from the Santa Marta Formation, a marine succession which has abundant marine invertebrates.

Elements of the ankylosaur included fragments of the skull; these dinosaurs were protected by thick dermal co-ossifications, which on the skull diagnostically bear marked lateral projections. A fraction of the left dentary (lower jaw) is fitted with alveoli and one lingual tooth (typical leaf-shaped and cylindrical stem) – all consistent with the family Nodosauridae. Also preserved are vertebrae (including fused sacral vertebrae), ribs, limb fragments, dermal plates and part of the tail club. The beast was approximately 3–4 m long and probably fell victim to the 'float and bloat' *post mortem* circumstance whereby a near-shore carcass (or at least what was left of it) made its way into the shallow marine Antarctic sea where the decomposition of the animal created gas which kept it afloat for a time. Eventually, it sank to the sea floor, to be covered by soft sandy and silty sediment. The completeness of the skeleton would then depend on what it was subjected to after death (scavenging, disarticulation from movement, etc.) and the time it took before it was buried.

On subsequent trips the Argentinians found more dinosaur bones as well as the skeletons of giant marine reptiles such as mosasaurs and plesiosaurs. The discovery of dinosaurs in Antarctica was not a total surprise as the land bridge across the Gondwanan supercontinent would have given access to a wide geographic range for dinosaur groups – as indicated by the relatively rich record in southern South America and the growing record in Australia and New Zealand (including the Chatham Islands). However, the only record of these Antarctic Cretaceous non-avian dinosaurs comes from the marine geologic record. What this means is that dinosaur bones must have been washed down from the higher levels along the Peninsula via rivers and streams into the shallow seas of the James Ross Basin.

The carnivorous theropods form another group with a long history in Antarctica. They were discovered in the Early Jurassic of Mt Kirkpatrick and persisted in Antarctica until the Late Cretaceous. An indeterminate member of this group was recovered from Coniacian to Santonian sediments of the middle part of the Hidden Lake Formation near Cape Lachmann on James Ross Island. The fossil, found in a calcareous sandy concretion with associated trace fossils and plant matter, represented a distal portion of a primitive theropod left tibia, and when compared with similar-sized tibias of *Coelosaurus antiquus* or the more robust *Piatnitzkysaurus floresi*, the size of the dinosaur can be estimated to have been two to three metres in length. The environment of deposition seems to have been a fan delta system, and the climate was mild. This tibia indicates that theropods were present in Antarctica until the Late Cretaceous.

In late 2003 Judd Case, James Martin and their colleagues found the remains of a bipedal meat-eater *Carnotaurus* on James Ross Island. About 70 million years old, *Carnotaurus* stood about 1.8–2.4 metres high, so was rather puny compared with the terrible giants such as *Tyrannosaurus* or *Allosaurus*. The bones discovered include fragments of an upper jaw, teeth and bones from the animal's lower legs and feet, pointing to a primitive predator that somehow survived in the higher southern latitudes shortly before the world saw the disappearance of all terrestrial dinosaurs at the close of the Cretaceous Period.

Other specimens of dinosaurs have been located on nearby Vega Island in the López de Bertodano

A reconstruction of a Cretaceous scene in Antarctica. Illustration: William Stout.

Formation. In the uppermost Maastrichtian Sandwich Bluff Member, a single cheek tooth of a duck-billed dinosaur (Hadrosauridae) was collected from the so-called 'Reptile Horizon', along with many mosasaur and plesiosaur marine reptile remains, and rare bird remains. The birds belong to at least four species and can be referred to modern avian orders, including charadriiform and gaviid birds (such as *Polarornis gregorii)*, based on the morphology of their respective tarsometatarsals.[3]

Much work needs to be done on the avian fauna. The morphology of the non-avian dinosaur tooth suggests that it is a hadrosaurine species with affinities to the teeth of the widespread North American genus *Edmontosaurus*. Another probable hadrosaur bone was collected from the latest Cretaceous of Seymour Island, but is not as diagnostic as the cheek tooth from the Sandwich Bluff Member.[4] Hadrosaurs are among the most common Late Cretaceous dinosaurs and are quite popular because of their amazing head crests.

A more complete skeleton of a dinosaur named *Biscoveosaurus* was found in the lower Cape Lamb Member, dated as Late Campanian to Early Maastrichtian, in deep shelf deposits below storm wave base (but adjacent to land). The fossils belong to a hypsilophodontid and comprise complete left and right dentaries and maxillae (top and lower jaw), cheek and palatal elements and a partial braincase. Other parts of the skeleton known include vertebrae, scapulae, coracoids and humeri. This hypsilophodontid attained a relatively large size of 4–5 metres and is one of the largest recorded of all the family. The fossils are associated with the abundant ammonite *Gunnarites,* other macroinvertebrates and wood. This Antarctic record expands greatly the goegraphical range of this widespread group that is now known to have a more or less cosmopolitan distribution, also being found in northern Alaska.

By the year 2000, the remains of six types of dinosaurs had been collected, spanning 20 million years from about 85–65 million years ago.[5] The biggest Antarctic beast of this time was the six-metre-high megalosaur. Those Antarctic groups with a long geological history include the nodosaurids, theropods and hypsilophodontids (i.e. ornithopods), all belonging to groups with origins extending far back into the Early Jurassic to Early Cretaceous, and with broad geographic ranges during the Mesozoic. During this time in their history there were very few barriers to impede their dispersal. The theropods and nodosaurids should be regarded as the residents of the supercontinent Gondwana and the hypsilophodontids and nodosaurids were in Australia by the end of the Early Cretaceous. A surprise is the record of hadrosaurs, as this occurrence in the Maastrichtian represents a late dispersal event for the group. They are a 'northern hemisphere' dinosaur, as their oldest record is North American (such as the Arctic on Bylot Island in Greenland) and Eurasian, with a subsequent dispersal to South America.

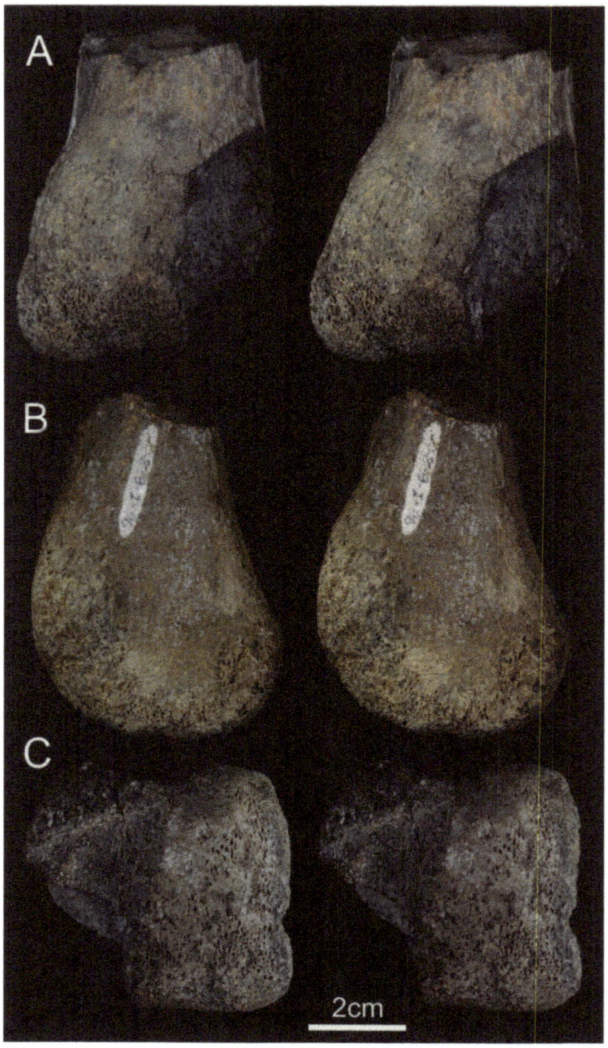

Stereo pair views of a limb fragment of a hadrosaur from Seymour Island. Image: Pat Vickers-Rich and Tom Rich.

How could hadrosaurs have arrived in Antarctica? The idea, based on current palaeogeographic reconstructions is that they must have entered the Antarctic Peninsula through a land connection, between Patagonia and the Antarctic Peninsula in the latest Cretaceous. These connections were generally present from the Late Cretaceous to the Oligocene, when they were severed by the emergence of the circumpolar current. So, if dinosaurs made their way to the south polar regions, so must have their diminutive companions of the Late Mesozoic, the mammals. Indeed, they did. The oldest record of mammals in Antarctica is from the Jurassic, and there is a very long gap until they turn up again in the Eocene. Judd Case has found from his work that some of the dinosaurs discovered in Antarctica had already disappeared from the fossil record in other parts of the globe. He believes that this is due to the late arrival in the high southern latitudes of flowering plants, which had already colonised warm regions of the globe earlier in the Cretaceous. These dinosaurs evolved adaptations to changing vegetation, but not Antarctic ones, which was one of last regions on earth to be populated by angiosperms and the animals that ate them. Antarctic fossil evidence shows that there was a high diversity of plants in the Late Cretaceous ecosystem, and it is likely that there was a diversity of dinosaurs as well, even though the record is still quite patchy.

The relatively depauperate dinosaurian record of Antarctica will undoubtedly grow as scientists trek through the inhospitable conditions and terrain of the south polar regions. However, this meagre record already provides a tantalising glimpse of a variety and range of both meat- and plant-eating polar dinosaurs.

The rise of modern birds

The very few bird remains from the Cretaceous that have been collected have been found only in the James Ross Island Group and include a loon-like bird and also a relative of ducks and geese.[6, 8, 9.]

One of the most important questions in avian evolution is the long-term controversy of whether or not extant bird lineages emerged after non-avian dinosaur extinction at the K–T boundary or, alternatively, if avian groups co-existed with other dinosaurs and were able to survive the mass extinction event. Major discoveries of Late Mesozoic bird remains from the Antarctic Peninsula are assisting to reconstruct the picture of early avian evolution, especially since some of the oldest known groups are now recorded from Antarctica. Thus, the high southern latitudes have much to tell us about the origin of birds.[7]

The Cretaceous avian fauna is widely varied, with many different kinds of birds from toothed, clawed, and other oddities that lived side-by-side with more familiar types of birds such as those related to loons and ducks. It was thought that the mass extinction at the K–T boundary exterminated not only all of the dinosaur groups, but also decimated most bird lineages such that only one group of beaked, clawless birds survived the calamity. Then, very slowly in the Early Cenozoic, birds diversified in all corners of the globe until these feathered creatures were kings and queens of the skies living in more woodland or forest environments. Other forms became highly cursorial, inhabiting more open plains areas. This view of bird evolution may be too simplistic, especially since bird remains do not preserve well as they are so delicate. Maybe there were more 'modern' birds than we realise back in the Cretaceous Period.

In the late 1980s, parts of a supposed loon-like bird were discovered in the late 1980s in the López de Bertodano Formation of Seymour Island.[8] The parts included part of the bill and the adjacent part of the cranium, parts of the otic region, four vertebrae, a small fragment of sternum, a femur and a proximal tibiotarsus. It wasn't until 2002 that this bird received the name *Polarornis gregorii*. Reconstructions abound of *Polarornis gregorii* as an Antarctic equivalent of *Hesperornis* species – thus, a diving, fish-eating bird that acted like a loon by sitting on top of the water, then taking a dive sporadically to catch a fish or crustacean. More fossils are needed to confirm these assessments of this enigmatic bird.

In early 1992 a more complete Cretaceous bird was discovered also along the Antarctic Peninsula, this time on nearby Vega Island. Jorge Noriega and Claudia Tambussi of Argentina first reported the Vega Island bird as a presbyornithid bird, comprising a disarticulated partial postcranial skeleton preserved in two halves of a concretion, dated as middle-to-late Maastrichtian.[9] Presbyornithids were living until the

Eocene or Early Oligocene and are in the extant group Anseriformes. The name *Vegavis iaai* was given to this bird which is believed to be most closely allied to Anatidae, which includes true ducks. Thus, at least duck, chicken and ratite bird relatives were coextant with non-avian dinosaurs. Julia Clarke and the original describers provided a full a description along with new analysis and illustrations based on CT-scan data. The age is no younger than Late Maastrichtian (c. 68–66 mya), and is the oldest well-confirmed fossil member of the group of modern birds, Neornithes, from which all modern birds were derived.

Because the evolutionary split leading to anseriforms is not the first in the avian family tree, several splits must have occurred earlier in their history. These include Galliformes (chickens, etc.) and Palaeognathae (ostriches, emus, etc.), but these are not yet known as Cretaceous fossils.[10] Thus, there is a general consensus within the bird scientific community that there is no longer any serious query that crown group birds in the lineage Neornithes were present in the Cretaceous. The question is now: For how long were these birds co-extant with non-avian dinosaurs? Further, there remains a problem with the time of origin of the next branch of modern birds, Neoaves, that makes up the vast majority of living birds. Most of the fossils that look like neoavians are very fragmented pieces of bone, and many specialists on bird evolution argue that Neoaves did not originate until the Early Cenozoic.

Vegavis iaai takes its unusual species name from the Instituto Antártico Argentino (IAA) expedition that collected the specimen. It is sad to note that the holotype received rudimentary preparation that degraded the delicate bones which were originally exposed. Fortunately, new preparation, x-ray computed tomography (CT) and recovery of latex peels of the specimen before its original preparation

Seymour Isand in the James Ross Island Group. Photo: Jeffrey Stilwell.

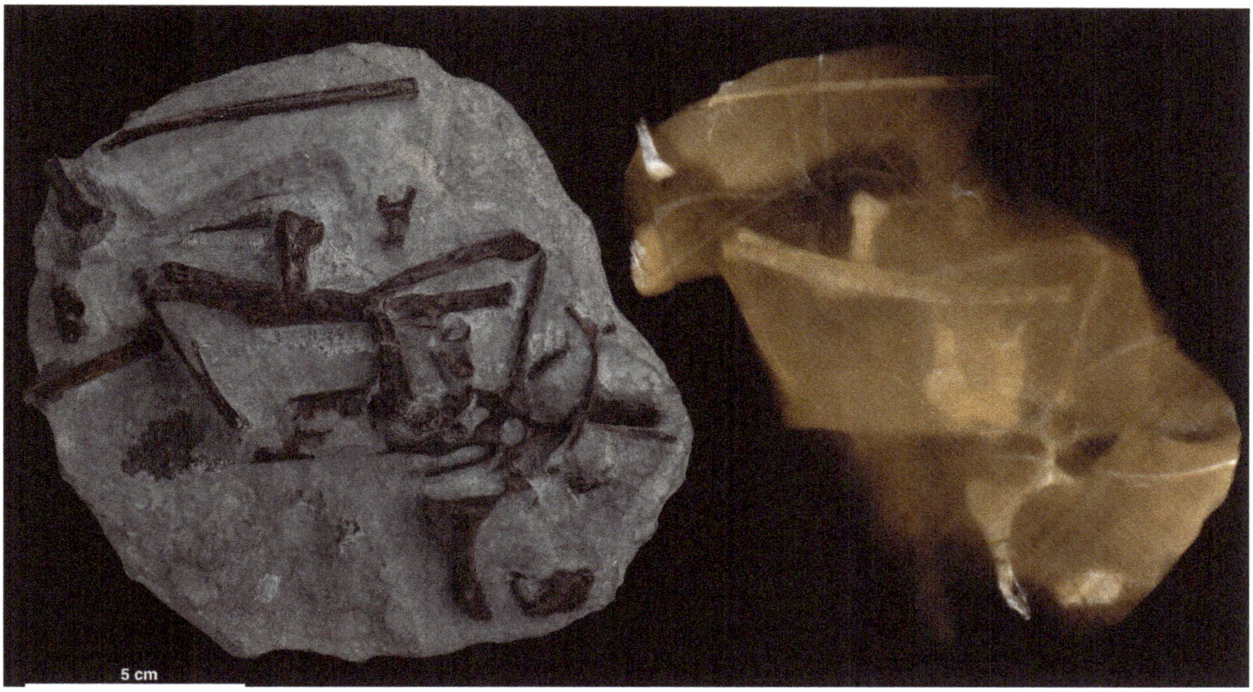

The original specimen and a CT-scan (right) of the anseriform bird *Vegavis iaai*. Photo: Julia Clarke.

revealed numerous, previously unknown bones and further significant anatomical details. Thus, these two birds and a possible few other indeterminate taxa attest to the importance of the Antarctica fossil avifauna in deducing the evolutionary history and ancestry of modern bird lineages.[11] Much more work in Antarctica in Cretaceous deposits and new discoveries of sub-Antarctic Cretaceous species such as those possibly to be found on the Chatham Islands will add more information about the radiations of late Mesozoic birds in the southern hemisphere.

Bill Zinsmeister at Purdue University with the remains of a Late Cretaceous Antarctic bird. Photo: Purdue University Media Unit.

But the modern bird story is set to get more interesting with the 2006 discovery, by Jeffrey Stilwell and his team in the Chatham Islands, of birds in the latest Cretaceous to earliest Paleocene, dated from around 65–63 million years ago, in sub-Antarctic palaeoenvironments. Several species have been recognised in one horizon and these birds are all quite distinct, the smallest being a diminutive, gracile species about 30 cm in height, and the biggest more than one metre. Will this new evidence show that modern birds were much more diverse than thought previously in the Late Cretaceous, but were drastically reduced in numbers at the K–T boundary, or that the post-K–T vacant ecospace resulted in accelerated diversification? Will these new finds provide the earliest evidence to indicate when the rest of Neoaves radiated? Some hypotheses identify the high southern latitudes as the centre of avian origins.

Early Cretaceous plants of Alexander Island

In a small part of Alexander Island, on the west coast of the Antarctica Peninsula, fossil forests 100 million years old, with logs up to seven metres high can be found still standing upright, frozen in time since the day they died. The trunks of coniferous trees, with their roots attached to the ancient carbonaceous soil deposits, testify to the ancient environment in which they thrived and died. Their roots show that they preferred to grow on the rich plains alongside the large meandering rivers, but the coarse sandstone deposits that bury the trunks are evidence that the power of floods would eventually blanket the whole plain with thick sediments.

The many fossil leaves that have been found in association with the logs show that a huge diversity of plants once lived here in what was then a predominantly evergreen, temperate rainforest at about 75°S. Such latitudes today are smitten by extremes of climate – devastating cold and blanketed with snow and ice. Indeed, nothing grows in such conditions at 75°S today in Antarctica. Yet 100 million years ago, despite seasonal fluctuations of light where 70 days of darkness may have been an annual event, forests were thriving in much warmer conditions. Such warmth came about because the large size of the landmass, the southern supercontinent of Gondwana, kept at bay the closure of cold currents that would later envelope and strangle the Antarctic continent after the break-up of the other continents was complete.

The braided flood plain here was dominated by taeinopterid ferns, small stand of podocarps and taxodioid conifers.[12] On the coastal plain, open canopy forests dominated by conifers would grade occasionally into broken conifer woodlands with fern thickets established here and there on immature substrates.

A major distinction between the polar forests of mid-Cretaceous Australia and Antarctica, and those of the far northern hemisphere, is that the winter in the south was much colder (with mean temperatures as low as –8°C), as opposed to the far north (+6°C cold month mean temperature) and dark. The milder northern winter would have favoured more deciduous vegetation, whereas the colder, harsher southern winter favoured an evergreen habit.[13] Under such conditions the evergreen habit permits more rapid photosynthesis as soon as light begins to return to the region, without the need to expend energy growing new leaves. Plants in these Antarctic forests such as *Nothofagus* and *Gingko* have been shown by modern experiments to be unaffected by 70 or so days of darkness.

The areas of the Weddellian Province, including the Antarctic Peninsula, were blanketed by conifer-dominated rainforests characterised by *Nothofagus*

Fossil trees such as this large trunk section attest to warmer conditions during the Late Cretaceous. Photo: Eduardo Olivero.

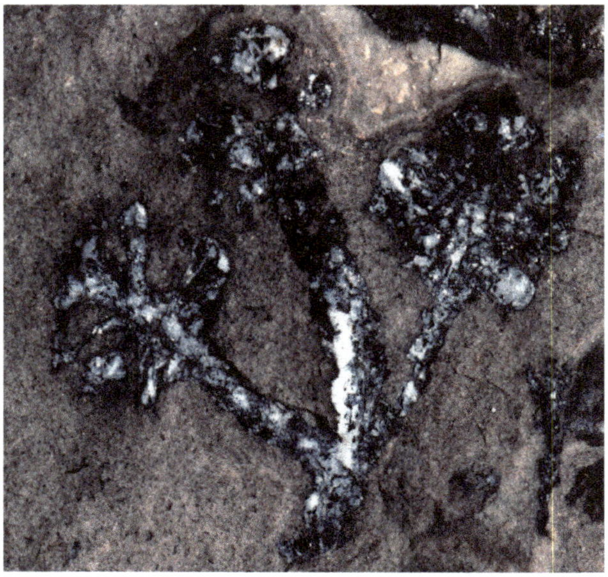

The conifer *Athrotaxites ungerii* from the Cretaceous of Alexander Island. Photo: David Cantrill.

Unnamed fern foliage from the Early Cretaceous of Alexander Island. Photo: David Cantrill.

A superbly preserved species of *Aculea*, an Aptian-aged fern from Livingston Island, Byers Peninsula. Photo: David Cantrill.

and podocarps. The Cretaceous Period was also the time of origin and diversification of angiosperms, with Australia having one of the earliest records. The origin and dispersal of some of the angiosperm taxa during the Late Cretaceous may have been related to opening and enlargement of oceans and fragmentation of Gondwana.[14] Palynological studies have indicated that there was an increase in endemism of angiosperm pollen from the Late Campanian through to the Paleogene. By the Late Maastrichtian, almost 50 per cent of angiosperm pollen was endemic to Seymour Island–James Ross Island Basin or to Antarctica. Most gymnosperm species were found to have Weddellian to austral distributions while cryptogam spores showed Weddellian to cosmopolitan distributions.

In the early 1970s, palynologist Rosemary Askin from the University of California, Riverside, collected many samples of fossil pollen and spores in the Transantarctic Mountains. She attributed the pattern of increased endemism of terrestrial floras, especially of angiosperms, to the continued drifting of the southern high-latitude landmasses. For planktonic marine dinoflagellate cysts, she suggested changing palaeogeography, ocean currents and the greater geographic isolation of Antarctica as causes for increased endemism.[15]

Amazingly, it seems that Cretaceous Antarctic trees thrived in the mid-winter Antarctic palaeoenvironments in complete darkness. By studying the tree rings of collected fossil specimens,

Standing tree in fluvial deposits of the Triton Point Formation, Coal Nunatak, Alexander Island. Photo: David Cantrill.

The conifer *Araucaria chambersii* from the Early Cretaceous of Alexander Island. Photo: David Cantrill.

The fern *Sphenopteris sueroi* from the Early Cretaceous of Alexander Island. Photo: David Cantrill.

Jane Francis from the University of Leeds has found that the trees exhibited no sign of frost rings or drought. Very early flowering angiosperm fossils, dating to about 85 million years in the early part of the Late Cretaceous, attest to equable environments. Flowers reminiscent of the Siparunaceae that live today in the Amazon, and also plants similar to the Australian *Eucalyptus* and Winteraceae such as the Tasmanian mountain pepper, have been found as well by Francis and her team. Thus, both poles were ice-free and warm during the Cretaceous, and it seems that CO_2 levels rather than palaeogeography were much more important in the story of Antarctica's ancient hothouse environments, as seen by animals and plants of both high latitude regions.

OPPOSITE: The ammonite *Maorites densicostatus* collected on Seymour Island was given its name 'jewel-of-the-sea' because of the densely ribbed pattern on its shell. It is one of the most abundant and best preserved fossil ammonites of Antarctica. Photo: Steve Morton.

10
WHEN GIANT REPTILES SWAM IN SOUTHERN SEAS

The numbers of these animals [marine reptiles] inhabiting the ancient seas must have been enormous, as the remains of their bones and fossil dung are found in such quantities … Another monster … the Plesiosaurus, which, with a shorter body, was also furnished with four paddles, and a long neck, like that of a serpent, and a small head; a structure which fitted it to find its prey in shallow waters, or for seizing such as flew above them. Both of these animals [ichthyosaur and plesiosaur] are altogether different from any now existing on our planet …

Outlines of British Geology … London, 1850, pp. 145–146.

The seas of the Late Cretaceous were home to many forms of life. Some were familiar – like the sharks, fish and some marine invertebrates of today, but others like the giant coiled ammonites and the huge marine reptiles were products of the dinosaur age. Diverse shore birds flew over the shallow Antarctic waters and waded along the forested coast. Although similar marine creatures have been found in marine deposits of this age around the world, what makes Antarctica's faunas so interesting is their southern location, cooler water conditions and their links to neighbouring Gondwana regions. The Late Cretaceous, although still warm compared with today, saw the gradual decline in sea surface temperatures and eventually, at the end of the period, witnessed one of the greatest mass extinction events in the history of Earth. The K–T boundary event (see page 132) saw the demise of dinosaurs (apart from birds),

Humps Island on the Antarctic Peninsula, where richly fossiliferous Cretaceous exposures have been discovered. Immediately west of Humps Island are cliff sections of James Ross Island, and glimpses of ice-covered peaks of the Antarctic Peninsula. Photo: Jeffrey Stilwell.

An unusual concretion with eroding fossils poking out, all of which belong to calcareous annelid worms called *Rotularia*, collected by Jeffrey Stilwell on the northern slopes of Humps Island in early 1987. Photo: Steve Morton.

marine and flying reptiles, and a host of plants and animals on both land and sea. Some groups, such as marine reptiles and ammonites, are encountered in the Antarctic fossil record within a few metres of the K–T boundary, are among the last members of their kind known from the fossil record. These Cretaceous marine fossils were first collected on Seymour Island by Otto Nordenskjöld's team in 1901–1903.

Marine reptiles

The Cretaceous Period witnessed a great diversity of marine reptiles with a worldwide distribution – they were kings of the seas, filling similar niches as do today's sharks, whales and seals. Marine reptiles are known to have eaten everything from molluscs on the sea floor to large sharks, and evidence suggests that they may have eaten birds and dinosaurs as well. Throughout the Cretaceous the marine reptiles indeed filled the role of the large predators, and Antarctica's record is diverse, including the sea-gliding plesiosaurs and the ferocious mosasaurs. The demise of these successful reptiles coincides with the K–T mass extinction event and the proliferation of fish, including sharks, in the Paleogene.

During the 1984–85 austral summer Jeffrey Stilwell discovered the skeleton of a plesiosaur embedded in a small hill on Seymour Island – it was one of the highlights of his college career. Unfortunately, the find was during a snowstorm at the end of a long day with his college advisor, Bill Zinsmeister of Purdue University, and the marine reptile team was unable to find it after the storm, the skeleton having been buried under snow.

The plesiosaurs were the largest marine reptiles of all time, some attaining a length of 15 metres with necks of over seven metres. They were highly adapted aquatic animals, having evolved rapidly into several body morphologies. There were two general types which reflected a divergence in feeding and locomotion: the 'pliosauromorph' with short necks and large heads, and the 'plesiosauromorph' with a long neck and small head. The short-necked species were fast and could rapidly change direction, ambushing their prey, whereas the long-necked species cruised the oceans with efficiency.[1]

Historically, plesiosaurs have been considered to be specialist hunters of Cretaceous seas, seeking out agile fish and cephalopod molluscs such as ammonites, belemnites and nautiloids. However, new work on fossilised stomach contents of two Austalian elasmosaurids sheds light on the feeding habits of these monsters of the sea. The specimens were collected in upper Aptian shallow marine deposits of the Great Artesian Basin in Queensland, Australia. The elasmosaurids were about five to six metres in length and would have weighed about 1000 kg. The contents of one individual contained 92 per cent bivalves and gastropods and eight per cent of belemnite guards.[2] The second specimen had an intact decapod carapace, several crustacean fragments and a fish scale, with also about 135 gastroliths. The provenance of the gastroliths was some 300 km away from the burial site. The large proportion of epibenthic prey in these specimens suggests that some elasmosaurs were not nekton specialists, but preferred to eat bottom-dwelling

The skeleton of a plesiosaur from the López de Bertodano Formation on Seymour Island, discovered on the way back to our camp in early 1985. A snow storm covered the skeleton and it was not relocated on that expedition. Photo: Jeffrey Stilwell.

An ancient shallow sea floor on the western coast of Seymour Island with Bill Zinsmeister looking for fossils in this concretionary lag horizon. Many of these concretions contain beautifully preserved fossils, which have been protected by the ravages of time in these mud balls. A good whack with a hammer can reveal a fossil fortune! Photo: Jeffrey Stilwell.

invertebrates, although their diet may have also varied with life cycle. Thus, the specialised elasmosaurid neck did not necessarily indicate a narrow ecology, and the success of these long-necked denizens may have been partly due to a versatile feeding apparatus that could be used both on the sea floor and in the water column.[3]

The Upper Cretaceous López de Bertodano Formation of Seymour Island has yielded abundant plesiosaur remains.[4] The plesiosaur now assigned to the sauropterygian genus *Aristonectes*, is also found in Patagonia. It had dental alveoli that were outwardly directed and long, slender interlocking teeth that suggested that it strained soft invertebrates, such as small swimming crustaceans, as it glided through the water. Another plesiosaur species is also present on Seymour Island, but it is too poorly preserved for accurate assessment.[5]

Mosasaurs were huge marine lizards closely related to todays monitor lizards (varanids). They appeared suddenly in the fossil record around the mid-Cretaceous and rapidly became a top predator, competing only with the plesiosaurs. Evidence of mosasaurs in the seas surrounding Antarctica is relatively scarce, limited mostly to fragmentary bones and teeth.[6] Despite this, a significant increase in the number of taxa has occurred since the early work in the 1980s. No fewer than six genera have been recognised, including new discoveries of mosasaurines, plioplatecarpines and tylosaurines. Five genera are identified in the Marambio Group alone: *Leiodon*, *Plioplatecarus*, *Mosasaurus*, cf. *Hainosaurus*, and cf. *Moanasaurus*. Although there are certainly taxa that existed in both hemispheres (e.g. *Tylosaurus*,

OPPOSITE: A reconstruction of a Cretaceous Antarctic shallow marine ecosystem with a huge mosasaur capturing a *Diplomoceras* ammonite, and eating its immense fleshy mantle. In the background, other ammonites such as *Pachydiscus* and a shark become alerted to the assault. Illustration: Tony Windberg.

A major discovery of a plesiosaur skeleton on Vega Island, Antarctic Peninsula. The plesiosaur died and sank mostly intact to the muddy Late Cretaceous sea floor, hence the wonderful preservation of the individual. Photo: Eduardo Olivero.

A cast of the cranium and jaws of the tylosaurine marine reptile, *Taniwhasaurus antarcticus,* described by Martin and Fernandez in 2005. Image: Jeffrey Stilwell, courtesy Monash Science Centre.

Hainosaurus, Mosasaurus), evidence suggests that there may have been some sort of latitudinal barrier with endemics present in the southern hemisphere, and possibly restricted to the Weddellian Province.

In 2002, a discovery on James Ross Island of the tylosaurine marine reptile, *Taniwhasaurus antarcticus*, is the most complete mosasaur skeleton found thus far. The Dinosaurio River, part of Santa Marta Cove on James Ross Island, has yielded diverse invertebrates, shark teeth, marine reptiles and the ankylosaur mentioned previously.[7] The bulk of evidence from this rich mosasaurian record in Antarctica suggests that the diversity of genera was higher near the high southern latitudes in comparison to boreal palaeogeographical oceanic realms.

One of the latest marine reptile discoveries came to light in December 2006, with the announcement of one of most complete plesiosaur skeletons ever found on the planet and arguably the most complete fossil skeleton ever found on the Antarctic continent. This particular skeleton from an elevation of 200 metres on Vega Island along the Antarctic Peninsula was special for many reasons. As typical for most plesiosaurs, erosion had disarticulated the skull from the rest of the body and the skull had been lost to the elements of time. The 1.5-metre-long petrified body of this young elasmosaurid plesiosaur was extracted from 70-million-year-old shallow marine sands of Vega Island, which were covered by volcanic ash. Did a volcanic event have anything to do with the creature's demise? Study and retrieval of the specimen by the team, led by James Martin and J Foster Sawyer (South Dakota School of Mines), Judd Case (Eastern Washington University), Marcelo Reguero (Museo de La Plata, Argentina), and others, was hazardous due to 110 km/hr winds that hammered the excavation site. The baby marine reptile rescue mission, pushed the boundaries of Antarctic fossil collecting in extreme environments, with water turning into slush before plaster could be mixed to 'entomb' the specimen for its long journey to the USA, permafrost making the ground so

A large tooth belonging to a mosasaur (left) and the needle-like tooth of a plesiosaur (right). Both were found by Jeffrey Stilwell on Seymour Island in 1985. Photos: Brian Pump.

Vertebrae of marine reptiles collected by Jeffrey Stilwell on Seymour Island in early 1985; a vertebra belonging to a moderate-sized, long-necked plesiosaur (left) and a large mosasaur (middle and right). Photos: Steve Morton.

frozen that metal tools snapped during use, and the resultant fossil payload being so heavy that it took five men to life the specimen onto the helicopter, which then delivered it to the tent camp along the shores of Herbert Sound until it was hoisted onto the ice-breaker *Laurence M Gould*.

The juvenile elasmosaur, probably related to *Mauisaurus*, a characteristic New Zealand genus known since the late 19th century, was so young that the bones had not yet been able to develop. But astoundingly, the animal's abdomen holds clues to its lifestyle. The stomach ribs span the stomach cavity, and are curiously forked and three-pronged. Preserved within the stomach are gastroliths (or stomach stones) that not only aided in digestion of food, but assisted to maintain buoyancy of the body. Associated sediments and chunks of ash with plant material inside indicates that a volcanic event resulted in the destruction of forests nearby, and the blast dumped ash into the ocean, which was inhabited by a diversity of marine vertebrates and invertebrates. High levels of silica in the system from the volcanic material had affected the exquisite preservation of the baby elasmosaur. Whether or not plesiosaurs gave live birth is another puzzle, but perhaps may still be determined with future discoveries.

Fish

The fossil fish fauna is undoubtedly the richest vertebrate fauna from the Cretaceous and Paleogene–Neogene of Antarctica, and many new groups remain to be described. As the majority of successions on the southern continent are shallow marine in origin, this is not at all surprising, since shark teeth are extremely resistant to *post mortem* burial processes, and shark and teleost vertebrae are abundant as well.

The first account of fish was in 1908 by Sir Arthur Smith Woodward, who described isolated fish remains from Cretaceous–Paleogene strata of Seymour and Snow Hill islands discovered during the Swedish South Polar Expedition. Some 72 years elapsed before new work would provide strong evidence of just how rich the record is on the Antarctic continent.[8] Much of this fish record has been exhumed from Cretaceous and Paleogene deposits of the James Ross Island Group, with only a sparse account from East Antarctica.

The oldest Cretaceous forms are from the Spartan Glacier Formation of Alexander Island, and represent a single shark tooth of the family Palaeospinacidae, possibly *Paraorthacodus*, an extinct group of neoselachian sharks. The Upper Cretaceous

A small concretion collected by Jeffrey Stilwell in 1985 from the López de Bertodano Formation of Seymour Island, when split open, revealed a single vertebra, belonging to a relatively large Late Cretaceous Antarctic shark. Photo: Steve Morton.

Ammonite-rich concretions in the López de Bertodano Formation of Seymour Island. Photo: Rodney Feldmann.

López de Bertodano Formation has revealed the presence of *Isurus* sp., *Notidanodon* and possibly *Sphenodus* as part of the primitive hexanchiforms.[9] Endemic Cretaceous teleost fish include *Antarctiberyx*, and appear to be less diverse than Late Jurassic assemblages from the Ameghino Formation of Longing Gap and James Ross Island, which include *Ameghinichthys*, *Vinctifer* and *Antarctithrissops*.

The fossil record of chimaeroids or rabbit fish (holocephalans) in Antarctica appear to also follow pelagic faunas with evidence suggesting that their evolution and diversity were controlled by shifting environmental conditions and changes in habitat. Despite the depauperate account of actual specimens found in Antarctica, the identified taxa are uniquely diverse. The extant holocephalan fish are characterised by hard tooth plates, which are fused to the braincase and comprise areas of hypermineralised tritoral tissue that act as a grinding medium for the crushing and/or cutting action of the tritoral pads. The chimaeroid group reached their highest worldwide diversity by the Late Cretaceous. Three species have been recognised in the Late Cretaceous of Antarctica, including *Chimaera zangerli* and *C. seymourensis* from the López de Bertodano Formation; and the long-ranging species *Ischyodus dolloi*.[10]

Fossil invertebrates of Seymour Island

The Mollusca classes Bivalvia, Gastropoda, Scaphopoda and Cephalopoda are represented in the thick Upper Cretaceous to lower Paleogene sequence on Seymour Island.[11] At present 34 bivalve species, at least 32 gastropods (many still not formally described), one species of scaphopod, at least 18 ammonites, one nautiloid, and one belemnite species have been recorded in the López de Bertodano Formation.

Gastropoda

Marine snails represent a moderate component of the molluscan assemblage. The early 20th century reports of Cretaceous Antarctic gastropods identified 18 species, many of which were new, from the islands of the James Ross Basin, including Seymour Island. However, the oldest Cretaceous molluscs are from the Lower

A superbly preserved fossil whelk *Taioma charcotianus,* collected during the Swedish South Polar Expedition of 1901–03, and described by Otto Wilckens in 1910. Photo: Brian Pump.

Cretaceous sediments of south-eastern Alexander Island, and many species remain undescribed.[12]

The gastropods of Seymour Island are very well preserved and many have their original shells with little or no alteration over a long time. Most fossils exhibit little alteration from diagenesis, so the exquisite nature of preservation of the Antarctic material allows detailed study of the diverse macroinvertebrate groups. The 32 gastropod species belong to at least 16 families and carnivores dominate, followed by in descending order of abundance: ectoparasites, deposit feeders and herbivores. The specimens exhibit little to no abrasion that indicate only a small to moderate degree of destructive erosive and/or transport processes, implying relatively quiescent conditions. Moreover, some of the shells reveal evidence of biologic activity, such as many small borings. These features, interpreted to have been made by epibionts such as algae, necessitate that, either in life or in death, the shell had to be relatively stable on the sea floor for a long enough period of time to allow the colonisation of its surface. One cannot discount the possibility that some shells, even while colonised, could have been transported and rapidly buried to the final site of deposition. The absence of other indications of transport from the study area such as sedimentological structures (e.g. current structures, etc.), suggest that *post mortem* transport of the material was at a minimum.

One account of a gastropod limpet (a species of *Acmaea*) in living position was published by Bill Zinsmeister in 1990. This is one of only a handful of known occurrences of limpets preserved in living position in the fossil record. As the herbivorous alga-feeding limpets live attached to a firm surface today, upon death the shells become detached, are incorporated in the nearby sediments, or are transported to the site of final deposition. A specimen of *Acmaea* was collected from Seymour Island that was tightly affixed to the large limid bivalve *Acesta shackletoni.*[13] We can surmise that when the bivalve rolled over in the soft sediment, it pinned the limpet between the shell and the sediment surface. The poor limpet may not have been able to escape because the weight of the bivalve shell pressed it firmly into the muddy sea floor, where it eventually saw its fate, preserved exquisitely in a medium-sized concretion.

The evidence of predation on several specimens is demonstrated by circular borings on the gastropods, probably made by naticids present in the fauna or even cephalopods such as the octopus. Damage to the outer lip of irregular growth scars in some gastropods indicates also the presence of 'lip-peelers' in the assemblages such as crabs – well known for their predation on molluscs.[14]

As Seymour Island was an integral part of the Weddellian Province during the latest Cretaceous,

An exquisitely preserved Late Cretaceous marine gastropod, *Amberleya spinigera*. Photo: Brian Pump.

this moderately diverse faunule provides important, additional insight into the character of the province. The accumulated data indicate that the Weddellian Province was inhomogeneous with respect to gastropod faunal composition. It seems likely that Antarctica, southern South America, New Zealand-Chatham Islands, New Caledonia and Australia had started to develop endemic centres by the end of the Cretaceous, especially given the final fragmentation of Gondwana and segmentation of oceanic circulation patterns, which led to increased faunal provincialism. The gastropods were characterised by moderate endemism at genus level (34 per cent) and high endemism at species level (nearly 100 per cent). Also, at genus level, gastropods are more cosmopolitan than thought previously. Bivalve relationships in the Austral Realm, in contrast, were higher, reflecting the different range and diversity modes of the two classes, which presumably arise in turn from their differing modes of life. Bivalves are primarily infaunal filter and deposit feeders, while gastropods, primarily epifaunal, have radiated into a wide variety of habitat and feeding types. With this variation in mode of life, gastropods have become more diverse and generally more specialised than bivalves, as reflected in the fossil record and modern assemblages. Two of the most abundant and characteristic gastropods are the large noded trochid *Amberleya spinigera* and the winged *Austroaporrhais larseni*.[15]

Bivalvia

The first Cretaceous bivalve was collected during Captain Larsen's visit to Seymour Island in January 1893 on the Dundee expedition.[16] Although Sharman and Newton first described fossil shells from Seymour Island, apparently this one Cretaceous specimen was misplaced and not made available to Sharman and Newton until after their first publication on the fauna. It is known that no Cretaceous rocks occur along the beach where Larsen made his landing, and there is no evidence that he ventured inland. Thus, he probably collected the specimen of *Lahillia* on the beach, especially as specimens of this group are extremely abundant in the upper part of the Cretaceous sequence and it seems unlikely that he

Postage stamps commemorating the British Antarctic Survey and depicting a range of Cretaceous bivalves: (from left to right) *Aucellina*, a species of *Pinna*, and *Trigonia* – members of a worldwide Mesozoic group of marine clams.

would have just collected one specimen. The highly abraded nature of the specimen indicates that it was well worn from beach abrasion. This description of *Lahillia* would be the only record of Cretaceous clams from the 19th century as the next account would not be until 1910, when Otto Wilckens described the bivalve fauna, along with gastropods, collected during Nordenskjöld's 1901–03 Swedish South Polar Expedition. Some 17 species of bivalves were recognised and a few of these were found to be the same as those described previously from southern Argentina.[17]

Fieldwork during the 1970s on Seymour Island revealed a much higher diversity than thought previously and this work culminated in a major revision of the Cretaceous bivalve fauna of Seymour Island.[18] More recent work by Jeffrey Stilwell revealed that Cretaceous bivalve ties, in terms of palaeobiogeography, were strongest between New Zealand and the Antarctic Peninsula during the Maastrichtian, a somewhat surprising result given that these faunas were separated by more than 90 degrees of longitude during the Late Cretaceous.[19] Bivalve links between southern South America and New Caledonia were moderate.

Ammonites

There are strong links and similarities between the ammonite faunas of Patagonia, Quiriquina (Chile), Antarctic Peninsula and the Carnarvon Basin in Western Australia.[20] The first reports of ammonites in the Antarctic Cretaceous was by Stuart Weller in 1903, followed by the popular account of the Swedish South Polar Expedition by Nordenskjöld in 1904, and the 1909 cephalopod monograph by early 20th century experts Kilian and Reboul. The

Dubbed the 'ammonite trap', these are concentrations of many types of cephalopod, large and small, from the Cretaceous of the James Ross Island Group. Photo: Eduardo Olivero.

A spectacular specimen of the ammonite *Pachydiscus*, showing its richly purple mother-of-pearl shell. Photo: Steve Morton.

The strongly ribbed and nodded ammonite, *Grossouvrites gemmatus*, from Seymour Island. Photo: Steve Morton.

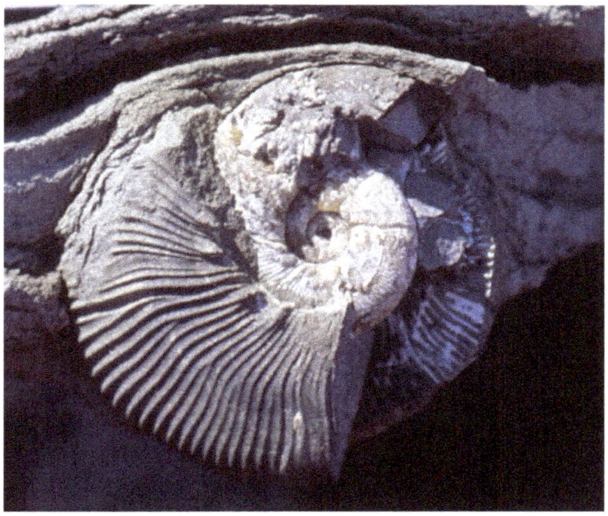

An unusual preservation of a vertical specimen of the ammonite *Maorites* and horizontal rock strata. The sediment is draped around the ammonite, and lithified. Photo: Eduardo Olivero.

dominant ammonite family, the Kossmaticeratidae, occurs only within this region.[21]

Some Antarctic ammonites attained very large sizes and are among the biggest recorded. An immense heteromorph ammonite (*Diplomoceras maximum*) collected during the 1994–95 austral summer by Bill Zinsmeister and his colleagues reached an uncoiled length of four metres. When coiled, this gargantuan ammonite resembles a rather mighty paperclip! The cephalopod occupied one-half

A superb *in situ* heteromorph ammonite *Ainoceras zinsmeisteri* from the Late Cretaceous. It was named after Bill Zinsmeister of Purdue University. Photo: Eduardo Olivero.

A James Ross Basin outcrop jam-packed with many individuals of the small ammonite *Neograhamites*. There are so many fossils of the same species that the deposit can be termed a 'coquina'. Photo: Eduardo Olivero.

of the shell, the rest being filled with gas and fluid within septate chambers to provide buoyancy for the shell. It is likely that this particular species could not swim actively, unlike most ammonites, and probably just drifted along near the sea floor and grabbed any food that crossed its path on the sea bed.

The ammonites of Seymour Island record and reflect the marked environmental changes at the close of the Cretaceous Period and are among the youngest species in the world, having gone extinct some two metres of section before the K–T boundary event, representing only a smidge of geologic time before the asteroid impact. Of particular recent interest is the documentation of ammonoids surviving the K–T boundary event for at least 200 000 years in Denmark and Netherlands, and probably in North America as well.[22] Research work on this 'Dead Clade Walking' phenomenon has revealed cases of vertebrates and invertebrates that cross (barely) the K–T boundary, but only to meet their final demise in the dawn of the Cenozoic Era.

Nautiloids

Nautiloids are poorly represented in the Antarctic fossil record, compared with ammonites, but this successful group of cephalopods survived the K–T boundary mass extinction and are alive and well today in warm Indo-Pacific waters. Two unnamed species of *Eutrephoceras* and *Cimomia* have been recorded from the Cretaceous of Seymour Island. *Cimomia* became extinct at the K–T boundary, but *Eutrephoceras* survived on Seymour Island into the Danian Age, that is, the first four million years of the Paleocene Epoch.

Belemnites

The belemnite family Dimitobelidae arose during the Aptian and was widely distributed among the Gondwanan landmasses. After Albian time, the family was represented only by the genus *Dimitobelus*, which by the Campanian–Maastrichtian, was represented by species endemic to portions of the Weddellian Province, including New Zealand-Chatham Islands, James Ross Island, and Seymour Island. As the Dimitobelidae were believed to have been restricted to shelf environments, it seems that the break-up of Gondwana and the subsequent southward movement and isolation of Antarctica-Australasia were the primary factors, which controlled the geographic distribution of this family. The development of a circum-Antarctic watermass may have also further contributed to this isolation. Belemnites are present on Seymour Island at only one locality, appropriately called the 'belemnite battleground' as they occur there in a monotypic concentration of scores of specimens with little else in the deposit.

Elsewhere in the James Ross Island Group, belemnites are locally common but disappeared by the Late Campanian. It is surprising that this locality, some 620 m below the K–T boundary, is the only horizon where they have been found on Seymour Island, especially when belemnites are relatively common elsewhere on Alexander Island and the islands of the James Ross Island Group. The material consists of several hundred specimens that were collected from just below the crest of a low ridge extending for a distance of about 2.5 km.[23] This single occurrence on Seymour Island suggests that a short-term oceanic perturbation allowed the belemnites to re-enter the James Ross Island Basin for a brief interval during the Early Maastrichtian, after which they apparently became extinct. This short-term event may be associated with changes in thermohaline circulation during the Early Maastrichtian.

Cnidarians

Corals have a relatively scant fossil record in Antarctica.[30] Most of the fossil corals described in the literature are from Cretaceous to Paleogene strata of the James Ross Basin, Antarctic Peninsula and King George Island. The fossil record of Seymour and Snow Hill islands include stylasterid hydrozoans and scleractinians.[31] An Eocene erratic from McMurdo Sound was found to include probable scleractinian corals.[32] Scleractinian corals are present in the Oligocene beds of the Cape Melville Formation of King George Island and those of the same age from the Victoria Land Basin.[33] Probably the most abundant coral in the Antarctic fossil record is the Cenozoic scleractinian *Flabellum rariseptatum*,

An extraordinary find of a complete belemnite fossil called a 'guard'. This belemnite, *Dimitobelus*, is shown here in pieces, next to coiled annelid worm tubes of *Rotularia*, which lived on the floor of the shallow Antarctic sea. Photo: Eduardo Olivero.

The K–T boundary mass extinction

The catastrophic extinction event that took place around 65 milion years ago wiped out about one-third of genera and two-thirds of species globally. Despite the fact that the Permian–Triassic mass extinction (~251 mya) was much more severe in terms of loss of life and environmental upheaval, the K–T (Cretaceous–Tertiary) boundary event receives more media attention, undoubtedly because the dinosaurs met their doom at this time after reigning the world for 160 million years. The event exterminated them all (apart from birds), as well as all marine reptiles, pterosaurs, a host of marine invertebrates, many plants and micro-organisms, etc. But was the extent of extinction the same all around the planet or was there a gradient in extinction intensities? Can Antarctica provide clues in the extinction debate?

The Cretaceous Period was a time of a 'greenhouse Earth' with little to moderate sea surface temperature gradients both in the sea and on land. Climatic conditions were warm and humid with only a slight temperature gradient between the equator and poles. Sea levels were generally high with large parts of the continents being covered by shallow seas. Flowering plants were evolving and destined to become incredibly diverse across the globe. Moreover, the final break-up of Gondwana resulted in the effective isolation of the southern continents by opening of the Southern Ocean. The isolation of major ocean basins increased shelf area around the dissected continental fragments, but at the same time created increased biotic provincialism, and thus, endemism. The global thermal decline at the end of the Cretaceous had ensuing environmental and biotic changes to both terrestrial and marine realms.

Seymour Island contains the most complete, well-exposed section of Upper Cretaceous to lower Paleogene rocks in the southern hemisphere and is composed of highly fossiliferous marine sandstones and siltstones reflecting a spectrum of environments along the shallow to mid-shelf. The sequence is exposed over the southern two-thirds of the island, an area of approximately 70 km^2. The extensive exposures of these rocks include one of the best K–T boundary sections in the world, providing an opportunity to examine in detail the events at the close of the Cretaceous.

The K–T boundary interval is located at the base of a glauconitic unit and crops out in a snake-like patterns for approximately 7 km, forming a belt about 2 km wide. No other K–T boundary section has aerial exposures that approach those on Seymour Island. Faunal diversity is relatively high with an admirable record of molluscs, corals, brachiopods, annelids, decapods, echinoderms and marine reptiles. Substantial tree fragments up to one metre in diameter are abundant at certain levels in the succession.

Analysis of the geological range data of species recovered from this boundary interval reveals that at least 20 species (55 per cent) disappeared through a 16-metre interval below the K–T boundary. Two extinction steps occur 1.5 m and 1 m below the Lower Glauconite. Two species (6 per cent) last occur in the K–T Glauconite and 14 (39 per cent) Maastrichtian species pass through the boundary event into the Paleocene. This survivorship of some 40 per cent is among the highest in the world and suggests that the magnitude of the event in the high southern latitudes was buffered to a degree.

The survival interval is represented in the 'Fish Bone Layer' and is believed to represent the 'crisis zone' immediately following the boundary event. The significant occurrence of fish debris over an interval of around two to three metres indicates multiple local kill events, which suggests that this layer represents an interval of unstable conditions following the extinction (e.g. plankton bloom, etc.). Based on radiometric dating of the deposits, this period of unstable conditions probably lasts some 20 000 years, and the recovery interval commenced some 300 000 years after the event, reflecting the influx of opportunistic species that filled the ecospace vacated by the extinction. Research in 2010 provided further strength to the hypothesis that the asteroid impact was the 'final straw'.

occurring also in the sub-Antarctic record and Tierra del Fuego, southern South America, indicating a circumpolar distribution for the group.

Crustaceans

In terms of distribution of decapod crustaceans, lobsters of the family Glypheidae are moderately well represented in the Antarctic Cretaceous, with at least three recorded species: *Glyphea alexandri* and *G. georgiensis*, both from Lower Cretaceous rocks of Alexander Island, and *G. australensis* from the Campanian–Maastrichtian of James Ross and Vega islands. Distortion of the carapace material of *G. australensis* indicates that the carapace was relatively supple and soft, and there is additional evidence of moulting stages. The number of decapod crustacean species from the southern hemisphere rivals that of the northern hemisphere.[24]

A Cretaceous mud-ball concretion collected in early 1987 from Humps Island, contained the well-preserved remains of a lobster, *Hoploparia stokesi*. Photo: Steve Morton.

Echinoderms

The first published account of the fossil record of echinoderms in Antarctica was based on the collections from the Swedish South Polar Expedition.[25] Echinoderms form a small, but nevertheless important component of the Mesozoic and Cenozoic marine invertebrate faunas of Antarctica, and at some localities, one finds little else. These monotypic concentrations, such as those in the La Meseta Formation of Seymour Island, comprise extremely abundant remains of echinoids (many of the sea urchins with spines in tact), asteroids (both complete starfish and layers of 'arms'), crinoids (sea lilies), and delicately preserved ophiuroids (brittle stars). The oldest Antarctic echinoderms are echinoids from the Early Cretaceous (Aptian) of Alexander Island.[26] The richest assemblages are from the lower two-thirds of the Marambio Group in the James Ross Basin, where no fewer than 17 taxa have been recognised from the Santonian to the Early Maastricthian.[27] The fossils comprise complete or partial tests (i.e. exoskeleton) and abundant spines. These species encompass those

The discovery of *Metanephrops jenkinsi* from the López de Bertodano Formation extended the range of a widespread extant genus into the Cretaceous. Photo: Rodney Feldmann.

Hoploparia stokesi was the first decapod and possibly the first fossil named from the Antarctic. Photo: Rodney Feldmann.

To give an idea of scale, Alistair Crame of the British Antarctic Survey lies next to the gargantuan heteromorph ammonite *Diplomoceras*. These ammonites evolved into weird and wonderful shapes at the end of their time on Earth, but how did they manage to swim? Photo: David Cantrill.

of regular echinoids, holasteroids, and spatangoids. Most of the species reveal affinities with those taxa of Australia, New Zealand, south-east Africa and South America. One group, however, *Micraster* aff. *regularis*, is morphologically more like species from Spain and France. These assemblages were buried along the deeper part of the Antarctic shelf during the Late Cretaceous. The absence of cassiduloids suggests that the fauna was not in a shallow water environment on the shelf. Moreover, no marsupiate species (those with brooding pouches) have been found in the older echinoid assemblages, but there are a marsupiate cidaroid and spatangoid species from the Maastrichtian and Eocene of Seymour Island, respectively.[28] Perhaps the deep water habitats of the older Cretaceous forms is related to this anomalous lack of marsupiates on James Ross Island, or the lack of fauna data.[29]

Sponges

In 2011, the first fossil sponge was described from Antarctica. *Laocoetis piserai*, from the mid-Cretaceous of James Ross Island, lived in relatively deep Antarctic waters on the margins of an active volcanic arc.

OPPOSITE: Mike Woodburne at a remote fossil site on Seymour Island, on 5 March 1982, the day fossil mammals were discovered in Antarctica. Photo: Bill Zinsmeister.

11
KILLER BIRDS, GIANT PENGUINS AND EARLY MAMMALS

The destruction which the ancient fauna of the Antarctic Gulf has endured and the length of time which has elapsed since its expulsion, deprives us of much hope of reconstructing it ... But palæontology must be chiefly called on to relate the story of the decline and fall of the Antarctic marine fauna.

Charles Hedley, *Considerations on the Surviving Refugees in Austral Lands of Ancient Antarctic Life*, 1895.

The organisms living at the beginning of the Cenozoic Era, which followed the K–T boundary mass extinction event, heralded the onset of the modern fauna. The pioneers of this 'brave new world' were made up of sturdy survivors, opportunistic and migrant taxa, and newly evolved groups that filled the ecological vacuum left by the mass extinction. The marked changes in composition of the biota after the Cretaceous reflects a real phenomenon, a response relating to myriad complex factors associated with extinction and survivorship processes. As communities stabilised throughout the Early Paleogene, the diversity of life forms climbed in both terrestrial and marine realms, albeit with some relatively smaller fluctuations due to more minor extinction episodes. The extremely rich fossil records of Seymour and Cockburn islands, along the Antarctic Peninsula, and the newly discovered fauna and flora in the McMurdo Sound region, attest to this spectrum of life and habitats which marked the dawn of the modern biota.

The summer campsite on Seymour Island in 1982. Photo: Mike Woodburne.

A tidal channel deposit in the La Meseta Formation of Seymour Island abounding with thousands of shells and many round concretions. This site also contains many fossil bird bones. Photo: Jeffrey Stilwell.

Whereas the Paleogene Period is represented by fossiliferous sedimentary rocks found in only a few areas, strata of the Neogene Period crop out in several areas dotted around the periphery of the continent. Detailed assessment of climatic changes in Antarctica throughout the Cenozoic Era is made possible by thick marine sequences preserved on land and extensive core samples from long-term, scientific drilling programmes.

The dawn of the 'brave new world' in the earliest Paleogene marks a major change in both fauna and flora, reflecting a turnover in the biota resulting from the decimation of scores of terrestrial and marine organisms affected by the K–T boundary crisis. The so-called 'disaster taxa' that survived the extinction event represent the winners and are found in the fossil record in very low numbers of individuals. The only place in the high southern latitudes to see the after-effects of this calamity in great detail is Seymour Island, where the rocks tell a story of great upheaval, even some 10 000 km from the asteroid impact site in Mexico. The fossil record relays also the pattern of recovery of organisms in the shallow marine environment of Antarctica. In particular, the Seymour Island record assists in unravelling some global questions concerning the K–T boundary event. These include:

- Why were the survivors spared from the mass extinction and what made them so special, compared with other groups?
- Why were some groups extinction resistant and others not?
- What mechanisms caused the great disparity in diversity of various groups across the boundary?

It is only since late 2004 that the Antarctic record has provided some clues to these major queries. Discoveries on Seymour Island have provided exciting new information on the magnitude of the extinction event in the high southern latitudes and also the complex processes of biotic recovery and faunal rebound with changing environmental conditions. Studies such as this are extremely valuable and relevant to today, as we are in the midst of another mass extinction event, comparable to the K–T boundary tragedy, with accelerated extinction rates that indicate that, at current estimates of rates of extinction, half of all species on the planet may be extinct by 2100! To compound matters, scientists are uncertain as to just how many species there are on the planet – approximations are from some 3 to 100 million, but no one knows for sure!

A spectacular accumulation, probably tidal channel, of a shell coquina from the middle section of La Meseta Formation of Seymour Island, dominated by large, disarticulated clams of smooth-shelled *Lahillia* and coarsely ornamented *Cucullaea*.
Photo: Jeffrey Stilwell.

It is important to bear in mind that, although the K–T boundary event was a prime mechanism of change, subsequent biotas of the Paleogene were shaped also by palaeoceanographic events, including the final break-up of Gondwana and ensuing changes in oceanic circulation and climate. The northern hemisphere record indicates major diversity changes in both terrestrial and marine realms, resulting from this event.

Paleogene molluscs

Paleogene macrofaunal remains are relatively scarce and distributed sporadically in the stratigraphic record of Antarctica; where they occur in many instances the record is quite rich and diverse. The paucity of palaeontological data for the Paleogene renders the recovery of fossil materials from almost any stratigraphical level a significant discovery.

The glacial moraines of Mt Discovery are the focus of many expeditions by the University of Nebraska-Lincoln team to East Antarctica, in search of Eocene life. The Royal Society Range is in the background. Photo: Jeffrey Stilwell.

A densely packed accumulation of the bivalve *Cucullaea raea*, representing a tidal channel deposit in the La Meseta Formation of Seymour Island. Photo: Mike Woodburne.

Further, this is true even the specimens occur in sediment clasts that have been removed from their stratigraphical context, such as the prolific fossil record from the Eocene glacial erratics from McMurdo Sound that yield our only known window into the life of that time from the Transantarctic Mountain region of Antarctica. The glacial erosion and subsequent transportation of Paleogene sedimentary materials and their inclusion in glacial sediments provide a useful means to increase the palaeontological record from Antarctica by providing a window into subglacial basin stratigraphy.

Molluscs dominate the shallow marine assemblages of the Antarctic Cenozoic Era.[1] However, our knowledge of molluscan fossils from the mid-Cenozoic, especially from the Late Oligocene to Middle Miocene interval is very scant and most assemblages are of relatively low diversity.[2] By far the acme of diversity for shelled molluscs is in the Early Paleocene of Seymour Island, and the Eocene shells from Seymour and Cockburn islands,

A large block of fossiliferous sandstone from the La Meseta Formation of Seymour Island, revealing a wealth of fossils, mostly macroinvertebrates. Photo: Steve Morton.

and the McMurdo Sound region. Interestingly, the Danian shells of Seymour Island are the only well-dated species of molluscs of this age anywhere in the southern hemisphere and are extremely important in they assist us to decipher the series of events at the end of the Cretaceous and also the post extinction rebound and recovery interval after this major biosphere upheaval 65 million years ago. The only other Early Paleocene fossils recorded are from New Zealand (although dated as Late Danian), and those of southern Patagonia, which are also rather poorly constrained biostratigraphically.[3]

The molluscan record from the earliest part of the Paleocene Epoch on Seymour Island is the only shelf fauna of this age from Antarctica, and provides the critical insight into the emergence of marine faunas of the southern hemisphere following the mass extinction at the end of the Cretaceous. In fact, Seymour Island has the best preserved record of macrofaunal turnover at this major extinction boundary compared with other austral sections. There may be others, especially in Patagonia, that remain unstudied, but Seymour Island remains the most detailed.

During the Swedish South Polar Expedition of 1901–03, Otto Nordenskjöld, Gunnar Andersson and their colleagues collected fossil material on Seymour Island. The only trained geologist was Andersson but his fieldwork was limited to the last days of the expedition. At the same time, Nordenskjöld's work was slowed down by several mishaps, which prevented these early explorers from recognising the remarkable fossil record of this small Antarctic island. Although they spent two-and-a-half years on the islands of the James Ross Island group, only a very limited amount of time was devoted to collecting fossils. They made seven sledding trips to Seymour Island and, except for the last one made by Andersson during late October and early November of 1903, each visit lasted only three days or less.

A beautiful example of the protobranch, deposit-feeding bivalve, *Australoneilo rossi*, from the La Meseta Formation of Seymour Island. Coincidentally, this locality is famous for the discovery of the first fossil mammal remains from the Antarctic continent. Photo: Brian Pump.

Fossil macroinvertebrates collected from Seymour and Snow Hill islands during the Swedish South Polar Expedition of 1901–03, described as Cretaceous. Source: Nordenskjöld, 1904.

Moreover, most of the sledding trips were restricted to the east coast, which is the least fossiliferous part of the island. An exception was the Bodman party of 21 to 25 November 1902, which forayed along the west coast of the island. They passed through Cross Valley to the east coast to collect penguin eggs at the Adélie penguin rookery. During this trip the members passed through some of the most richly fossiliferous strata of the island, but only a few fossils were collected without specific locality data. As a result of this haphazard collecting, the first major works on the mollusc fossils of Seymour Island by Otto Wilckens lacked locality details. It was only during the last visit to Seymour Island by Andersson that adequate field data were compiled. At this time little was known of the stratigraphy of the island, and some of Wilckens' taxa were collected from different stratigraphic horizons and ages. Furthermore, the common belief of the time was that the Danian belonged to the uppermost part of the Cretaceous instead of the Paleogene, as it is today. Gunnar Andersson showed great foresight as the absence of ammonites in the Danian sections suggested to him that these fossils were, indeed, Paleogene in age, but Wilckens nevertheless included them with the Cretaceous taxa that he described in his monograph.

To date, 60 species of Antarctic molluscs have been recorded from this interval, and two-thirds have been described.[4] Of these, 20 species of bivalves, 37 species of gastropods, two species of nautiloid, and one species of scaphopod have been recognised – all collected from a spectrum of temperate beach to tidal-dominated environments. This has given us, unlike anywhere else in the southern hemisphere for macroinvertebrates, details of the changes in the environment over thousands of years and how these perturbations affected the biota.

Suspension-feeding bivalves dominate the Danian clam fauna at 55 per cent, followed by 45 per cent of deposit feeders. Epifaunal bivalves are surprisingly rare during the Danian. Epibyssate clams are also rare and comprise the rare species, *Acesta webbi*, *Pinna freneixae* and *Seymourtula antarctica*. Deposit feeders are members of the Nuculidae, Nuculanidae and Mallettiidae – all infaunal burrowers. This is consistent with the idea that epifaunal forms were particularly affected by the phytoplankton crash at the K–T boundary, leading to the demise of many suspension-feeding groups, compared with deposit feeders. In the Austral Realm, this pattern of selective extinction is corroborated by the K–T sections in Antarctica and also Early Paleocene molluscan faunas of New Zealand.

Lahillia larseni is a relatively common member of the Maastrichtian bivalve community, but despite its survival across the K–T boundary into the Danian, it remains a rare member of the Early Danian assemblages, where it usually represented by fewer than five individuals at any given locality. However, in the Upper Glauconite, it is locally abundant, occurring in floods of thousands of individuals. The number of articulated specimens of these bivalves and other species is relatively high, indicating that

This glacial moraine along Mt Discovery is a treasure trove of fossil remains, among a sea of disparate rock types including sandstones, dolerites, diorites and granites. In the middle is a circular piece of fossiliferous sandstone, which contains macroinvertebrates either similar or the same species as those from Seymour and Cockburn islands from the Eocene. Photo: Jeffrey Stilwell.

the environments in which they lived were below fair-weather wave base. The absence of shallow-water sedimentary structures in these units corroborates this interpretation.

The Paleocene marine gastropods lived in sandy, shallow- to middle-shelf environments in temperate coastal waters of normal salinity. In descending order of species-level diversity, the gastropod fauna consists of five general groupings: carnivores, epifaunal grazers, deposit feeders, epifaunal browsers and a single ectoparasite. Some forms inhabited both epifaunal and infaunal habitats during their life cycle. Most of the Danian snails were carnivores (60 per cent), and all were considered to be highly mobile taxa. Carnivores represented so-called 'bloom taxa' that flourished in post K–T environments. None of the recorded gastropods is particularly abundant at any locality, apart from the aporrhaid *Struthiochenopus hurleyi* and cerithiid *Bittium (Zebittium) brooksi*. Their abundance and association

The very large, suspension-feeding clam *Lahillia* from the Eocene is one of a few bivalves to actually flourish after the K–T boundary mass extinction event. Although the last Antarctic *Lahillia* bivalves are recorded in the La Meseta Formation of Seymour Island, the genus persisted into the Miocene in Patagonia. Photo: Steve Morton.

One of the most characteristic invertebrates from Antarctica, *Antarctodarwinella,* is an early member of the Struthiolariidae, and is recognised by its normally heavily callused, globose shell with sinuous growth lines. These marine snails are so prolific on Seymour and Cockburn islands that they have been used for local range zones in the biostratigraphy of the La Meseta Formation. Photo: Brian Pump.

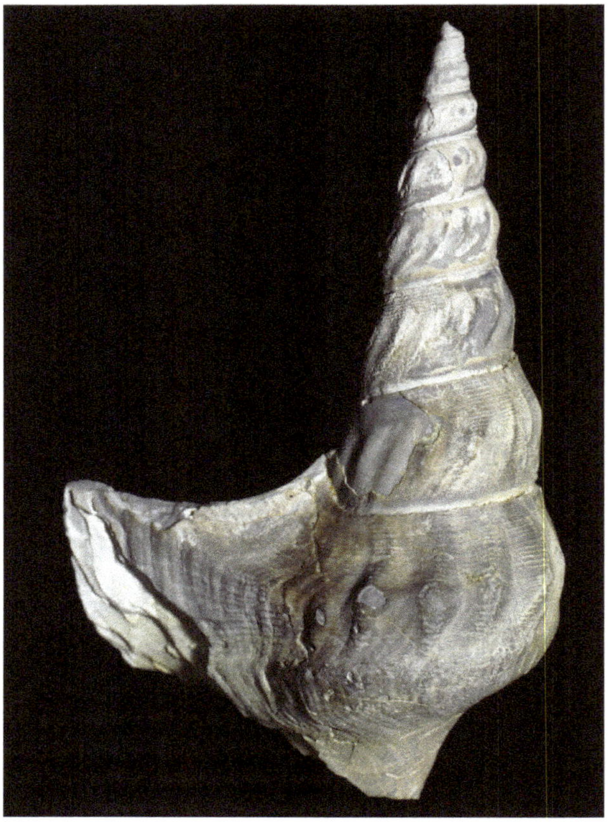

The majestic fossil gastropods of Antarctica, *Struthioptera camachoi,* was named in 1977 after Argentine palaeontologist Horacio Camacho. This winged family of Aporrhaidae is represented only by this species and one other in the La Meseta Formation on Seymour Island. Photo: Steve Morton.

A survivor of the K–T boundary mass extinction event, *Taioma* lived on in the Antarctic region after the extinction event for some 20 million years before its eventual demise, probably due to rapidly cooling conditions in Antarctica at the end of the Paleogene. Photo: Steve Morton.

The bivalve *Panopea philippi,* preserved in the upper part of the La Meseta Formation, died in life position in the Late Eocene with its posterior pointing upwards. It belongs to a group of deep-burrowing, suspension-feeding clams called hiatellids. Photo: Jeffrey Stilwell.

Fossil echinoids (sea urchins) from the La Meseta Formation of Seymour Island. Some are so well preserved that hair-like spines are still present in exceptional specimens. Photos: Steve Morton.

with the mass accumulations of *Lahillia* suggests that they were opportunistic forms and appear in great numbers following the boundary event. The overall rarity and low diversity of gastropods attests to the significant impact of the K–T boundary event and is a reflection of the decimation and general low diversity of macroinvertebrates, in general, after the mass extinction. As the diversification with the faunas began to increase resulting from the stabilisation processes in the marine realm, the isolation of Antarctica also became more pronounced as the other southern continents moved northward. These major changes associated in oceanic circulation reflecting the final break-up of Gondwana enhanced this isolation of Antarctic faunas.

There is a rather substantial gap in the Paleogene record of molluscs. This hiatus from the Danian to the Early to Middle Eocene represents approximately more than 10 million years of the Antarctic record. However, the possibility remains that the unstudied and poorly dated faunas of Cross Valley on Seymour Island may fill in part of this gap, although the mediocre preservation and low diversity would not bolster the record to any large degree perhaps even with detailed collecting in the field. The Eocene Epoch is an intriguing time where climatic conditions deteriorated rapidly from the so-called 'greenhouse' conditions that dominated Earth from mid-Mesozoic time through the Cenozoic to the 'icehouse' conditions that have dominated the

One of the best preserved accumulations of Eocene macro-invertebrates with bivalves such as *Miltha* and *Panopea* and struthiolariid gastropods, collected just east of Mt Discovery campsite in November 1995. Photo: Jeffrey Stilwell.

A fossilised Eocene reef environment in the lower section of the La Meseta Formation, characterised by a diversity of benthic, bottom-dwelling organisms such as mobile asteroids (starfish), attached brachiopods (lamp shells), large pectinid bivalves, and hemispherical and lacy bryozoans (moss animals). Photo: Eduardo Olivero.

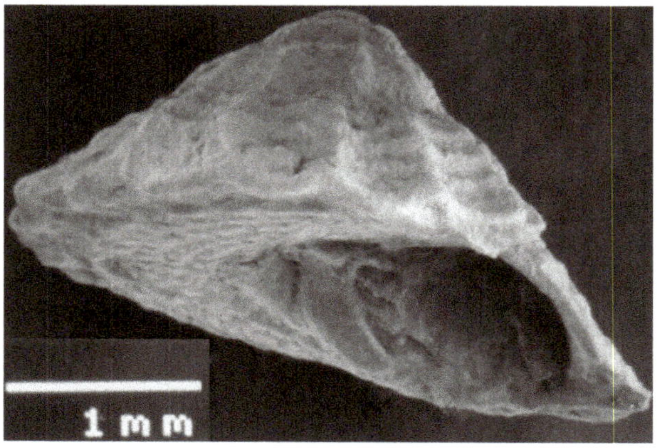

A closer look (SEM) at some sedimentary deposits from the Eocene fossiliferous glacial erratics of McMurdo Sound reveals a world of small invertebrates, such as this – the only known Antarctic record of *Astraea lilliputia*. Image: Jeffrey Stilwell.

climate since that time. The record of Eocene rocks is sparse, but this is no surprise given the small area of rock not covered by ice and snow and available for study.

The discovery of Eocene rocks in East Antarctica provides only the second major site on the entire continent where we can study the palaeontology of this important interval.[5] In most places around the planet, the study of fossils from erratics, and thus out of context from their original site of deposition, would warrant little study and thus, not a lot of attention by scientists. But, when most of the continent is covered, this study becomes much more important and exciting. There can now be a comparison of the biota between East and West Antarctica during the Eocene, and some palaeontological surprises that enabled a large group of scientists to interpret the climatic and palaeoceanographic setting of this high latitude environment some 50 to 40 million years ago.

Our record of Eocene macroinvertebrate is astounding, with more than 200 species from Seymour and Cockburn islands, and the McMurdo Sound region. On Seymour Island alone, 172 species of Eocene molluscs have been recorded from 196 fossil localities, and many more taxa await formal description. The next most diverse assemblages of Eocene molluscs is from McMurdo Sound, East Antarctica, where highly fossiliferous sandstone and siltstone erratics abound in the glacial moraine deposits which blanket the flanks of Mt Discovery and Minna Bluff. There are fewer Eocene fossils from the surrounding islands, including Ross Island. A single modiolid mussel recovered from sediments may be as old as Late Eocene in age.

Although shells dominate the shallow shelf marine assemblages during the Eocene, other macroinvertebrate groups are quite common and some not usually encountered in the field, are exquisitely preserved such as ophiuroids or 'brittle stars'. The La Meseta Formation is known worldwide for its diversity, too, of brachiopods, echinoderms (echinoids or sea urchins, asteroids or starfish, crinoids or sea lilies), cirriped crustaceans or barnacles, arthropods (crabs, lobsters, etc.), calcareous annelid worms, cyclostome and cheilostome bryozoans, and corals. Worthy of mention here are the crustaceans, which comprise 13 species, mostly crabs, which is the most diverse assemblage of fossil decapods known from high-latitude sites.[6] They revealed that the shallow waters surrounding Antarctica during the Eocene played an important role as an area of origin and a dispersal pathway in the biogeographical distribution of decapod taxa and other macroinvertebrate groups. Poriferans (sponges) are conspicuously absent, despite being a dominant group of Pleistocene and Holocene macroinvertebrates in subfossil assemblages today, such as the beautifully preserved sponges (some 10 000 years old) in the 'Ice Pinnacles' of McMurdo Sound. Nearly all of the Neogene records of echinoderms are from the Miocene of King George Island (South Shetland Islands), Cockburn Island, McMurdo Sound and the Vestfold Hills. Microfossils are not as abundant in the Peninsula Eocene, but are represented by marine palynomorphs, benthic foraminifera and ostracods. However, an abundance of microfossils have been recovered from the Eocene of East Antarctica, where rich assemblages of marine

The large, articulated bivalve *Eurhomalea antarctica* lived in great abundance in the shallow sands along the Antarctic Peninsula in the mid-Eocene. This specimen highlights the morphology of the shell with growth lines providing scientists with data on how long it may have lived and what the sea-surface temperatures were like at the time. Photo: Brian Pump.

diatoms, ebridians, silicoflagellates, spores, pollen, and their palynomorphs, have been studied in detail.[7]

The important record of marine macrofossils identified in ocean drill cores should also be mentioned, as many fossils of Mollusca, Annelida, Bryozoa, Cnidaria, Brachiopoda, Echinodermata, Porifera and Vertebrata do provide some information on the make-up of the mid-Cenozoic Antarctic biota – however scant it may be.

Paleogene and Neogene plant fossils

Until recently, the Cenozoic flora of Antarctica was restricted to surrounds of the Peninsula – a single Late Oligocene leaf from the CIROS-1 borehole in McMurdo Sound, and younger, Late Neogene floral assemblages from the Sirius Group of the Transantarctic Mountains.[8] The discovery of abundant plant material in Eocene glacial erratics of the McMurdo Sound region during the 1992–93 and 1993–94 field seasons are the most southerly Eocene plants known in the world and the oldest Cenozoic plant remains from East Antarctica.[9] The *Nothofagus* or southern beech fruits described from East Antarctica are the first records from the Cenozoic, and confirm that *Nothofagus* was reproducing sexually, and associated *Araucaria* leaves are also a first record from East Antarctica.

The Antarctica Paleogene flora can be divided into three dominate assemblages:

- A Paleocene assemblage of ferns and some *Nothofagus*, includung podocarp and araucarian conifers, and a variety of angiosperms;

MacRobertson Shelf

In 1995 and 1997, the Australian Antarctic Division provided logistic support on its ship MRV *Aurora Australis* for marine research cruises for Geoscience Australia. These cruises carried out seismic studies of sediments deposited over the last few thousand years, and also took shallow dredge and core samples in the MacRobertson Shelf. A major surprise came when some of the cores yielded charcoalified wood, foraminifera, and spores and pollen, much older than expected, and contained in material that had slumped down the side of submarine valleys onto the sea floor. These submarine valleys are quite spectacular and are up to 1400 metres deep across the continental shelf in this locality and up to 2000 metres deep near Windmill Islands (site of Casey Station). The microfossils gave new insights into the environment before Antarctica became glaciated in its modern form. The foraminifera came from three different ages, about 40 and 55 million years and something older than 65 million years. They included Eocene *Globigerinatheka kugleri* and *Asterigerina lornensis*. Even older, Paleocene, forms included *Planorotalites pseudomenardii* and *Subbotina triloculinoides*. The spores and pollen told of Antarctic vegetation prior to the destruction of that vegetation by the advancing ice. Antarctica was forested with southern beeches, *Nothofagus* spp., now well known in the earlier Gondwana continents, plus many other groups now known in southern hemisphere cold temperate rainforests such as in Tasmania, New Zealand and South America.

A flowering plant leaf from the Eocene. Photo: David Cantrill.

One surprise was the presence of pine pollen known under the name *Dilwynites,* now known to belong to the ancestors of the Wollemi Pine, *Wollemia nobilis*. The discovery of this evidence of past vegetation led to the establishment of a garden at the Australian Antarctic Division in Kingston, Tasmania, displaying those Tasmanian plants known to have an Antarctic ancestry, including man ferns, huon pine, creeping pine, guitar plant, sundews, celery top pine, but especially the southern beeches – *Nothofagus* – myrtle and decidous beech.

Carbonised leaf remains from Cross Valley, Seymour Island, dated as Paleocene in age. These plants lived not long after the K–T boundary mass extinction event. Photos: David Cantrill.

- An Early–Middle Eocene assemblage dominated by two large-leaved *Nothofagus* species (among the largest nearly 70 mm long; two types of fruits present, two-flanged and three-flanged – one of which is from currently living temperate subgenera), two species of fern, an araucarian conifer, and three types of non-*Nothofagus* dicotyledons; and
- Late Eocene leaf assemblages of the La Meseta Formation, which consist almost entirely of *Nothofagus* leaves.

The Eocene assemblages of both West and East Antarctica were probably at comparable latitudes of around 65°–70°S, based on the palaeogeographic reconstructions and paleomagnetic studies. The presence of southern beech in the Sirius Group, sediments deposited at a much higher palaeolatitude of 85°S, indeed, reinforces the conclusions that southern beech formed a dominant part of the forest biomass for the Late Cretaceous and much of the Cenozoic.

The long-term diversity of Cenozoic Antarctic floral communities tells us that the deep freeze of today has come about relatively recently in terms of geologic time. The Late Paleocene–Early Eocene thermal event about 55 million years ago (known globally as the warmest part of the last 65 million years) was the last phase of globally warm greenhouse climates on Earth before the onset of a major cooling phase that resulted in the Late Cenozoic glaciation.

The trunk of *Araucaria marensii* from the Eocene La Meseta Formation, Seymour Island. Photo: David Cantrill.

A carbonised southern beech leaf from the lower part of the La Meseta Formation, dated as mid-Eocene. Photo: Steve Morton.

The Eocene plants of Antarctica potentially yield the first signs of global cooling, and especially, the deterioration of Antarctic climes.

The record of trunks, branches and twigs in the shallow marine environments represent forests that grew on adjacent land. The wood collected to date represents trees of the living Araucariaceae, Podocarpaceae and Nothofagaceae, all of which grow today in cool temperature forests of Australasia and South America. Cold winters probably occurred in these high latitudes of the Paleogene, but it had to be warm enough to prohibit the formation of significant ice at lower altitudes. But what about the last remnants of Antarctic forest – those southern beech that were severely stunted in growth and probably existed prostrate on the ground? These stressed-out plants would have had to tolerate several months of darkness during the polar winter, and no doubt became dormant, after which in the summer they grew continuously during the long hours of summer light. In fact, the Late Neogene record of southern beech in the highest south latitude only 5° away from the South Pole is incredible in itself, given the environmental stress that these hearty plants must have endured as Antarctica lapsed into a barren frozen world of snow and ice. In terms of tree survival in a harsh environment, these plants were surely 'extremophiles' of the plant world. Thus, *Nothofagus beardmorensis* from the Sirius Group would have had to be very frost tolerant to

Part of an Eocene tree branch, La Meseta Formation, Seymour Island. Photo: Steve Morton.

Black carbonised plant remains in coarse sandstone from the McMurdo Sound erratics of Mt Discovery. Their discovery in the mid-90s indicates that temperate forests flourished around Antarctica during the mid-Paleogene. Photo: Steve Morton.

have survived in these extreme conditions. Some of these twigs, roots, leaves and pollen are virtually inseparable from *Nothofagus gunnii*, the Tasmanian deciduous beech, and may only be 2–3 million years old, implying that Antarctica could not have been completely immersed in a deep freeze. Moreover, both light levels and temperatures were obviously not low enough to prohibit the growth of forests at this palaeolatitude. It was a Late Cenozoic day when the last Antarctic tree died as the temperatures plunged beyond the tolerance of the species.

Earlier in its geologic past, when Antarctica was part of Pangaea and later Gondwana, the waxing and waning of ice sheets would have greatly reduced floral diversity in the Antarctic region; but the rebound of plants was more possible as re-colonisation took place. As Antarctica finally broke off from Australia and became the isolated continent it is today, both thermally and geographically, the gradual deterioration of climate saw the complete demise of its forests.

The Paleogene record reveals that some of the Antarctic trees were deciduous, based on the vernation of the leaves or the manner of leaf folding while within the bud. Some Eocene specimens look 'fan-like'. It is highly likely that these leaves were dropped at the start of the polar winter, and hence, when there would have been several months of darkness. Today, both *Araucaria* and *Nothofagus* occur in regions with a low range of mean annual temperature along with an element of rainfall seasonality. A modern analogue would be the *Araucaria-Nothofagus* forest in the Valdivian Andes, where the climate is probably comparable to that of Eocene Antarctica, and where the *Araucaria* grows emergent over deciduous *Nothofagus*, such as that seen in Vulcan Llaima. Climatically, the presence of southern beech suggests cool, ever-wet conditions with a minimum temperature no less than –22°C and air temperatures during the growth season of at least 5°C for several weeks. All three Paleogene assemblages discussed above suggest that over time there was an increasing dominance of *Nothofagus* concurrent with decreasing taxonomic diversity as temperatures fell throughout the Cenozoic.

Paleogene bird fossils

More than three thousand bones (mostly isolated elements) of penguins (Sphenisciformes) have been collected from the Antarctic since they were first discovered during the Swedish South Polar Expedition in 1901–03. They constitute by far the most comprehensive and diverse fossil bird found on the continent, the majority collected from the richly fossiliferous La Meseta Formation as isolated elements (apart from one semi-articulated individual collected from a 1986–87 expedition). Apart from *Waimanu,* a 'proto-penguin', from probable upper Paleocene sediments of the Waipara Formation and a beautifully preserved skull from the Kaiatan Stage (Late Eocene) from South Island, New Zealand, the Seymour Island penguins and those of the Eocene of Peru are among the oldest-known formal records of these water-adapted birds and definitely the most diverse.

In 2006, Jeffrey Stilwell and his team discovered even older early birds from Chatham Island, east of New Zealand, which are inferred to be close to 65 million years old and lived alongside the described theropod dinosaurs on the Chatham Peninsula. These fossils represent at least two but probably four entirely new groups of birds, which is very important in the evolutionary story of modern birds, especially since they evolved in warm, 'greenhouse' palaeoenvironments of the Late Cretaceous.[10]

The majority of Eocene penguins belong to endemic groups and only *Palaeeudyptes* and *Archaeospheniscus* are identified also from New Zealand and Australia. In total, 15 species grouped into six genera have been proposed since the first reports at the beginning of the 20th century. Current evidence suggests that only 10 of these proposed taxa appear to be taxonomically distinct, and their type specimens are represented by foot bones called tarsometatarsi. Six of those species were probably no bigger than the emperor penguin *Aptenodytes forsteri,* the largest and heaviest extant species, and it is possible that perhaps most (if not all) species may have lived together or co-existed in West Antarctica. Their absence in East Antarctica

A glorious day at the foothills of Mt Discovery, McMurdo Sound. This was the campsite and area of several expeditions to East Antarctica to search for Eocene fossils in glacial moraine deposits. Photo: Jeffrey Stilwell.

A reconstruction of *Anthropornis* – the biggest Antarctic fossil penguin known. Some individuals reputedly grew to 1.7 metres and weighed up to 90 kg. Illustration: William Stout.

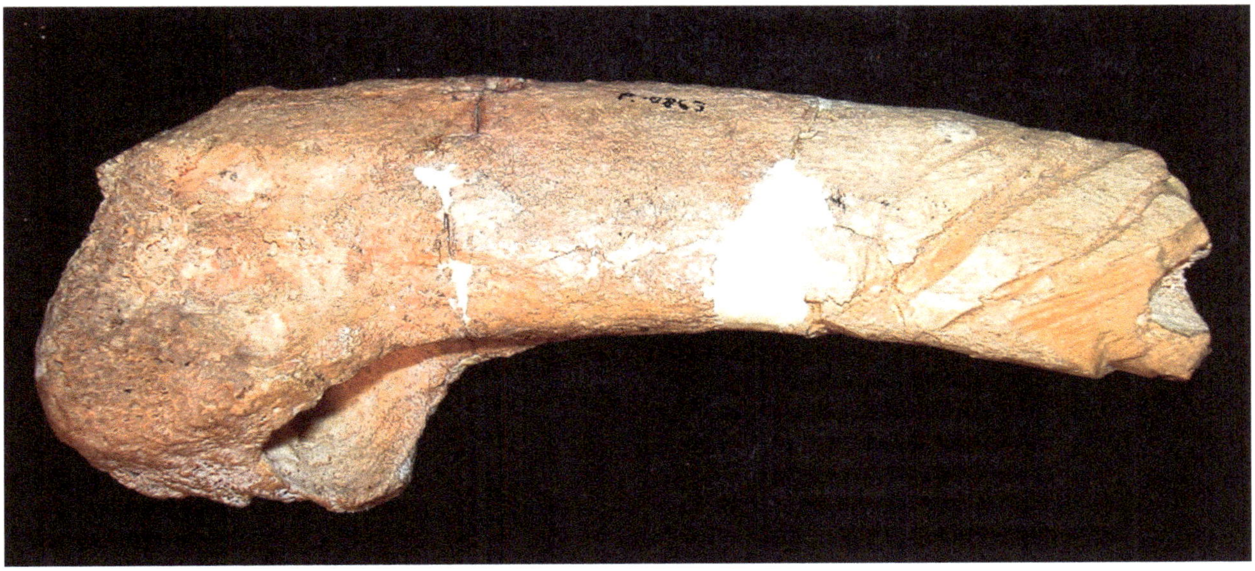

Large humerus of a fossil penguin, *Anthropornis,* from the La Meseta Formation of Seymour Island. Fossil penguin bones were first collected by Otto Nordenskjöld and his team during the 1901–03 Swedish South Polar Expedition, and studied by Carl Wiman, who published his research on the birds in 1905. Some of these penguins were immense, reaching well over a metre tall. Photo: John Long.

A reconstruction of a phorusrhacoid – a giant, flightless, carnivorous bird. Illustration: William Stout.

Avian footprints from King George Island provide the only recorded evidence of birds from the continent of Antarctica for the Oligocene Epoch. These trace fossils were originally researched by Pat Vickers-Rich of Monash University. Photo: Tom Rich and Pat Vickers-Rich.

is probably due only to the limited fossil record of coeval rocks of Eocene age, which are known only from glacial erratics and not *in situ* outcrops.

A single fossil bird bone was discovered from an Eocene erratic of McMurdo Sound and indicates that birds were members of the Eocene East Antarctic vertebrate community as well, despite the fact that we know very little of their diversity. As recently as 2008, reputed new penguins have come to light from the Seymour Island record, as described in *Antarctic Science* by Polish palaeoornithologist Piotr Jadwiszczak, who identified an 'intriguing' partial tarsometatarsus of a relatively small penguin, which may even represent a new genus and species.

The diversification of these birds in the austral Paleogene after the K–T mass extinction event was exemplary, compared with other known avians of the time, and their evolutionary success is told by the abundance of their remains in the high southern latitudes of the Antarctic Peninsula, and other mid- to lower-latitude regions of the southern hemisphere.

A surprising number of non-spheniscid birds have been collected from the La Meseta Formation, but all are isolated fragmentary elements. Major groups thus far identified include a phorusrhacoid, a falconid, a presbyornithid, procellariids, diomedeids, pseudodontorns, and two different charadriiforms including a possible phoenicopterid.[11] Oligocene–Miocene footprints from a large non-volant bird and a variety of smaller volant taxa represent the only known post-Eocene evidence of fossil birds from Antarctica. These ichnotracks have been noted on King George Island in the South Shetlands.

A severely crushed fragment of a long bone of a large flying bird is the only record from mainland Antarctica. Based on size and proportions this specimen has been identified as belonging to the pseudodontorn (Family Pelagornithidae), the second occurrence of these birds from Antarctica.[12] These 'false-toothed' birds are first recognised in the Early Eocene of England, and the Antarctic occurrences are the next oldest known, after which the groups had a widespread distribution and eventually became extinct in the Pliocene. These birds were relatively medium- to quite large, pelagic seabirds with large robust bills that supported rows of bony tooth-like projections, and probably looked similar to gliding birds like albatrosses, rather than those that used flapping for movement. The specimen recovered from an Eocene erratic at Mt Discovery suggests that these birds had a high-latitude circumpolar distribution. Out of hundreds of fossiliferous erratic boulders, only one has contained bird bones, and it is surprising that no penguin material has been found to date. Pseudodontorns are thought to have been oceanic birds and probably were associated with high areas of ocean productivity, as around the globe they are often associated with the remains of whales, other seabirds, leatherback turtles, teleost fish and sharks. Marked upwelling systems can often generate, strong steady winds, and this would have been ideal for large gliding birds such as the pseudodontorns.

Paleogene fish fossils

The high diversity of fish in the Paleogene of Antarctica attests to the probable marked levels of productivity in the proto-Southern Ocean. Most of the remains of fish represent sharks and are well-preserved teeth. Nearly everything known about the fish record of the

A nicely preserved fossil teleost tooth from the Eocene glacial erratics of McMurdo Sound. Interestingly, these same teeth are found on Seymour and Cockburn islands, indicating that these sharks had a circum-Antarctic distribution during the Eocene. Photo: Brian Pump.

Jeffrey Stilwell, John Kaser and Steve Bohaty (left to right), all then of University of Nebraska-Lincoln, moments after the discovery of the first fossil vertebrates in East Antarctica on 6 December 1993. The Eocene fossils were found in the glacial moraine deposits of Mt Discovery. Photo: Rich Levy.

continent is derived from Seymour Island and in the glacial erratics of McMurdo Sound.[13] The diversity of sharks is quite high while in contrast the fish record, mostly teleosts, is poor with representatives only of siluriforms and other unidentifiable fish bits. Chondrichthyans comprise *Carcharias*, *Galeorhinus* (only known record from Antarctica – McMurdo Sound), *Cetorhinus* (first record of a fossil basking shark from the continent is one of the earliest records of the genus), *Carcharhinus* (first record in the austral Eocene), sawfish *Pristis* (also first Eocene record) and an unidentified eagle ray (myliobatoid).

The La Meseta Formation hosts also the best record of chimaeroids in Antarctica, attributed partly to the varied shallow marine depositional environments. *Callorhincus*, *Ischyodus*, and two species of *Chimaera* occur in the lower part of the formation; however, in the upper section of the formation the only group present is *Ischyodus*, which actually increases in abundance up-section.

Mike Woodburne and Bill Daily at the fossil site on Seymour Island where the first fossil mammal was discovered on 5 March 1982.
Photo: Bill Zinsmeister.

Paleogene fossil mammals

One of the most spectacular finds in the history of Antarctic palaeontology was the chance discovery by Mike Woodburne and Bill Daily of primitive marsupial mammal fossils in a remote fossil site on Seymour Island in early 1982. Fossil terrestrial mammals are still to this day only recorded from Seymour Island in the La Meseta Formation, dated as Early to Late Eocene. Since the ground-breaking research by Charles Darwin in the early and mid 19th century, the strange distributions of animals and plants of the southern hemisphere had always puzzled scientists. Thus, why were marsupials, such as koalas, Tasmanian devil, the now extinct thylacine, kangaroos, and wallabies, etc., found in South America and Australia, but apart from the opossum, not in the north? Until 1982 there were no fossil mammals known from Antarctica, and until as late as 2007, fossil mammals (other than bats) were finally discovered in the Miocene of New Zealand.

The fossil mammal discovered unexpectedly on an austral summer day in 1982 was named as *Antarctodolops dailyi*. How did this mammal (now assigned to the genus *Polydolops*) relate to better known mammals of similar age in southern South America? Work on Seymour Island by both American and Argentine workers has turned up a variety of placental mammals and a polydolopid marsupial, all with South American affinities, and all of essentially mid-Eocene age. Several years later, another report on another species of mammal called *Polydolops seymourensis* was discovered, thus indicating another closely related species of the genus originally discovered in 1982.[17] Subsequently, more fossil mammals turned up with careful searching. These include two ungulate species (hoofed mammals), a litoptern, an astrapothere and a tardigrade sloth.[18] More recently, a non-therian mammal identified as a sudamericid gondwanathere has been discovered in the La Meseta Formation and another litoptern *Notolophus arquinotiensis*.[19]

In summary, the palaeomammalian fauna of the La Meseta Formation now includes seven species of marsupials from four different families, a commendable number of new taxa in 28 years of fieldwork; these include described taxa *Derorhyncus minutus*, *Pauladelphys juanjoi*, *Xenostylus peninsularis*,

A reconstruction of the small Eocene mammal, *Polydolops*.
Illustration: William Stout.

A few preserved teeth of a fossil mammal from Seymour Island, locality 456. Many mammalian remains have been recovered by sieving the loose sand from the near-shore tidal channel deposits. Photo: Mike Woodburne.

Marambiotherium glacialis, Polydolops dailyi, Polydolops seymouriensis, Perrodelphys coquinense, Trigonostylops sp., *Victorlemoinea* sp., Tardigradidae genus and species indeterminate, and *Notolophus arquinotiensis*. The latter species, which has recently come to light, tells us that we know very little about the early evolution of ungulates, and that Antarctica has played a significant role in the early evolution of ungulates despite a poor state of current knowledge.

Thus, there are two species of polydolopid mammals, but also three species of opossum-like derorhynchids: a prepidolopid and a microbiotheriid, and a sudamericid gondwanathere from the older beds of the La Meseta Formation. These are dated as early Middle Eocene, about 45 million years. Birds from these horizons, including the diverse penguin assemblages, and other birds such as ratites (ostrich-like) and a phorusrhacoid (cursorial carnivore) in younger horizons, are associated with the diverse mammals from this prolific geologic unit. The size range includes small marsupials and rodent-like gondwanatheres to large-sized ungulates, along with a relatively large sloth and big birds. These terrestrial organisms, as corroborated by plant and macroinvertebrate evidence, indicate a seasonal, if not relatively cool temperate climate with temperatures averaging –3° to 2°C. The fossil evidence suggests that the wide and bimodal range of mammal body sizes means that smaller-sized critters adapted to colder winters through physiological

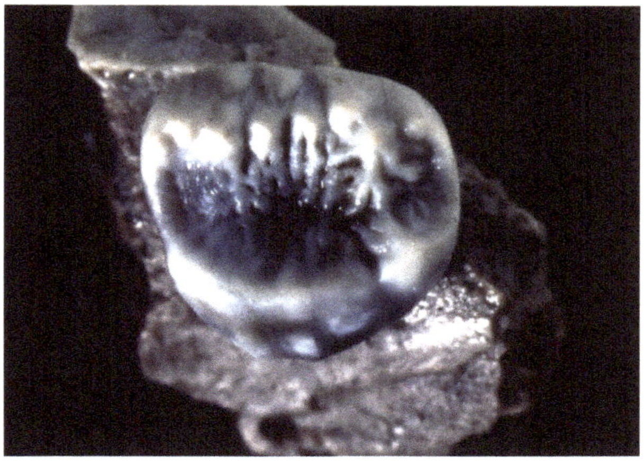

This small molar tooth, only millimetres across, represents a small Eocene mammal from Seymour Island. Photo: Mike Woodburne.

This small jaw with teeth from the mid-Eocene on Seymour Island was the first mammal fossil to be discovered from Antarctica. Found by Mike Woodburne in 1982, it was called *Antarctodolops dailyi* and made headlines around the globe. Photo: Mike Woodburne.

strategies such as torpor and hibernation, while the larger animals adapted to cool climes with larger body sizes and hence smaller surface-area-to-volume ratios. Just why middle-of-the-road-sized mammals are not present in the fossil assemblage may be a result of exclusion by selective disadvantage in colder winter settings along the Eocene Antarctic Peninsula.

Whether or not terrestrial mammals disappeared after the Eocene Epoch is not readily apparent from the Antarctic fossil record, but their absence in younger Oligocene and Neogene deposits may be a reflection of cooling temperatures during the Eocene and after, eventually exterminating most terrestrial animals as the eventual deep freeze approached and lack of terrestrial vegetation and other food sources.

Later work by Argentine researchers in Patagonia has turned up what are effectively Early Paleocene monotremes. The interpretation by Mike Woodburne and his colleagues is that they came to Patagonia from Antarctica–Australia, but he also suggests that the group, which dates back to the near mid-Cretaceous in Australia, very likely was widely distributed in the Cretaceous of Gondwana which includes Australia–Antarctica–Patagonia at least, if not the rest of South America. This means that the palaeogeographic reconstruction by Judd Case and Mike Woodburne, based on the current state of knowledge, is still valid and one would expect to see a faunal continuity between the three now separated places (South America–Antarctica–Australia) at least during the Maastricthian and into the earliest Paleogene. Unfortunately, there are no deposits of lowermost Paleocene age in Australia that have recorded mammal fossils, and this frustrates vertebrate palaeontologists studying the evolution of mammals. However, other high latitude, connected regions during this time such as 'Zealandia' (New Zealand-Chathams Islands, New Caledonia, etc.) may hold some secrets, which are slowly coming to the surface. For example, the late 2008 discovery of a fossil tuatara (a rhyncocephalan lizard called *Sphenodon*) in New Zealand after 150 years of searching illustrates nicely that there are still many fossil treasures to be found in the southern hemisphere.

Dan Chaney, Mike Woodburne and Judd Case on the snow-covered slopes of Seymour Island. The snow is ephemeral and the outcrops are blown clean of the white stuff in a day or two, or it melts quickly. Photo courtesy Mike Woodburne.

A reconstruction of the Eocene sea around the Antarctic Peninsula, which hosted a high diversity of marine vertebrates and invertebrates.
Illustration: William Stout.

The evolution of marine mammals

Whales have general similarities to the great radiation of hoofed mammals, or artiodactyls. Discoveries from the Eocene of Pakistan has filled in some of the vital gaps in the story of their ancestral group, the mesonychids, showing how they left the land and became fully aquatic creatures, eventually known as whales. The earliest known whales are the archaeocetes, meaning 'ancient whales'. Like their ancestors, the hoofed mesonychids, they had long skulls, with large triangular molar teeth. *Pakicetus*, the oldest known whale, was a small archaeocete with much smaller skull and auditory bulla (for hearing) than any of the other four known Early–Middle Eocene primitive whales. *Pakicetus* lacked also the specialisations for effective hearing in water found in the other whales, so it is generally regarded as the most primitive known whale.

The discovery in the 1990s of *Ambulocetus*, and more recently *Rodhocetus* from the Early to Middle Eocene of Pakistan, along with an earlier form, *Pakicetus*, of basal Eocene age, shows the transition from running terrestrial mammals through to cetaceans that swim with spinal undulation of the tail flukes. *Ambulocetus* had well-developed arms and legs and a long tail. It was capable of terrestrial locomotion, but in water was thought to have swum by flexing its vertebral column up and down, as in modern cetaceans. This is suggested by the shape of the lumbar vertebrae. Its feet were large and could have assisted in its swimming. Its back muscles primarily powered the rear limbs, as in phocid seals today. On land the animal was much like that of a seal, having spawling hands and feet. *Ambulocetus* is clearly a whale with several specialisations of the skull found only in early archaeocetes.

In 1994, the first discovery of an early whale *Rodhocetus*, with the complete lumbar, thoracic and sacral regions of the vertebral column preserved, gave special insight into the nature of how cetacean swimming began.[20] The high neural processes on the anterior thoracic vertebrae along with the pelvis which articulated directly with the sacrum are features otherwise only encountered in terrestrial mammals that support their body weight with their limbs, suggesting that this early whale retained some ability to move about on land. The neck vertebrae are short and the skull is very long, as in later whales. The unfused sacral vertebrae would have enabled the creature to flex its spine up and down more easily, also as in modern whales.

The remains of *Rodhocetus*, one of the most complete early whale fossils, came from deep-shelf sediments, suggesting it was the earliest form to venture fair distances into the open sea, and not be restricted to shallow, near-shore habitats. Also, as in modern whales, *Rodhocetus* has the functional adaptations required for major flexing of the body using the back and abdominal muscles, as seen in the presence of unfused sacral vertebrae; high neural spines on the dorsal vertebrae; short, stout anterior caudal vertebrae; and long ventral chevrons. Thus *Rodhocetus* is an almost perfect intermediate mammal between the mesonychids and the true whales.

As the whales continued to evolve through the Eocene Period, they increased rapidly in their size, reduced the hind limbs to just remnants, the tail was lost, and the front limbs developed more into powerful flippers for directional guidance and manoeuvrability.

Other whales of this time were experimenting with various means of locomotion. *Indocetus*, for example, had long hind limbs and a fused sacral region, as in fully terrestrial mammals, and it inhabited shallow marine environments. *Protocetus*, on the other hand, had a sacrum that did not articulate with the pelvis, and was thus thought to be fully aquatic, and incapable of supporting its weight on land. These remarkable fossils show us that the mesonychid ancestors of whales took gradually to the water, and through an amphibious, seal-like locomotory phase, eventually graduated into a wholly aquatic body form. Following the success of occupying this new niche, the diversification and radiation of different whale types ensued. Among these were the first of the giants of the animal world, the mighty baleen whales.

Archaeocetes, like our primitive Eocene whales, were primarily predators with large serrated teeth amd long, stout jaws. The most speciailised lineages of whales were the two modern lineages, one leading to baleen whales (filter-feeders), the other to the odontocetes, which include our sperm whales, killer whales, dolphins and porpoises. The archaeocetes came to an end at about the same time as these two modern lineages diversified, at the end of the Oligocene Period. During the Miocene, representatives of many of the modern whale genera, or at least their families, had appeared, including *Baleanoptera* – the great blue whale.

The origin of baleen whales from toothed archaeocete whales has been elucidated in recent years from new finds and recent research by Ewan Fordyce of Otago University, and Erich Fitzgerald of Monash University and Museum Victoria. Both of these researchers have worked on one particular whale, *Mammalodon colliveri*, a small form known from the Late Oligocene of Victoria, Australia.

Mammalodon was first described in 1939, and was then thought to be an archaeocete whale, as it had

Paleogene reptiles

The crocodilian record of Antarctica is extremely poor with the only records from Eocene deposits in McMurdo Sound. The isolated tooth fragment discovered by Jeffrey Stilwell is only tentatively identified based on the size, shape and, surprisingly, the fracture patterns on the specimen.[14] The physical attributes, associated with their proportions, confirm it as crocodilian, and probably a piscivoran reptile, one that ate fish. These authors speculate that it may belong to a South American gavial crocodile, which are the only group to be geographically related to Antarctica at this time.

Cretaceous crocodiles are present on all Gondwanan continents except Antarctica, but there is no reason to indicate why they have not yet been discovered except that it is a matter of time before the appropriate sequences are covered. Crocodilians are among the few reptiles to successfully live through the Cretaceous–Tertiary boundary extinction event.

Two Eocene turtles have been exhumed from the La Meseta Formation. The first was described as tentatively belonging to *Psephophorus* (Dermochelyidae) from isolated platelets and a small portion of the shell of epithecal nature.[15] The second turtle is provisionally referred to a congeneric species found in the mid-Eocene of New Zealand.[16] Identified from bony ossicles and associated ribs, these are the only records of marine turtles known from the Antarctic continent.

One of the Miocene predators of the early Southern Ocean, this spectacular skull of the cetacean, *Janjucetus,* from the Jan Juc Formation of Victoria, exemplifies the bizarre, top carnivores of the southern polar regions. Photo: Erich Fitzgerald.

typical archaeocete teeth. Ewan Fordyce prepared the original skull and lower jaws, and had several vertebrae to work on. He came up with a new picture of this strange whale as being a sort of intermediate form, a missing link between baleen whales and archeocetes. His evidence for this was that although *Mammalodon* has teeth, they were widely spaced apart and had high gum lines, suggesting there was a large amount of lip or skin tissue between the tooth gaps. Furthermore, the lower jaws in toothed whales meet at strong symphysis to give strength to the bite, yet in filter-feeding baleen whales the lower jaws are loosely connected to allow some rotational movement of the long jaws when catching krill or plankton. *Mammalodon* showed the latter condition, a weak median symphysis. The skull showed it had a more baleen whale style of echolocation, and thus it was deemed to be more of a primitive baleen whale (yet still retaining teeth) than a typical specialised archeoecete. *Mammalodon*'s widely spaced teeth may have functioned like that of a crabeater seal, which sieves food out of the sand using its serrated triangular teeth. Eventually the use of soft tissue between the teeth would have become more efficient for filtering smaller food items, and baleen would have replaced teeth entirely.

In September 2006 Erich Fitzgerald published his description of a complete fossil whale skull from the Late Oligocene of Victoria, named *Janjucetus.* Although *Janjucetus,* (about three metres in length) was closely allied to *Mammalodon,* its skull was more complete and thus it provided a more complete suite of characters for analysis. Fitzgerald believes that early forms like *Janjucetus* were sifters and suction feeders that ate crustaceans, similar to modern crabeater seals.

The enlargement of their body size seems to accompany all filter feeders throughout time, such as the giant whale and basking sharks, the huge titanichthyid and homosteid placoderm fish, even the huge 16-metre-long Jurassic teleostean fish *Leedsichthys.*

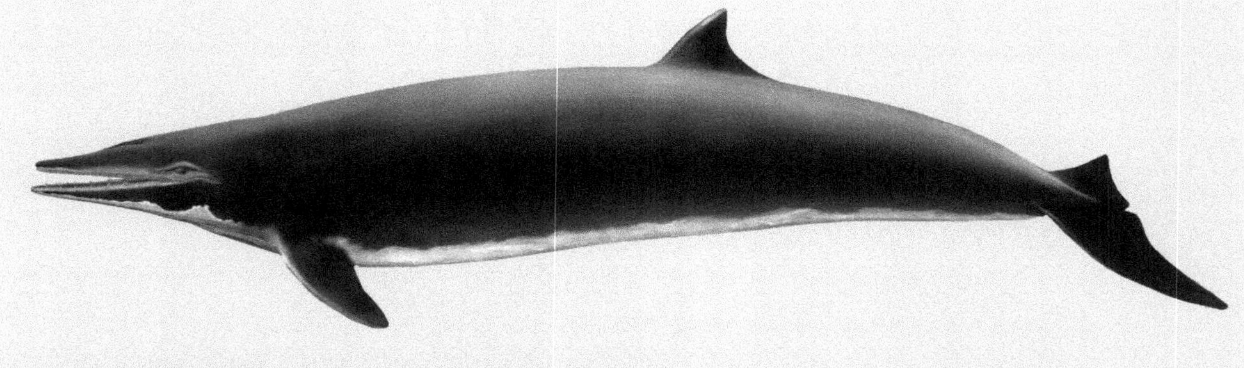

Llanocetus, the oldest Antarctic whale, is an early toothed mysticete. Illustration: Mary Parrish.

According to research by Ewan Fordyce, cetaceans expanded into the proto-Southern Ocean in the later Eocene from their northern Tethyan origins. The Cetacea is the only group that is known to have a fossil record that is before the Holocene Epoch, and there are many accounts of subfossil seals that are thousands of years old around the continent. The authors recall finding relatively well-preserved mummified seals far from the coast in isolated regions in the Transantarctic Mountains. These seals travelled up the Dry Valleys and became confused and eventually perished – possibly seeing the snow and ice in the mountain highs as coastal. However, despite the record of cetaceans in the Antarctic fossil record being rather patchy with the oldest report of whales from Cross Valley, whales must have been in the Antarctic Peninsula region earlier than 34 million years ago, as has been suggested by Ewan Fordyce.[21] These older cetacean bones indicate clearly that whales were in Antarctica no later than the Middle Eocene.[22]

Apart from fragmentary specimens of a small *Dorudon*-like animal and possibly a larger *Basilosaurus*-like species, the most complete whale known from the fossil record is a toothed stem-mysticete from the uppermost levels of the La Meseta Formation. The first bones of this skeleton were considered to be of a small whale, later given the name *Llanocetus denticrenatus*.[23] This whale is estimated to have been more than nine metres in length and has been interpreted as a filter-feeder structurally intermediate between basilosaurid archaeocetes and more-crownward mysticetes. Close to 34 million years old, *Llanocetus denticrenatus* is now considered among the most basal named mysticetes and thus provides a lower age limit for the origin for the mysticete sister-group Odontoceti.[24]

Mammals have thus had a rich and varied evolution, resulting in forms as diverse as the gargantuan behemoths of the seas, the flying bats and the upright-walking thinkers like ourselves. Yet it is with great humility we must always remember that we started our humble lineage as poor second-raters in the wake of the dinosaurian ascendency, and if not for the dinosaurs' mysterious demise, we may still be crouching in their mighty shadows.

OPPOSITE: Fossiliferous strand moraines in McMurdo Sound, Southern Victoria Land. Photo: Jeffrey Stilwell.

12
THE BIG CHILL: FROM PARADISE TO POLAR ICE

On the 10th December I started on my last sledge journey into Robertson Bay, principally for the purpose of securing eggs of different birds … On this sledge journey a remarkable discovery was made by Dr. Klövstad … When he returned he was proud to show me several insects. The discovery is far-reaching in its information. The insects were found in the mosses, and their existence throws a satisfactory light on our meteorological work, and it is improbable that the temperatures about Geikie Land will go much below what we experienced, otherwise the life of insects would not have been possible … until we found that it was a dead stowaway in a jam-box from London!

Commander Borchgrevink, *First on the Antarctic Continent*, 1901.

The transition from 'greenhouse' to 'icehouse' conditions, which is estimated to have occurred some time between the Late Eocene to Oligocene, was thoroughly underway by the Miocene when temperatures plummeted around Antarctica and changed drastically the composition of its biota. The fossils reported from sediments deposited at the beginning of the Miocene reflect the 'big chill' around the Antarctic continent as the development of the circumpolar current and deep sea conditions saw the thermal deterioration of surrounding waters. However, there were episodes of warmer and wetter climatic pulses, which brought with it biotic stress resulting from severe change. The Miocene fossils also give credence to the idea to a dynamic nature and behaviour of the ice sheet for some time slices

The 'Ice Pinnacles' of McMurdo Sound. The veneer of glacial moraine deposits masks an extremely slippery, treacherous surface and large, deep cracks in the ice demand that caution is taken in this area. Photo: Jeffrey Stilwell.

during the early part of the Neogene, representing comparable modern sub-Antarctic to Arctic climes. The fauna and flora unable to cope with this massive environmental upheaval died out although many of these taxa had existed in the Antarctic regions for millions of years. Truly polar conditions commenced during the Early Pleistocene, with bottom water temperatures ranging from about –2° to 0°C. We must keep in mind, however, that there may have been a slightly warmer episode or interglacial during this time.

West Antarctica sponsors an important Neogene fossil record, fortunately due to extensive exposures around the Antarctic Peninsula. Climatic inferences from these fossils tell the tale of major changes in the Antarctic environment during the Late Paleogene and Early Neogene. Bear in mind that many factors influence climate and it appears that two of the most marked fluctuations in climate during the Cenozoic, which relate to temperature decreases, occurred at least in part through the influence of plate tectonics on ocean circulation. Thus, at the end of the Mesozoic and into the Early Cenozoic, the same land connections between Antarctica, Australia and South America that allowed terrestrial animals (e.g. dinosaurs, mammals, etc.) to spread across these continents, also prevented the circumpolar flow of water around the Antarctic continent. Frigid, cold waters flowed north into the Indian, Pacific and Atlantic oceans, mixed with tropical water, and the return flow of warm water to the south kept the polar region relatively warm – and ice free, until eventually Australia and Antarctica separated. When immense fragmentation of the continents occurred, the cold polar waters could flow around the continent, like they do now, which resulted in isolation of these waters from warmer masses to the north. Gradually, the Antarctic turned into a deep freeze, and a permanent ice cap developed. When this happened the result was a noticeable cooling effect on global climate. The actual onset of glaciation in Antarctica probably coincided with the sharp drop in sea water temperatures near the Eocene–Oligocene boundary. This is implied by oxygen isotopes from both macro- and microfossils, and also the type of fossils present like those in the uppermost horizons of the La Meseta Formation on Seymour Island.

In contrast to the decent fossil record for this time on parts of West Antarctica, East Antarctica does not produce as many fossiliferous deposits of this age, although some macrofossil-bearing deposits occur on the Vestfold and Larsemann Hills, Ingrid Christensen Coast, Amery Oasis of the Prince Charles Mountains area, and several localities in the McMurdo Sound region, including the Transantarctic Mountains.

Among the oldest Neogene deposits in Antarctica are those on King George Island, dated to the earliest part of the Miocene. They contain a modest, but very important, macropalaeontological yield, in terms of climatic significance. These exposures contain costate pectinid bivalves such as *Austrochlamys*, which are absent from the extant polar Antarctic settings, co-occuring with another group that may be allied with *Adamussium*. The presence of costate pectinids can be interpreted as an ecological indicator that bottom temperatures at the time of deposition of the Miocene units was probably warmer than today's temperatures, and possibly comparable to modern Arctic or sub-Antarctic domains that see temperatures of up to 5°–7°C.[1]

Bivalves and annelids recovered from the Lower Miocene Battye Glacier Formation in the Prince Charles Mountains suggest that the shallow- to mid-shelf environments were perhaps oligotrophic habitats of normal salinity with a soft, pebbly and sandy bottom that was mobile enough to host deep burrowing forms. Other Miocene localities include the Jason glaciomarine diamicton, that contains shell fragments and polychaete tubes, as well as the Prospect Mesa gravel, which are Late Miocene in age. Miocene faunas which are not that well

The 6000-year-old sediment in this horizon contains layers of the bivalve *Laternula* in living position. Photo: Pat Quilty.

A well-preserved fossil moss from the Meyer Desert Formation, dated as Pliocene. Photo: David Cantrill.

constrained biostratigraphically include DSDP sites 270 and 272 and some McMurdo Sound erratics contain macroinvertebrates of Miocene age.[2] Those paucispecific faunules from the Antarctic Peninsula comprise fossils from the Miocene–Pliocene volcanic rocks of the James Ross Island Volcanic Group, which unconformably overlie unconformable and faulted Cretaceous and Eocene sedimentary sequences on Cockburn Island. The Pliocene-aged faunas are seemingly richer than the older Miocene assemblages.

The Late Pliocene extinction of *Austrochlamys* in Antarctica is believed to be related to the loss of shallow water habitat, caused by the expansion of shelf ice due to accelerated cooling about 2.5 million years ago.[3] But, the genus did not die out completely, and there is a single species that lives today in southern-most South America. Oligocene to Recent *Adamussium*, also a pectinid clam, is endemic to the Antarctic and still survives today, perhaps due to its escape response and ability to move quickly away from approaching floating ice and predators such as fish (which are recorded to have preyed on juvenile *Adamussium colbecki*).

In recent years, unexpected fossil discoveries are providing us with proof for major global climatic shifts in the Neogene. Most of our knowledge of mid-Miocene events related to global cooling are derived from marine geologic and fossil records, so new evidence from Antarctic terrestrial environs is important. The overall cooling that has occurred over the last 50 millon years was punctuated by three steps that comprise the Late Eocene–Early Oligocene (c. 34 million years ago), the Middle Miocene (c. 14 million years ago) and Late Pliocene (c. 3 million years ago). In August 2008 it was reported by glacial geologist Adam Lewis of North Dakota State University and his colleagues that they had found freeze-dried fossils poking up out of the ground, which seemed well out of place. It turns out that the mostly aquatic *in situ* fossils from the western Olympus Range in the McMurdo Dry Valleys represent exceptionally well-preserved mosses from the last vestiges of a tundra community that lived around 14 million years ago, and are associated with prehistoric remains of insects, diatoms and tiny freshwater crustaceans. Studies of these remains indicate that the Antarctic cooled by at least 8°C over about 200 000 years, and that the biota lived during the last phase of wet-based alpine glaciation in deglaciated terrain. This means the climate must have been warm enough for glaciers to melt and create the conditions for these organisms to thrive. The lake probably persisted for thousands of years, but lake water levels apparently dropped at least once, which resulted in a lacustrine basin bottom composed of a bryophyte-rich mire.

In May 2007, a report at the International Congress of Palaeoentolomology in Spain by Allan Ashworth of North Dakota State University on Neogene insects in glacial moraines of the Beardmore and the Dry Valleys regions of the Transantarctic Mountains astounded the palaeontologic community as the fossil record of southern hemisphere insects has, at best, been meagre until now. The Antarctic fossil record of insects is particularly depauperate, so these new discoveries are more than welcome, and yield some surprises. These Miocene to Pliocene remains of insects comprise a head, pronota, elytral fragments and femora of listroderine weevils of the family Curculionidae, elytra of a trechine ground beetle of the Carabidae (Coleoptera), larval head capsule of a chiromomid midge (Diptera), the puparium and segmented leg of a cyclorrhaphan (Diptera), and even the eggs of an unidentified insect! Further specimens have been recovered adjacent to the Polar Plateau in the Dry Valleys, which contain abundant insect fragments from the Curculionidae and perhaps

Charcoalised wood, probably re-cycled Paleogene (or possibly Cretaceous), from sediment core, Nielsen Basin, MacRobertson Shelf, offshore a bit east of Mawson. Photo: Pat Quilty.

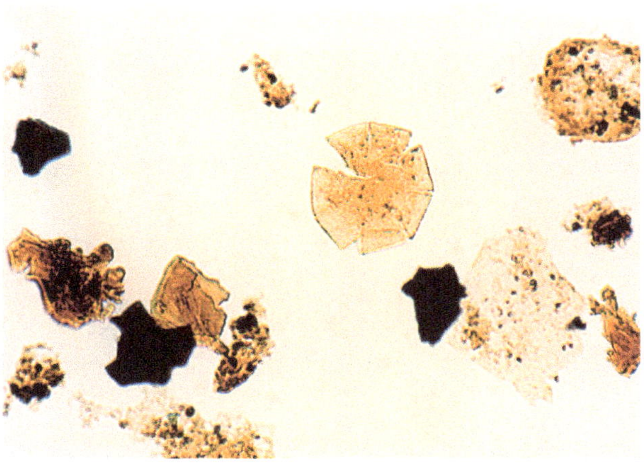

The pollen of the Antarctic *Nothofagus*. Photo: Pat Quilty.

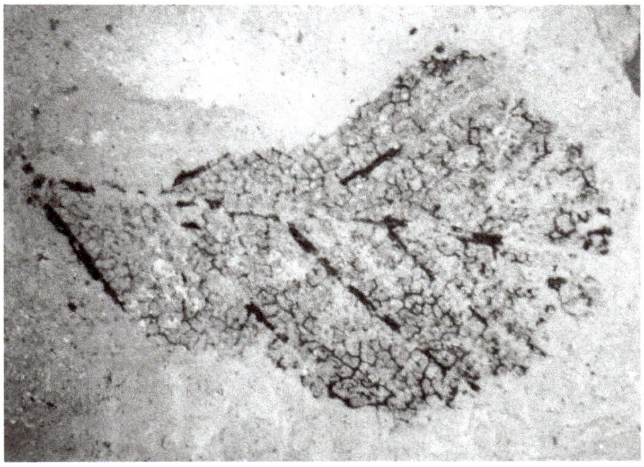

An early Oligocene beech leaf from Cape Roberts, CRP-3 core. Photo: David Cantrill.

the Hydrophilidae. Associated volcanic ashes have been dated radiometrically as mid-Miocene age, indicating that the insects are approximately 13.6 million years old. Preserved with the insects are fossil plants including the southern beech *Nothofagus*, and so-called cushion plants (mosses), ostracods, fresh-water molluscs such as clams (*Pisidium*) and lymnaeid gastropods, and a tooth and pharyngeal bones of a small freshwater fish, most of which are first records for the Antarctic fossil record and all confirming a tundra type of community of plants and animals. The small branches of *Nothofagus* that are only three centimetres in diameter are inferred to have belonged to trees that were 120–130 years old, and these trees and leaf mats, meant that the trees probably lived prostrate to the ground like Arctic willows today, due to wind and harsh conditions. The discovery of several organisms not thought to have ever occurred in Antarctica requires a reappraisal of the role of Antarctica and Gondwana as an evolutionary centre.

As the climate deteriorated, the ice sheets grew to their present size and the vast majority of the fauna and flora became extinct – it was too cold and too arid to support any organism apart from a depauperate assemblage of soil-dwelling creatures. The Pliocene Antarctic ice sheets are considered to have been both dynamic and polythermal, and fossil biotas add greatly to the picture of Neogene Antarctic evolution.

The Early Pliocene marks the second phase of the deterioration in Antarctic climate, whereby some three to four million years ago, another pulse of global tectonics affected ocean circulation. Around this time, the gap between South and North America was closed with the formation of the Isthmus of Panama, blocking equatorial waters of the Atlantic from flowing westward into the Pacific, as had happened previously. Because the world had already cooled down, the North Polar ice cap developed quite rapidly, and again the effect was the lowering of temperatures in other parts of the globe. Comparable to the big drop in temperatures at the close of the Eocene, this temperature change in the Pliocene created marked changes in the biosphere, and clearly, the Cenozoic record tells us that climate, plate tectonics and evolution are woven into a web of complexity.

The Pliocene fossil record, at least what is known about it, has received a great deal of attention in the literature, not only for its palaeontologic significance, but also because these shallow marine deposits record a probable Late Pliocene interglacial episode before Antarctica plunged into its present 'deep freeze' state. Among the most conspicuous fossils found in the Pliocene are an abundance of pectinid bivalves, mainly as thousands of pieces of predominantly umbonal fragments, that is, sections such as the less fragile part of the shell near the beak where the hinge is located on the clam. These pectens are represented by an extinct, but large and relatively thick-shelled clam that is quite unlike the small, fragile scallops that inhabit the frigid Antarctic waters today.

Experts have noted that most of the fragments of *'Zygochlamys' anderssoni* represent shelly umbonal parts of right valves, suggesting that the preferential preservation of this part of the clam means that the left valve was probably quite thin and fragile, compared with the right. Another poorly represented pectinid, *Adamussium colbecki,* is known only from internal and external moulds. The bivalve *Laternula elliptica,* one of the first Antarctic molluscs to be described in 1831, is the only infaunal clam that still inhabits its circum-Antarctic waters today. This clam is also the largest infaunal bivalve of the Southern Ocean and lives in water depths of less than 100 metres in a spectrum of substrates, and exhibits pronounced seasonality. Marine snails are rare in the fauna, but include epifaunal grazing and/or browsing limpets and carnivorous muricids. Other faunal groups include bryozoans, barnacles, brachiopods, echinoids, and very rare vertebrates (penguin and dolphin bones). Fortunately, microfossils such as diatoms, silicoflagellates, ostracods and foraminifera, provide biostratigraphic control to age the deposits and provide information on climate change in the southern high latitudes. The geologic and palaeontologic evidence compiled thus far suggests that a climatic amelioration occurred in the Antarctic Peninsula region some three million years ago with marked marine incursions into East Antarctica during the Late Pliocene.

Marine Plain – a unique site for fossils

Marine Plain, 10 km south-east of Davis Station in the Vestfold Hills of East Antarctica, is emerging as one of the most significant fossil sites in that vast continent, and it has consequences beyond the location itself.

The principal fossils recovered so far are dolphins and whales, of which only *Australodelphis mirus,* meaning 'the wonderful southern dolphin', has been described formally. Others of this group, yet to described, include other new species of *Australodelphis* and a fossilised right whale. No seals or birds have been discovered to date despite dedicated searching. Other fossils include the youngest fossil crustacean from the Antarctic (representing an important family no longer living in the Antarctic), abundant molluscs (bivalves and gastropods which have been studied but not yet published, including groups that have since moved away from the Antarctic), echinoderms, serpulid worms, sheet and stick bryozoans, abundant

The jawbone of the cetacean *Australodelphis* showing typical fracturing caused by freeze/thaw action. Photo: Pat Quilty.

A reconstruction of *Australodelphis* – a Pliocene Antarctic cetacean. Illustration: William Stout.

The original discovery of *Australodelphis mirus* consists of about 300 fragments of bone shattered by freeze/thaw action. Sørsdal Formation, Marine Plain, Vestfold Hills, East Antarctica. This was the first discovery of post-Eocene bone in Antarctica. Photo: Pat Quilty.

The Marine Plain site of first cetacean skull found in 1985 and described as *Australodelphis mirus*. Photo: Pat Quilty.

The skull and shoulder girdle of an *Australodelphis* specimen in the position the fossils were found in the field. *Australodelphis* had no teeth and seems to have eaten squid. Photo: Pat Quilty.

Ewan Fordyce (left) with the late Andrew Grebneff and the reconstructed skull of *Australodelphis mirus*. Photo: Pat Quilty.

Piece of original skull (holotype) of *Australodelphis mirus* showing the internal bone structure and veins of opaline silica reworked from the diatomite. Photo: Pat Quilty.

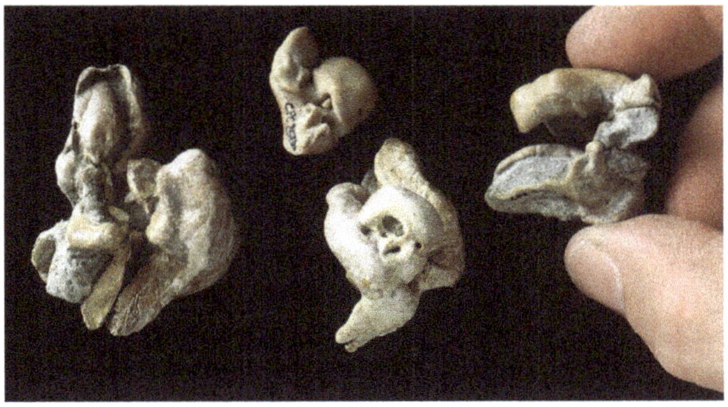

Earbones of some specimens of *Australodelphis mirus*. Photo: Pat Quilty.

microfossils in the form of diatoms (a major component of the sediments and responsible for estimates of age and environment of deposition), few foraminifera and worm jaws (scolecodonts), and innumerable minute spherical bodies (spores?) of an unidentified group of organisms, probably algae of some sort. The molluscs comprise at least nine species of gastropods from five families and ten bivalve species from eight families in two primary horizons in the Early Pliocene. These include snail genera such as *Parmophoridea, Falsimargarita, trochids, Euspira, Chlanidota, Trophon* (many with a long fossil record in Antarctica), and clams such as *Leionucula, Neilo, Limopsis, Lissarca,* pectinid taxa, *Limatula, Cyclocardia, Laternula* and a hiatellid. These molluscs are a blend of cosmopolitan to widespread groups, and others which are endemic to Antarctica. The invertebrates belonged to a typical shallow, soft, sea-bottom community in slightly warmer waters and reduced ice cover and volume compared with present.

The sediments at Marine Plain are unique in East Antarctica in being *in situ*. They are about 7.5 metres thick and about four million years old (mid-Pliocene). They are the only known source of vertebrate fossils since Antarctica first became glaciated some 34 million years ago. The information derived from study of the fossils is clear evidence of a warmer, but still glaciated continent at the time.

Yet this location, so close to a large Antarctic facility that had been operating for almost 30 years, was discovered, like all the best scientific discoveries, by accident. Pat Quilty, then Chief Scientist for the Australian Antarctic program, became aware of the age of the sediments and went there for a few days early in 1985 simply to collect sediment samples to search for the microfossils in which he has specialised called foraminifera. On the first day in the field, he spied at some distance an assemblage of 'rock' fragments which he thought might be schist – a type of rock that splits into platey pieces – which would have been interesting as such rocks are unknown in the region and may have been transported to the site in glacial ice. He picked up a piece, examined it with hand lens and recognised it as bone. He knew immediately that this was an important discovery as there was a 40-million-year gap in our record of vertebrate fossils in Antarctica.

Ewan Fordyce (left) with the late Andrew Grebneff at University of Otago with the collection of vertebrate material from Sørsdal Formation, Marine Plain, Vestfold Hills, East Antarctica. Photo: Pat Quilty.

Back in Australia, he then had to determine what he had found and who was the best person to work with in order to study the material. That turned out to be Ewan Fordyce at the University of Otago in New Zealand who had the pieces restored by his technician Andrew Grebneff, and identified the skull of the dolphin. It is a bizarre dolphin, lacking teeth, with a very asymmetrical skull and features at the back of the skull indicating it had a neck. It seems to have echolocated food (probably squid) by emitting a low and a high frequency sound pulse. Its body shape has converged on that of unrelated dolphins. Further field seasons have yielded many other specimens of cetaceans (whales and dolphins) and much remains to be done to complete description of what has been recovered so far.

The discovery was considered significant enough that the area is now designated an Antarctic Specially Protected Area (ASPA) under the conservation provisions of the Antarctic Treaty. The discovery highlighted the possibility of further discoveries in the region and the message went out for field parties elsewhere to be alert for sediments that may contain fossils. Almost immediately, Jim Burgess of the Australian Defence Force Academy (Canberra), who was working about 100 km farther

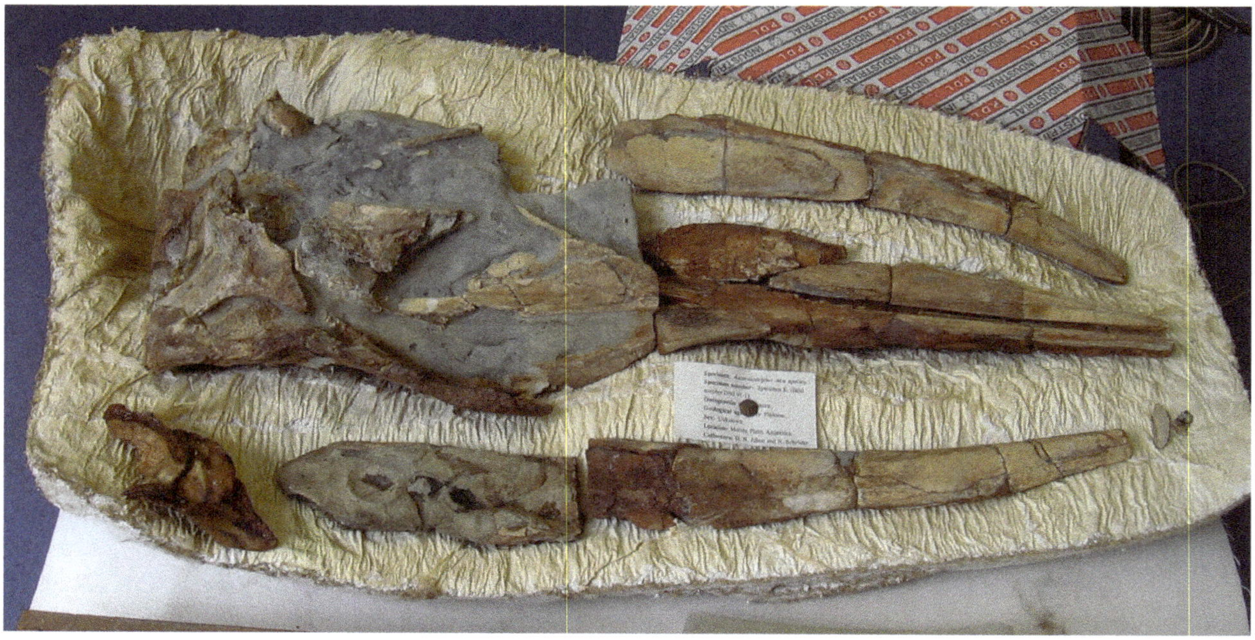

The skull and lower jaws of an undescribed *Australodelphis* species. The affinity of this species is uncertain because the lower jaws seem so large compared with lower jaws of other species. Sørsdal Formation, Marine Plain, Vestfold Hills, East Antarctica. Photo: Pat Quilty.

The foraminifer *Ammoelphidiella antarctica* from the Larsemann Hills, in sediment dumped by glacial action about 100 km south of Marine Plain. Photo: Pat Quilty.

A planktonic foraminifer, *Subbotina triloculinoides*, from a sediment core from the Nielsen Basin, MacRobertson Shelf, a bit east of Mawson. Photo: Pat Quilty.

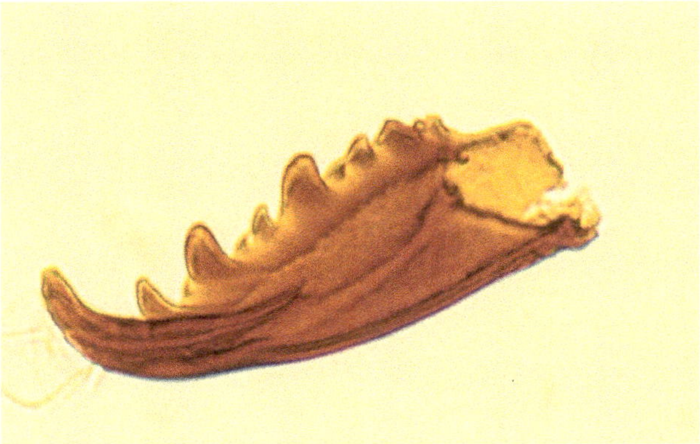

The jaw of a microscopic polychaete worm of Pliocene age, from the Sørsdal Formation, Marine Plain, Vestfold Hills, East Antarctica. Photo: Pat Quilty.

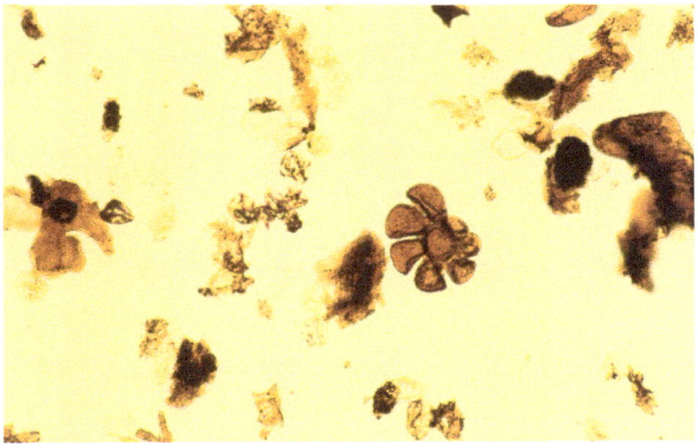

Agglutinated foraminifera (species unknown) and palynology residue, from the Sørsdal Formation. Photo: Pat Quilty.

south in the Larsemann Hills, contacted Pat Quilty to say he had found a thin (40 cm) layer of sediments on Stornes, a peninsula towards the western end of the Hills. He sent some to Tasmania where they were found to contain abundant and very well-preserved foraminifera including the characteristic Antarctic Pliocene foraminifer *Ammoelphidiella antarctica* and diatoms indicating an age identical to that of the material at Marine Plain. Unfortunately, this material is not *in situ* but seems to have been deposited as the ice melted. The time at which this happened is unknown but was in the last four million years.

It was beginning to seem that the region could be an important source of fossils to provide evidence of past environments as Antarctica evolved. More recent deposits of the Pleistocene and Holocene relay the complexities of the advances of the Antarctic ice sheets during recent times. Several coastal regions contain marine sediments and associated fossils are preserved in emergent deposits above sea level reflecting the influence of the prevailing glacial, marine and climatic conditions. These emergent sediments have been studied since the early 20th century and are present around the entire continent. Because most of the Antarctic coastline is occupied

A field camp at Marine Plain. Snow sublimates rather than melts and is gone in an hour or so. Photo: Pat Quilty.

by ice shelves, ice streams, outlet glaciers and ice walls, these glacial systems are influenced heavily by the relative dynamics of the adjacent ice sheets, varying from the marine-based West Antarctic ice sheet to the land-based East Antarctic ice sheet. As the Antarctic coastal areas are directly impacted by these different ice sheets, any coastal environmental shifts which are synchronous around the continent would suggest an overriding imprint of climate change. Species that are incorporated into emerged marine deposits are derived from different habitats and represent distinct ecological conditions in the Antarctic coastal zone, including saline lakes, sites on top of ice shelves, ornithogenic soils from avian rookeries and raised beach sediments. Emerged marine species include those which are restricted to the marine environment during their lifetime (e.g. fish, whales, marine flora, benthic invertebrates) as well as those which can move onto land independent of sea level (birds, seals). Most of the fossils below the Late Pleistocene–Holocene marine limit occur in low-coastal environments.

Middle Pleistocene assemblages have been studied from ocean drill cores in Victoria Land Basin, and comprise up to 80 species of molluscs, bryozoans, polychaetes, octocorals, echinoids, brachiopods and scalpellid cirripeds (barnacles).[4] Every taxon that has been identified in the drillhole is found to be living in present-day Ross Sea. This record of the Pleistocene is the most diverse benthic one in the Antarctic. By analogy with the distribution of similar modern assemblages in this region, no proximal ice shelf was present, so that it seems that conditions were completely open-marine. As these communities live today in near-zero temperatures under sea ice, it would suggest that these benthic organisms are quite hearty and the presence/absence of seasonal sea ice does not seem to be a factor constraining these communities. Carbonate-rich layers in the same drillcore have been interpreted as forming from bank-type sedimentation occurring under summer open-marine conditions. This time of prolific carbonate production over the bank marks a period of glacial retreat and optimal climatic conditions and the complete shut-down of the carbonate factory was apparently quite abrupt and thought to be linked to a glacial re-advance.[5]

These species, which are distinct from Pliocene deposits that also occur in Antarctic coastal areas have been identified from at least 150 different taxonomic groups.[6] Around McMurdo Sound, the most abundant, widespread and best-preserved marine fossils are benthic molluscs and sponges, including the circum-Antarctic scallop *Adamussium colbecki* and burrowing bivalve *Laternula elliptica*, and poriferan *Rossella villosa*. Subfossil animals such as mummified crabeater seals have been reported more than 50 km from the coastline at elevations above 1000 m, and the authors have seen several mummified carcasses in the most remote areas of the Transantarctic Mountains. Curiously, even snow petrels have been found in rookeries that are about 300 km from the ocean! Some of these deposits, such as the 'Ice Pinnacles' of McMurdo Sound contain rich assemblages of macroinvertebrates, comprising diverse molluscs and abundant sponges, and provide a relatively accessible means of assessing changes on a continental scale during the Holocene. The 'Ice Pinnacles' are essentially raised ice pressure ridges that have scraped off the sea bottom, exposing the benthos to the Antarctic elements and for palaeontologists – at least to those who are willing to scurry over the slick blue ice to collect the samples. Broad areas of sediment-laden ice are quite conspicuous along the shelf, and this debris is concentrated into long ridges and ice-cored mounds. The spectacular preservation

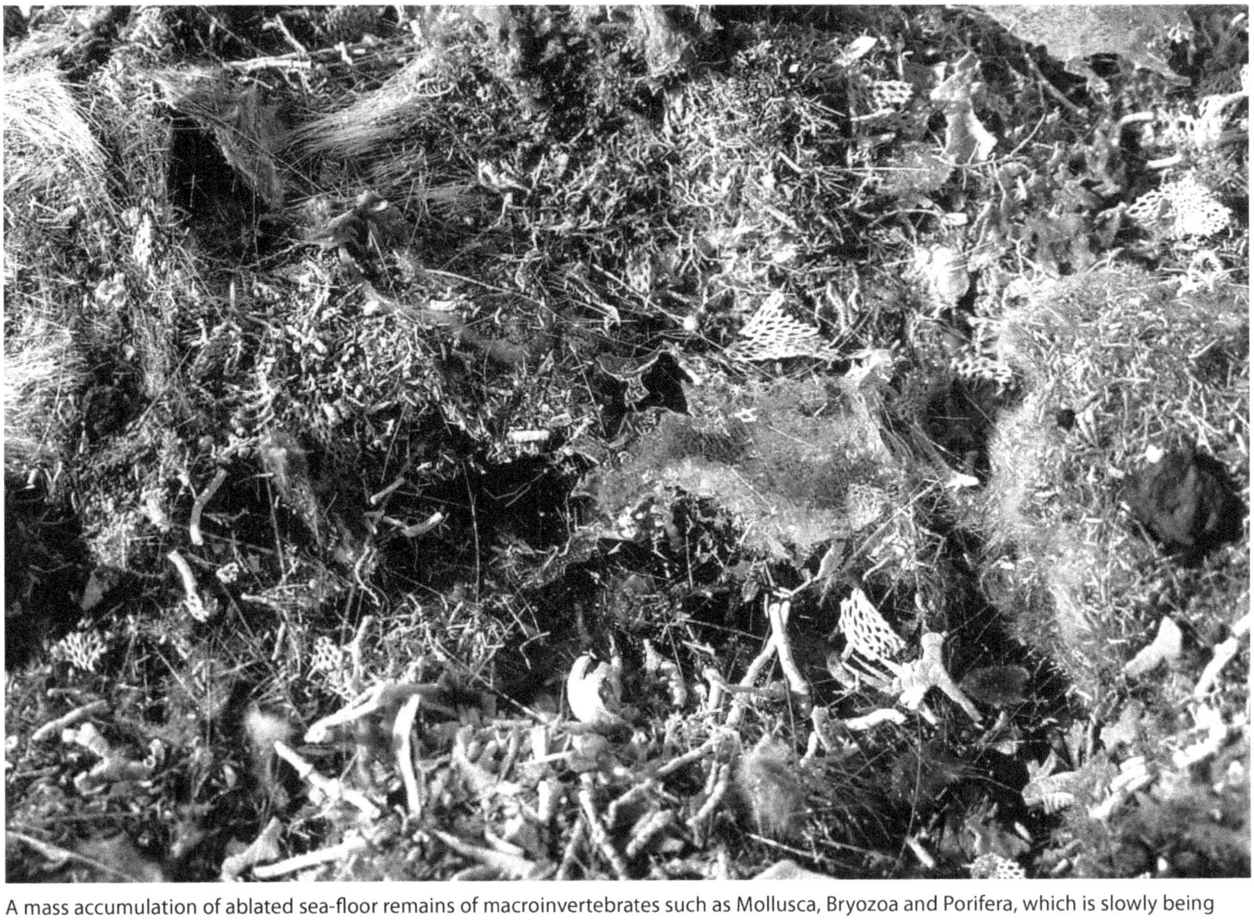

A mass accumulation of ablated sea-floor remains of macroinvertebrates such as Mollusca, Bryozoa and Porifera, which is slowly being exposed to the elements. Movement of the ice pressure ridges of the 'Ice Pinnacles' of McMurdo Sound has literally scraped the sea floor bringing the 7000–10 000 year old subfossil treasures to the surface. Photo: Jeffrey Stilwell.

Frozen in blue ice for thousands of years, these subfossil macroinvertebrate remains of molluscs, brachiopods, bryozoans and sponges from the 'Ice Pinnacles' of McMurdo Sound provide researchers with data on the benthic life of the Antarctic seas during the early Holocene. Photo: Jeffrey Stilwell.

of delicately preserved features of the benthos in the ridges has been of interest to researchers since the members of Captain Scott's team encountered them during the 1901–04 *Discovery* expedition.

Importantly, these emerged deposits surrounding the continent tell an important tale of the dynamic nature of climate through time over the last several thousands of years. The configuration of the Antarctic ice margins has oscillated approximately in phase with the fluctuating climate conditions that have occurred around Earth. The pool of information gathered indicates that marine species inhabited coastal environments from at least 35 000 to 20 000 years ago, at which time extensive iceberg calving created a 'meltwater lid' over the Southern Ocean. The general absence of these marine species from 20 000 to 8500 years ago coincides with the subsequent advance of the Antarctic ice sheets during the Last Glacial Maximum. The concurrent re-appearance of the subfossils in emerged beaches around the continent, all of which have Holocene marine-limit elevations an order of magnitude lower than those in the Arctic, shows that there was minimal isostatic rebound as relative sea level rise decelerated. What this means is that Antarctic coastal areas were responding to a mid-Holocene climatic shift associated with the hydrological cycle. However, more research is needed to understand these environmental connections between the circum-Antarctic coastal system and Earth's climate.[7]

OPPOSITE: A group of very curious and noisy emperor penguins at Cape Evans, Ross Island. Photo: Jeffrey Stilwell.

13
LIFE TODAY IN THE FROZEN SOUTH

I am at the very centre of the great white continent, Antarctica. The South Pole is about half a mile away. For a thousand miles in all directions, there is nothing but ice. And, in the whole of this continent, which is about one-and-a-half times the size of the United States and larger than Europe, there is a year-round population of no more than 800 people. This is the loneliest and coldest place on Earth, the place that is most hostile to life. And yet, in one or two places, it is astonishingly rich.

Sir David Attenborough's opening words to the TV series *Life in the Freezer.*

The thought of actually living in Antarctica today, in an environment of body-numbing cold and ferocious katabatic winds howling off the South Polar ice cap, is enough to make anyone shudder. Even writing this description of Antarctic climate makes the authors get a chill, despite some six expeditions between us! With frigid wind speeds clocked at up to 252 km per hour who wouldn't get a shiver?

Even though Antarctica seems like a vast, barren wasteland that is both inhospitable and static, it hosts not only a surprising diversity of among the extreme creatures known on the planet, but also the ice sheet itself holds some surprising forms of life. The Antarctic ice sheet is complex and dynamic, and is crucial to the world's climate. In both Antarctica and Greenland, deep cores drilled through the ice cap are revealing how mankind has affected the climate in recent times, more than ever before.

For land-based creatures or plants, Antarctica represents one of Earth's most inhospitable

Dr Herlof Klövstad was the first scientist to notice insects living on the Antarctic continent. His discovery was made on 10 December 1899, when the 1898–1900 British Antarctic Expedition members were on the last sledge journey into Robertson Bay in search of bird eggs. Reproduced from Borchgrevink (1901).

Specimens of large Antarctic lichens called 'reindeer moss', with its antler-like branches, from Southern Victoria Land.
Reproduced from Borchgrevink (1901).

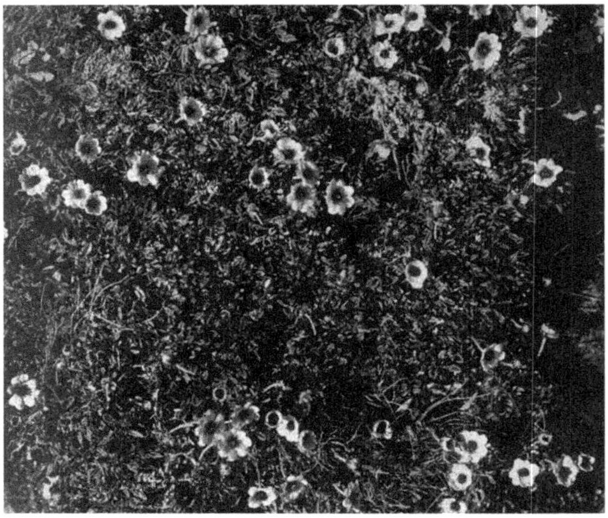

The only flowering plants are found in the most northern latitudes of Antarctica. Reproduced from Nordenskjöld (1904).

Plant life is relatively scarce in Antarctica, but includes moss and lichens and shown here in an early photograph.
Reproduced from Nordenskjöld (1904).

environments in which to live. The last tree in Antarctica lived only a few million years ago – this lonely and very stressed *Nothofagus* saw the end of the great southern beech in the highest polar southern latitudes, where it evolved many millions of years before, in the Late Cretaceous. There were times of stress through this long stretch of deep time, and the southern beech trees were able to survive despite the waxing and waning of ice as Antarctica dropped further into a deep freeze state. Today, no trees exist on

Carrie Schweitzer collecting fossil crabs in the La Meseta Formation on Seymour Island. Notice the splash of green vegetation on a mostly brown and white landscape. Photo: Rodney Feldmann.

the continent, but scores of lichens, bryophytes, fungi and algae along both maritime Antarctica (consisting of the western side of the Antarctic Peninsula and adjacent islands), and continental Antarctica, (which largely comprises the continental landmass and other regions of the East Antarctic Peninsula). Even a few animals live here, surprisingly!

Though an abundance of animals have adapted to life associated with the sea in this part of the world, few animals have adapted to the rigours of a terrestrial existence. One exception is the flightless midge *Belgica antarctica* which, at 2 mm in length, is the largest free-living animal on the continent.[1] Not only must the midge survive the lengthy austral winter as a larva, it also must survive the austral summer, which is characterised by wildly varying and surprisingly high temperatures in the microclimate that the midge inhabits both as a feeding larva and as an adult. A few other arthropods call Antarctica home – these include the mite and a springtail, which may be washed from nestling under rocks and mosses by melting snow and end up floating in small pools of freshwater by the thousands in astounding masses.

Occasionally, splashes of colour from red and green snow algae and lichens brighten some terrestrial habitats in the frozen wasteland. Most of these plants must be carefully protected today, as people trampling them causes much detriment and some may take 100 or more years to regenerate after an assault from stepping on them. Terrestrial plant life exists in three main habitats: permanent snow free regions with scarce precipitation, areas with winter snow accumulation and water from summer melting, and exposed nunataks and mountain peaks There are plants living on virtually any exposed ice free rock, in lacustrine environments, in near coast snow and ice, and even in snow banks and volcanic rocks. Yes, life finds a way!

More precipitation falls in the maritime region – even a bit of rain in the summer with milder temperatures – and, as such, the climate despite its extremes is still more favourable for terrestrial plants and the tiniest of animals. The continent of Antarctica itself has been too cold and dry to support virtually any vascular plants for millions of years. Its flora presently consists of around 250 lichens, 100 mosses, 25–30 liverworts, and around 700 terrestrial and aquatic algal species. Two flowering plants, *Deschampsia antarctica* (Antarctic hair grass) and *Colobanthus quitensis* (Antarctic pearlwort), are found on the northern and western parts of the Antarctic Peninsula. No flowering plants have been recorded south of 70°S. Species of moss endemic to Antarctica include *Grimmia antarctici*, *Schistidium antarctici* and *Sarconeurum glaciale*.

We still know relatively little about the Antarctic flora, as sampling is conducted relatively infrequently, and the number of researchers studying the flora is limited. Compare this diversity of flora with that of the Arctic, where there are some 2000 lichens, up to 700 bryophytes and 900 angiosperms – even some flowering plants living as far as 83°N in northern Ellesmere Island!

The most important factor affecting the distribution of Antarctic flora is the availability of free water. Antarctic flora living in the deep south must be able to withstand severe physiological stresses – for example, countless freezing and thawing, desiccation and other extreme factors. It may be surprising to learn that another extreme organism lives in Antarctica, but only in the hottest of places – yes, there are Antarctic 'hotspots' as well. These are the so-called 'fumarole' communities that exist solely on volcanic ground, which are openings in the earth's surface and emit gas and steam. Temperatures often reach 60°C only 10 cm below the surface of the volcanic substrate. The steam and melting snow/ice provide a continuous sprinkler system for plant colonies thriving in this special environment. These include the communities of soil microbes at over 3000 metres along Mt Erebus, dormant Mt Melbourne with fumaroles still active, and a few localities in the maritime Antarctic and Deception Island (South Shetlands), where the volcano last blew its top in 1985.

A changing climate

Over millions of years, the level of carbon dioxide in Earth's atmosphere changes because of slow geologic processes such as weathering of rocks, the plunge of continental crust into subduction zones, and the release of gases into the atmosphere by volcanoes. These natural processes are much more gradual and slower that the current rapid increase of gases into the atmosphere by the burning of fossil fuels. Human-induced climate change has the potential to trigger large-scale changes in Earth systems. This change may have severe consequences at regional and/or global scales. Despite the estimate that plausible abrupt or irreversible Earth system events could take more than a century to come about, the probabilities of triggering the events in the first place should not be ignored, given the dire consequences. What this means is that we urgently need the information to submit to governing bodies worldwide that climate change as a result of anthropogenic forcing could increase the risk of a large, abrupt event. How do we know if slowing the rate of human forcing of the climate system would ensure that humanity could avoid crossing the thresholds that would lead to abrupt climate change? Fortunately, the Antarctic rock and glacial record is paramount in telling us that we do affect the world's climate globally, and now we have the quantitative evidence to back this up.[2]

Adding fuel to the global concern about greenhouse warming and its causes, was an event in Antarctica that sent warnings around the planet that probable major environmental upheaval is definitely starting to happen. Floating ice shelves, notably the Wordie and Larsen A and B shelves, broke up very rapidly during the 1990s, following regional warming of some 2.5°C over the previous 50 years. Composite satellite images of the collapse from 1995 to 5 March 2002 are, indeed, startling. Research indicates that the observed rapid regional warming around the Antarctic Peninsula is not well predicted by climate models, and its cause as of 2003 remained contentious. In 2009, a thin strip of ice, just 6 km wide, was all that is holding back the collapse of a huge ice shelf in Antarctica. According to glaciologists, the Wilkins ice shelf – previously some 16 000 square km in area – was disintegrating very

Calm before the storm – beautiful cloud formations and Mt Discovery Quaternary moraine deposits with the Royal Society Range of the Transantarctic Mountains in the background. Photo: Jeffrey Stilwell.

quickly. On 28 February, an iceberg 41 km long and 2.5 km wide broke off the ice shelf. This triggered the runaway disintegration of a further 570 square km of ice. Ted Scambos of the University of Colorado postulates that the ice shelf may not survive more than a couple more melt seasons. Other researchers, including David Vaughan of the British Antarctic Survey, believe it could be gone within weeks.

Will future warming in the region indicate a further penetration of the temperature rise into even higher southern latitudes? The consensus is that whatever caused the regional warming along the Peninsula, the break-up of the Wordie and Larsen ice shelves was probably due in part to basal melting, and to summer meltwater draining through crevices, leading to major fractures in the shelves themselves. In the instance of the Larsen ice shelves this has been followed by the acceleration and active surging of a number of former contributory glaciers, leading to an older hypothesis that ice shelves tend to hold back contributory glaciers. What this also means is that the Larsen ice shelves are already making a small but significant contribution to sea level rise. Recent developments indicate that these new observations should cause researchers to rethink about the stability itself of the much larger West Antarctic ice sheet. Fortunately, the Ross and Filchner-Ronne ice shelves, which stabilise the West Antarctic Ice Sheet, are not immediately threatened as it would take a further 10°C warming in their region before the critical –5°C mean isotherm reaches their ice fronts. But increases to basal melt rates under the ice shelves due to ocean warming and changing circulation beneath the ice shelves are also possible. The initial rise in sea levels would disturb ocean currents worldwide, making it likely that more warm water would penetrate the polar regions, causing more melting and setting up a cascading chain of its own cycle of melting and warming. It is also important to bear in mind that it is not only a sea level rise by melting of the ice caps, but also there is a component of increase due to the ocean water expanding as it warms, thus contributing to increased chance of flooding worldwide as the polar regions heat up.

As far as it may be to fathom, melting of ice at the poles will affect those in low-lying regions and also increases the vulnerability to others to tropical storms. By the year 2100, the world will see, at present rates of change in the Antarctic, a major rise in sea level worldwide. But by how much? There have been qualitative estimates given of a rise on the order of as high as 50 cm, but data are wanting at this time to give an accurate estimate of sea level rise. The year 2005 may go down in history as a turning point in the controversy surrounding global warming, and Antarctica is a key player in quantifying for the first time that mankind is modifying the planet in a drastic way by the burning of fossil fuels.

There are few parts of the world's oceans that are not affected by the effect of currents from or around the Antarctic region. For many years measurements on the absolute temperatures of Antarctic were based on a few scattered bases mostly along the edge of the continent. These seemed to show that Antarctica was not subject to the global warming effects measured elsewhere on the planet and highlighted in the Assessment reports of the Intergovernmental Panel on Climate Change in 2005 and again in 2007. However, work by a team of US scientists led by Eric Steig of the University of Washington and published in *Nature* on 22 January 2009 showed that Antarctica has been warming slowly over the past few decades at an alarming rate. They measured a large number of data points across the continent using satellites and found that Antarctica has been warming by 0.1°C per decade for the past 50 years. This means the stability of the ice sheets will be threatened and eventually reach a point where they break away and cause dramatic sea level rises around the globe. If all the ice held in these ice sheets melted, seas would rise by six to seven metres around the world, causing forced widespread displacement of large populations of people currently living near the coast.

The longest and oldest ice core ever drilled beneath Antarctica, nearly 3.2 km long, has fuelled the controversy regarding the effects that humans have on Earth's climate. Scientists now have proof that human activity has dramatically altered the

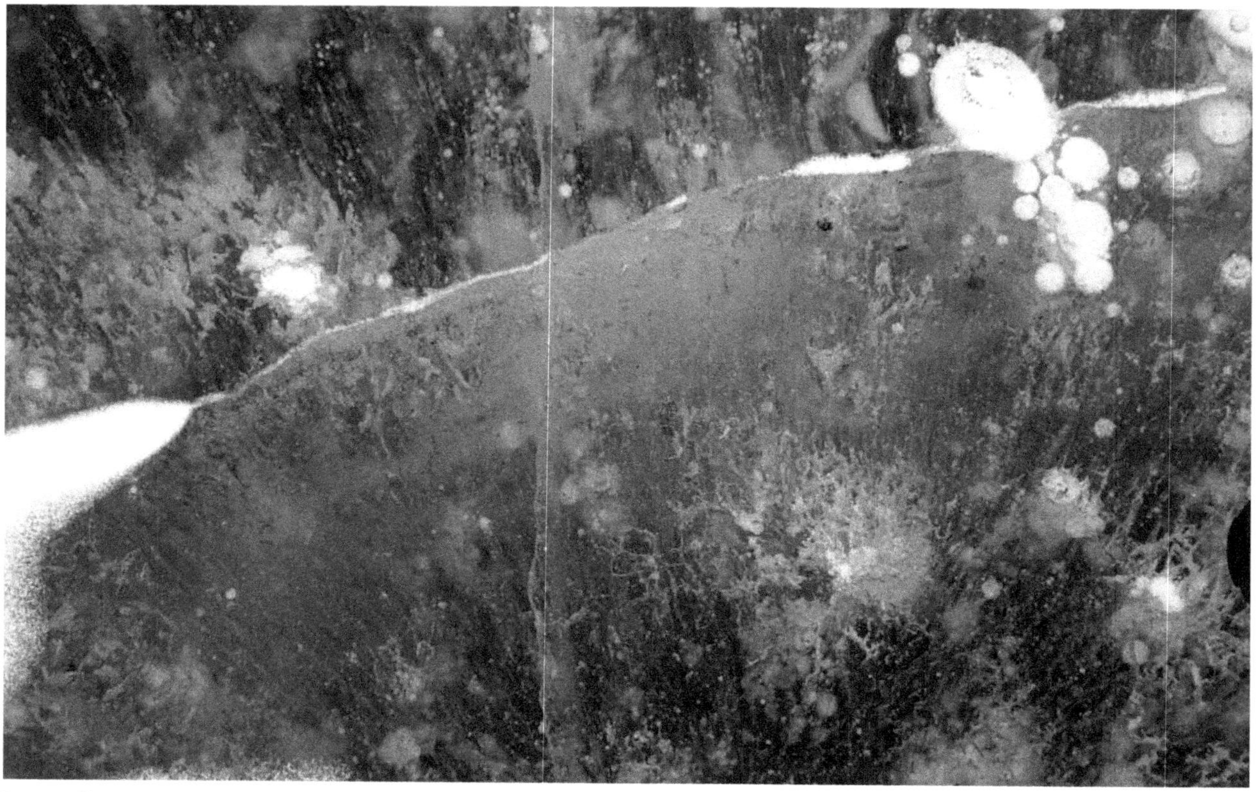

Recrystallised and very old blue ice in a small frozen lake on the flanks of Mt Discovery, McMurdo Sound, East Antarctica.
Photo: Jeffrey Stilwell.

Looking for Eocene age fossils in the Mt Discovery glacial moraine deposits. A partially cloud-covered Mt Erebus is in the background.
Photo: Jeffrey Stilwell.

planet's climate system, especially since the start of the Industrial Revolution. Startlingly, a report from November 2005 in the journal *Science* shows that the levels of two greenhouse gases, carbon dioxide and methane, have not been as high as they are today for 650 000 years! In fact, ice cores taken from a high plateau called 'Dome C' by the European Project for Ice Coring in Antarctica indicate that the last time that carbon dioxide levels were as high or higher was probably tens of millions of years ago. Thus, carbon dioxide levels are 27 per cent higher than they have been since the Pleistocene and methane gas levels are 130 per cent higher.

Ice cores are invaluable tools to delimit climatic responses to environmental change. They contain tens of thousands of trapped air bubbles that have become compressed over time from overburden stress and pressure. We can think of these bubbles as minuscule time capsules that capture samples of air and greenhouse gases from times long past, while the chemistry of the ice reveals temperatures from long ago. These new data are extremely important as the best previously available record for recent Antarctic climate was one extending back 440 000 years, drilled at the Russian Antarctic Vostok station. One of the major reasons for the excitement in the scientific community is that the ice core comprises eight full glacial cycles.

Even in the freezing Antarctic, the winter-to-summer changes in temperature and precipitation are enough to leave annual layers in the accumulating ice, similar to varves seen in lake deposits. These cores can be dated quite precisely by painstakingly counting layer by layer, extending the yearly record in the Antarctic back some 650 000 years. The terrestrial record of oxygen isotopes in the ice cores gives scientists working on the history of Antarctic glaciation a complement to the sea water record from cores taken from deposits laid down over vast amounts of time – at least back to the Late Eocene.

What else can be gleaned from the ice cores? Dust collected from the cores can also provide us with information on the actual composition of the atmosphere in recent times by tapping into ice-trapped gas bubbles, which can can relay to us data on past variations in atmospheric composition.

The big debate today about climate change in scientific and public circles is heating up, as temperatures are on the rise. Thus, the fluctuations of two greenhouse gases, carbon dioxide and methane, have the ability to cause temperature change, but quantifying the amounts of these gases with respect to natural processes such as volcanic outgassing and anthropogenic processes such as burning fossil fuels, is at the forefront of scientific research on the reasons for global warming. There isn't one person on the planet who wouldn't be affected one way or another if melting of the ice caps on Greenland and Antarctica happens, although the melting of the floating ice in the Arctic would have less effect of rising sea level of around eight metres compared with the Antarctic ice cap. If the collapse of the ice sheets did occur, millions of persons living in coastal regions would be quite unhappy about the ensuing rise in sea level on the order of 65 to 70 metres! Quantifying just how long it would take for communities to be affected by the sea level rise is not an easy task, but people would definitely notice changes in the short term.

There are other possible outcomes to the increase of global temperatures from greenhouse warming. The possibility remains that the increase in temperatures will certainly result from added atmospheric carbon dioxide, but this will counteract the tendency for Earth to plunge into another glacial era, which based on long-term records, could be just around the corner. However, there is little likelihood that the two effects will exactly balance one another. Most scientists who have studied the question believe that the CO_2-induced warming will win out, and that we are in for a 'superinterglacial' period that will last until most of our fossil fuel resources are consumed.

A dune field is one of the last sights you would expect to see in Antarctica. Here Chris Rudge is marvelling at the sand dunes of Bull Pass.
Photo: John Long.

A stream flowing in Antarctica? Yes, but for only some 50 days a year during the austral summer. The Onyx River is shown here as a veritable 'raging torrent', well, at least for Antarctica! Photo: John Long.

What can we expect from global warming? Desert environments and their biota are very responsive to changes in climate, given their extreme nature. Temperatures, CO_2 and UV levels are all set to increase from global warming. This will not have a serious effect on plant life, except that the make-up of communities may shift and change due to the increasing levels. As would be expected, new plants may be able to establish themselves in certain regions where this was impossible before – either through accidental introduction by us or on purpose. The predominant expectation is that climate change will increase the surface habitat area for colonisation.

The seas surrounding Antarctic are highly productive. This complexity of the Southern Ocean is due to a thick layer of warm, salty, nutrient-rich water which flows as the Antarctic Circumpolar Current, swirling endlessly around Antarctica. Although the waters of the Southern Ocean are the coldest in the world, they are still much warmer than the super-cooled Antarctic atmosphere, and are filled with enough nutrients such as phosphate, nitrate and silicate, to sustain the growth of phytoplankton necessary to breed an estimated 100 million tonnes of krill a year, which is greedily consumed by some mighty mammals and birds. The massive abundance of krill, in turn, attracts bigger species that feed on them, including whales, pinnipeds, crabeater seals, leopard seals, fur seals, and birds, such as petrels and penguins. It also feeds many other polar vertebrates and invertebrates such as the native fish, squid, krill and other marine species, some of which have developed a special 'antifreeze' component to their body fluids enabling them to survive in waters as cool as –1.8°C.

Response of invertebrates to climate change

Recent research on shallow marine polar invertebrates tells an unusual tale of a combination of very ancient Paleozoic features of benthic communities and those of the modern deep sea. If you go deeper than 10 metres along the Antarctic shoreline, there is a permanent sub-tidal community of sponges, starfish, crustaceans and molluscs that never encounter solid ice, and can be considered to be a quite stable community as such. An important question to ask is: Why are the Antarctic assemblages so strange in terms of community structure? It is no surprise that polar marine organisms show a predisposition to a physiological adaptation to cold-water conditions, and also that as the world warms from the combination of natural and anthropogenic causes, that they would be at great risk from the direct effects of temperature increase. Living Antarctic invertebrate assemblages are similar to Paleozoic ones in that skeleton-breaking predation is quite limited, as attested by the dense ophiuroid and crinoid populations, other slow moving bottom-dwellers, limited numbers of skates and rays, and the absence of crabs and sharks – very different from the Antarctic ancient past of the Eocene.

Most molluscs (bivalves and gastropods) and brachiopods are quite thin-shelled. The bivalves and gastropods are susceptible to predation by the shell-drilling carnivorous gastropods, and it is common to collect clams and lamp shells with small holes that were drilled into them by snails such as naticids and muricids. Research by Richard Aronson of the Florida Institute of Technology, USA, and his colleagues has shown that the endemic nature of the Antarctic near-shore faunas are the products of the cooling event at the Eocene–Oligocene boundary about 33.5 million years ago.[3] This cooling trend, starting about 41 million years ago, essentially eliminated many predators (e.g. modern bony fish (teleosts), sharks and rays, decapod crustaceans (crabs and lobsters), regressing the benthic assemblages into an archaic state similar to more than 250 million years ago. As such the top predators of Antarctic shallow seas are not the subject of scary movies, but slow-moving starfish, worms and other slow-moving critters that cannot break hard-shelled prey similar to common Paleozoic environments – most unlike the dominant assemblages elsewhere around the planet living today. What happened was that the disappearance of the sharks, skates, rays, etc. also meant a cessation in predation pressure, and the slow-moving groovers of the Antarctic sea bottom took over in mass abundance, forming extraordinarily dense accumulations. The tables are now turned, and the increases in temperatures (at least 1°C in the last 50 years along the western Antarctic Peninsula) are seeing a return of crabs to the Antarctic Peninsula – thus running the Eocene cooling event in rapid-reverse. What this means is that exotic species will be introduced with the climatic warming and many echinoderms that are currently flourishing in Antarctic shallow waters will be reduced or exterminated. Ultimately, climate change will promote 'biotic invasions' and acutely affect the endemic nature of the Antarctic shelf fauna, and make it more akin to near-shore faunas elsewhere on the globe, not the Antarctic shallow benthos we know today.

Another fascinating area of study is the diversity of marine invertebrates in the Southern Ocean. A central paradigm in biology is the deep-seated notion of increasing species richness from the poles to the tropics. New research has questioned the old data collected in the late 1950s and early 1960s. Detailed studies are showing very different diversity patterns of invertebrates in both hemispheres, and as such, a marked asymmetry, where the gradient of diversity is more regular in the north than in the south. In the 1990s it was found that Antarctica holds a high diversity for many taxa, and of course, many are endemic, and found only in the Antarctic region. As such, the data available do not reveal a clear cline of increasing diversity from Antarctica to the tropics – interesting, indeed! Yet, many variables are at play over evolutionary time, such as extinction and speciation, and more detailed investigations are required to get to the bottom of the biodiversity of marine organisms. Research on this topic is extremely important in the conservation debate in polar ecosystems.

Antarctica – the future

As has happened over the last 2.5 billion years on Earth, the continents will continue to move, sea floors will spread, mountains will be built and eroded, sediments will be deposited wherever they can find a space, volcanoes will still erupt and blow their tops, and earthquakes will keep Earth rocking and rolling. Antarctica will continue to separate from Australia at an accelerated speed of 7 cm per year. For at least the next 2 to 3 billion years, Earth will be able to support life, although the mind boggles if one considers what evolutionary milestones might lay ahead in the next few billion years, if we can as a species in the short-term are able to curtail the whirlwind of environmental change projected. When the heat of Earth finally runs out, circulation in the outer core stops and then, because of the lack of a magnetic field, harmful radiation will reach the planet's surface. Convection in the mantle will also creep to a stop, and fewer plates will cover the surface crust. Then, the mantle will solidify and the crust will form a thick solid plate, atmospheric gases and seawater will disperse, and Earth will become geologically dead, a piece of rock floating through space and time. Millions of years later, the Sun will expand into a red giant for a 'short spell' (at least a few million years) and then collapse into a white dwarf. It will shine only barely, and then will fizzle out to become a black dwarf. With the weakening of the gravitational pull of the Sun, Earth and the rest of the Solar System will have dispersed. What happens next is pure speculation. Perhaps the universe may expand forever, or at some point may contract leading to the 'Big Squeeze' or 'Crunch', and then maybe even a 'Big Bang' after that.

However, for at least the next 50 million years, geologists predict that the continental movements will continue on their present course. In the higher southern latitudes, Antarctica will rotate anticlockwise, but will remain over the South Pole, as

A faint wisp of smoke rises from the crater of Mt Erebus. Fang Ridge is visible on the left. Photo: Jeffrey Stilwell.

it has since the Late Mesozoic and all of the Cenozoic. South America will also turn anticlockwise, but how much it stays locked together with North America depends on the plate configurations around the Caribbean Basin. Australia and New Guinea will move north by some 30° of latitude and nearby New Zealand will become an elongated island between 0° and 30° south that is sub-parallel to the Equator, all relating to massive shifts along the active Alpine Fault that splits this microcontinent obliquely. A major change will be seen in Africa, in which the Great Rift Valley of East Africa will form a great new ocean and, although new, this ocean's position will resemble that of the Tethys Ocean that existed from about 380 to 50 million years ago.

Meanwhile, in the very near future, the diversity of the Antarctic ecosystem is set to become quite interesting, with a concerted effort of hard data currently being collected by an active team of researchers. The first comprehensive 'inventory' of sea and land animals around a group of Antarctic islands reveals a region that is rich in biodiversity and has more species than the Galapagos. The study provides an important benchmark to monitor how they will respond to future environmental change. Reporting in December 2008 in the *Journal of Biogeography*, the team from British Antarctic Survey and University of Hamburg, described how they surveyed the land, sea and shores of the South Orkney Islands, near the tip of the Antarctic Peninsula, using scuba divers and trawled nets to catch creatures as deep as 1500 metres.

Animals recorded were then checked with a century of literature and modern databases and the team concludes there are over 1200 known marine and land species. These include sea urchins, free-swimming worms, crustaceans, molluscs, mites and birds. Five were new to science.

David Barnes from the British Antarctic Survey (BAS) has stated that this is the first inventory of

Cockburn Island, viewed from Seymour Island. Photo: Jeffrey Stilwell.

Ice-blue and white, the Neumayer Channel holds a special treat for those who pass through – minke whales, seals and birds are frequently seen. The Antarctic Peninsula is in the background. Photo: Jeffrey Stilwell.

its kind in the polar regions. The biotic inventory is part of the Census of Marine Life (COML) – an international effort to assess and explain the diversity and distribution of marine life in the world's oceans. These surveys of south polar life are extremely important if we are to understand how these animals will respond to future change – a starting point like this is critical. The researchers were quite surprised to learn there were so many different species on and around these islands. This abundance of life was completely unexpected for a location in the polar regions, previously perceived to be poor in biodiversity. The research team, consisting of 23 scientists from five research institutes, spent seven weeks on the BAS Royal Research Ship *James Clark Ross,* appropriately named after the British polar explorer, in 2006. It is heartwarming to know that we may finally, after so long, be able to present to the scientific and broader community the facts about the biodiversity of Antarctica, instead of guessing how many taxa there are, so that we are better armed to conserve these special organisms.

Antarctica, the frozen continent, is the last great wilderness on Earth and still holds many secrets. Biologists are still cataloguing new forms of animal and plant life both on land and in the cold waters of the Southern Ocean. In fact, old notions about biodiversity are being tested in the far south, and macroinvertebrates such as the molluscs are telling us that the long-held belief that an increasing cone of diversity from the poles to the equator exists may not necessarily be the case at all.

ENDNOTES

Chapter 1

1 Mt Erebus is an intraplate volcano, situated at the southern end of the Terror Rift within the Victoria Land Basin, a major sedimentary basin with over 12 km of fill and underlain by 21-km-thick crust. The basin is bounded to the west by the Transantarctic Mountains, with 40-km-thick crust.

2 The 'SWEAT' hypothesis (south-west USA–east Antarctica) developed by Moore (1991), Dalziel (1991) and Hoffman (1991). Alternatively, Li (1997a, b) suggested that this 'SWEAT' connection did not begin until around one billion years ago, based on several lines of evidence including the detrital zircon grains in the late Precambrian rocks of the Transantarctic Mountains rift zone that are dated at around 1.2–1 billion years ago.

Chapter 2

1 Ross 1847b, pp. 332–333.

2 Hooker's botanical collections on Cockburn Island were quite historical in that they were the first ever to be made from Antarctica proper, and would not be extended upon for some 150 years by the British Antarctic Survey (see Berry 2009).

3 Ross 1847b, p. 335.

4 The rock samples collected remained unstudied until the end of the century when G Prior wrote a short paper on the few samples collected, which included calcareous glauconitic sandstone samples.

5 Bull 1896.

6 Dr A E Ortmann from the Princeton Patagonian Expeditions monographs.

Chapter 3

1 James Eights not only found the first fossil from Antarctica, but also described the first extant flowering plant (angiosperm) and used for the first time the word 'tabular' to characterise flat-topped icebergs.

2 *Transactions of the Albany Institute*, 1833, p. 64.

3 The 'cockles' Larsen referred to were probably what we now know to be the large cardiid bivalve *Lahillia,* which abounds on Seymour Island.

4 Aagaard 1930, p.84.

5 Burn Murdoch 1894, p. 356.

6 Burn Murdoch 1894, p. 364.

7 A paper in *The Proceedings of the Royal Society of Edinburgh* published in 1898.

8 *ibid.* p. 61.

9 The fossils collected by Stokes were originally housed in the Walker Museum in Chicago, but all nine specimens of the Stokes collection described by Weller are now in the Geology Department, The Field Museum in Chicago, all labelled as from 'Admiralty Inlet, Louis Philippe Land, Antarctica. Stokes Collection' with UC numbers (Paul Mayer, Collections Manager, pers. comm. 2008).

10 Weller was assisted by T W Stanton of Washington.

11 William Zinsmeister of Purdue University has recognised through his research that the Stokes collection was undoubtedly taken from a fossiliferous concretionary layer about 30 m from Nordenskjöld's Hut (not yet built at the time of the collection), where the horizon follows a steep cliff face. A long forgotten controversy surrounds the publication of the 1903 paper in that Nordenskjöld was organising the scientific results of the Swedish South Polar Expedition.

12 Zinsmeister 1981.

13 In 1908, Ernest Shackleton's expedition had also stopped at the Beardmore Glacier and collected various geological specimens yet they had not been fortunate in finding any plant fossils.

14 It would be another 25 years before humans set eyes on the Pole again when a US aircraft made a brave landing on the ice near the Pole.

Chapter 4

1 Although acritarchs (enigmatic microfossils, possibly from single-celled algae) of latest Proterozoic age were once reported from the Marina Glacier region of Northern Victoria Land, the age of these fossils has now been discredited according to Dr Bert Rowell of Kansas University (ref. in Cooper *et al.* 1982). Similarly the presence of Late Precambrian stromatolites in Antarctica was reported by Russians (cited in Tingey 1991), but again no definite evidence for the dating has been confirmed. Another report by Buggisch and Kleinschmidt (1991) presented evidence of latest Precambrian microfossils from the Watts Needle Formation of the Shackleton Range, and may occur also within the basement complex of Northern Victoria Land. So to date no definite Precambrian fossils have been proven to occur anywhere in Antarctica, and in particular, no trace of 'metazoans' or advanced life forms that characterise marine habitats of the latest Precambrian assemblages elsewhere on the globe.

2 Plimer 2001.

3 Zhuravlev & Wood 1996.

4 Rees *et al.* 1989.

5 Wrona 2004.

6 Rode *et al.* 2003.

7 Described by Webers *et al.* (1992) and Shergold *et al.* (1976).

8 Laird *et al.* 1972.

9 Upper Cambrian sediments in central Australia have yielded fish scales described by Young and Laurie 1996.

10 Sturm & Carryer 1970; Laird 1991.

11 Laird 1991.

12 Burrett & Findlay 1984.

Chapter 5

1 The Neptune Group, in the Pensacola Mountains, may be of Silurian age but so far has not produced any fossils. 'Silurian' corals were identified by Jell (1981) from a moraine sample collected at the Cloudmaker, near the Beardmore Glacier, by Shackleton's expedition. However,

an expedition back to the locality by Bert Rowell and colleagues could not locate any further material, so they cast doubt on the locality of the specimen as coming from Antarctica (Rowell *et al.* 1987). However, the original determination of these specimens being Antarctica's only Silurian fossils received support from Cocks (1989), based on the fact that further specimens from the Cloudmaker moraine are housed in the Sedgewick Museum in Cambridge. Rocks of possible Silurian age in Antarctica might exist at the base of the Beacon Supergroup, where a series of sandstone, conglomerate and shale units are of indeterminate age, lacking body fossils but rich in traces of life, lie underneath the Middle Devonian richly fossiliferous Aztec Siltstone.

2 Boucot *et al.* 1967.
3 The geology of the Ohio Range was originally studied by Bill Long (1962, 1965) who was able to determine that the sedimentary succession here was of Devonian age, based on the occurrence of the brachiopod *Pleurothyrella antarctica* (Boucot *et al.* 1963). Long collected much fossil material which was first described by Doumani *et al.* (1965), and later by Bradshaw and McCartan (1983). Long had suggested that the sedimentary rocks of the Ohio Range represented a near-shore, beach and swampy lagoon environment with occasional river or delta sequences. Bradshaw and McCarten (1983) recognised nine different environmental divisions (called lithofacies) in the Horlick Formation. These ranged from intertidal to upper shore deposits in shallow marine habitats through to river channel deposits.
4 Spörli 1992; Webers *et al.* 1992.
5 Young 1986.
6 These fossils were lodged with the British Museum (Natural History) and studied by Arthur Smith Woodward, who published his paper on the material in 1921.
7 The first true soil deposits or paleosols are also recorded from this unit by Retallack *et al.* (1997).
8 Turner & Young 1992. Originally the first specimens were described from latex casts made of the scales, but new material found on Mt Crean by John Long in 1992 consisted of specimens encased within a calcareous concretion. These were acid prepared and freed from their rock casing as three-dimensional scales.
9 Turner & Young 1992.
10 Miller *et al.* 2004.
11 Brazeau 2009.
12 Turner & Young 1992.
13 Discovered on VUWAE 13 expedition of 1970–71.
14 Young 1982; Long & Young 1995.
15 Long & Young 1995.
16 Burrow *et al.* 2008
17 Young & Burrow 2004.
18 Burrow *et al.* 2008.
19 One other interesting finding from Burrow *et al.* is an isolated set of acanthodiform jaws from Mt Ritchie in the Warren Range in the uppermost 'phyllolepid' biozone of the Aztec Siltstone.
20 Young 1988.
21 Burrett *et al.* 1990.
22 Ritchie 1975.
23 Ritchie 2004.
24 Long 1984; Ritchie 1984.
25 Young & Goujet 2003.
26 Young 1988; Young & Long 2005.
27 The oldest complete fossil shark, *Doliodus problematica*, comes from the Early Devonian of Campbelltown, Canada, and was described by Miller *et al.* (2004).
28 Long *et al.* 2008.
29 Young *et al.* 1992.
30 Johanson & Ahlberg 2001.
31 Grindley *et al.* 1980.
32 Grindley & Mildenhall 1981. Another stem of a *Haplostigma*-like lycopsid was collected from the Patuxtent Mountains of western Antarctica at the Okanogan Nunatak. Schopf (1968) recorded it as *Haplostigma* preserved as a mould in quartzite, and from the carbonaceous layers he identified *Cyclostigma*-type stems. Edwards (1990) reassessed these specimens based on photos of them found in the Schopf collection and agreed with the identification of *Haplostigma*, but suggested the evidence was lacking for the presence of *Cyclostigma*. She thought the impressions more closely resembled the species *Colpodexylon cashierense* from Venezuela, although diagnostic features were still missing so a positive identification could not be confirmed. Similar impressions had also been collected from the Dover Sandstone in the Neptune Range by Dr John Isbell of the Byrd Polar Research Centre.
33 Kyle 1977a; Kemp 1972.
34 Findlay & Jordan 1984; Edwards 1990.
35 Grindley *et al.* 1980. Edwards (1990) doubts the validity of the *Drepanophycus* identity and suggests that the material more closely resembles *Haskinsia* sp., an herbaceous phyte with triangular leaves. The age of the assemblage is most likely Middle–Late Devonian.
36 McLoughlin & Long 1994.
37 McGhee (2001) claims that reefs shrunk in area by a factor of 5000 after the Late Devonian extinction event.

Chapter 6

1 Fitzgerald and Barrett 1986.
2 Bradshaw 1984.
3 Carpenter 1969.
4 Bradshaw 1984.
5 Tasch 1970.
6 Taylor & Taylor 1990.
7 Maheshwari 1982.
8 Jefferson & Taylor 1983.
9 Taylor & Taylor 1990.
10 Smoot & Taylor 1986.
11 Balme and Playford 1967; Kemp 1973; Playford 1990.
12 McLoughlin & Drinnan 1996.
13 McLoughlin & Drinnan 1997.

Chapter 7

1 Kitching *et al.* 1972, p. 524.
2 Colbert & Cosgriff 1974.
3 See review by Crame 1994.
4 Sidor *et al.* 2008.
5 Chaloner and Creber (1989) reviewed the problem of

high-latitude polar forests and suggested that the only factor that inhibits tree growth in the highest latitudes at the present day is temperature. It seems probable that the annual input of solar energy to the polar regions is sufficient to produce large annual growth rings and it has been demonstrated that many evergreen trees can tolerate long, dark winters. Studies on the composition of both living and fossil forests have shown that the low angle of sunlight does not cause excessive mutual shading of trees.

Chapter 8

1 Taylor *et al.* 1989a.
2 Hammer & Hickerson 1994.
3 Hammer 1997.
4 Smith *et al.* 2008.
5 Smith & Pol 2007.
6 Hammer & Smith 2008.
7 Hammer 1997.
8 Zeuner 1959.
9 Thomson & Tranter 1986.
10 Smith & Tranter 1985.
11 Smith & Tranter 1985.
12 Pirrie *et al.* 1992.
13 Doyle 1991.
14 Howlett 1989.
15 Kelly & Doyle 1991.
16 Crame 1982, 1983, 1985; Medina *et al.* 1983.
17 Crame 1985.
18 Thomson 1982. More recently, Dr Dan Hikuroa of Auckland University completed a monographic treatment in 2004 (his PhD dissertation) on the Late Jurassic macroinvertebrates of the Antarctic Peninsula.

Chapter 9

1 Long & Molnar 1998; Rich & Vickers-Rich 2000; Stilwell *et al.* 2006.
2 Gasparini *et al.* 1987; Olivero *et al.* 1991.
3 Case *et al.* 2000.
4 Rich *et al.* 1999.
5 Case *et al.* 2000.
6 Chatterjee 1989; Olson 1992; Noriega & Tambussi 1995.
7 Recent work on tracing the phylogeny (or 'fossil family history') of birds in concert with new information on palaeobiogeographic distributions of fossil and living birds, and also, importantly, molecular sequence data, have tested ideas about the timing of radiation of modern (crown-group) birds. Slack *et al.* (2006) provided evidence for a lower estimate of 62–61 mya for the divergence between penguins and other birds and thus establishes a reliable calibration point for avian evolution – and hence, a radiation then in the Late Cretaceous. This means that modern birds lived at the same time as more archaic birds such as *Hesperornis, Ichthyornis, Baptornis,* and diverse groups such as the Enantiornithiformes.
8 First reported by Sankar Chatterjee in a single page abstract in the *Journal of Vertebrate Paleontology* in 1989.
9 Noriega & Tambussi 1995. The initial description was very short and left room for arguments. Tambussi brought the specimen to a symposium on fossil birds at the 1996 meeting of SAPE at the Smithsonian, where it was widely seen by other palaeontologists, who generally agreed with the diagnosis. The first time one of us (JDS) saw the material was at the Museo de La Plata around 1999. Assoc. Professor Julia Clarke (now of University of Texas, Austin) had further fossil preparation done and re-evaluated the specimen utilising abundant new data from CT scans, digitially reassembled, for the second, more thorough description that was published in *Nature* in 2005 (see Clarke *et al.*).
10 S Hope, *pers. comm.* 2006.
11 Case *et al.* 2000.
12 Cantrill & Nicholls 1996.
13 Falcon-Lang & Cantrill 2001b.
14 Dettmann 1989.
15 Askin 1989.

Chapter 10

1 O'Keefe 2001.
2 McHenry *et al.* 2005.
3 See discussion in McHenry *et al.* 2005.
4 The first report of marine reptiles from Antarctica was by Del Valle *et al.* (1977), who described a few post-cranial elements of uncertain affinity, but the first formal desciption to species level was Chatterjee and Small (1989). This new species has had a tumultuous time in terms of taxonomic status or how it relates to others of similar ilk and age. It was first assigned to *Turneria seymourensis,* then redescribed as a species within *Morturneria* by Chatterjee and Creisler (1994), but has most recently been shuffled around again into the Elasmosauridae in the genus *Aristonectes* by Gasparini *et al.* (2003).
5 Chatterjee & Small 1989.
6 Gasparini & Del Valle 1981, 1984; Chatterjee & Zinsmeister 1982; Jeffrey Stilwell, pers. obs.
7 In 2002, Novas *et al.* described *Lakumasaurus antarcticus,* now considered to be *Taniwhasaurus,* from the upper part of the Gamma Member of the Santa Marta Formation, dated as latest Campanian to early Maastrichtian.
8 Welton & Zinsmeister 1980.
9 Grande & Eastman 1986; Grande & Chatterjee 1987.
10 Ward & Grande 1991; Stahl & Chatterjee 1999 (oldest global record of the genus); Stahl & Chatterjee 2002.
11 First described in Weller 1903; Nordenskjöld *et al.* 1904; Kilian & Reboul 1909; Wilckens 1910.
12 Thomson 1971a; Jeffrey Stilwell, pers. obs.
13 Zinsmeister & Macellari 1988.
14 Since Wilckens' (1910) classic study on the molluscan assemblages of Seymour Island, the unpublished work of Fricker (1999) remains the most up-to-date, comprehensive treatment of the latest Cretaceous gastropod fauna, along with recent papers by Stilwell and Zinsmeister (2002b, 2003), who described one of the largest opisthobranch gastropods ever recorded worldwide, *Acteon (Tenuiactaeon) antarctihadrum* from Seymour Island, and a new genus of perissityid given the name *Antarctissitys,* which curiously has affinities with a diverse suite of confamilial taxa from the Late Cretaceous of North America, rather than other austral groups.
15 Wilckens 1910; Zinsmeister & Griffin 1995.
16 Sharman & Newton 1898.

17 Wilckens 1905–07.
18 Zinsmeister & Macellari (1988), based on a substantial amount of material from 600 fossil localities.
19 Stilwell 1994.
20 Macellari 1985, 1988; Riccardi 1991.
21 Research has ascribed the endemism and diversification of the Kossmaticeratidae within the Weddellian Province primarily to the geographic isolation of this area as a consequence of the break-up of Gondwana, and postulated that surface oceanic circulation may have had an influence on the distributions of a few genera.
22 Eckart Håkansson, University of Western Australia, formerly Geological Institute, University of Copenhagen.
23 Zinsmeister 2001.
24 Feldmann & Gaździcki 1997.
25 Lambert 1910.
26 Taylor 1966; Kelly 1993.
27 Néraudeau *et al.* 2000, and references therein.
28 McKinney *et al.* 1988; Blake & Zinsmeister 1991.
29 Néraudeau *et al.* 2000.
30 The first fossil species were Cretaceous forms described by Felix (1909).
31 Stolarski 1998; Felix 1909; Bibby 1966; Macellari 1988; Stilwell & Zinsmeister 1992; Filkorn and Feldmann 1993; Filkorn 1994; Stolarski 1996.
32 Harwood & Levy 2000.
33 Birkenmajer *et al.* 1983; Roniewicz & Morycowa 1985, 1987; Stolarski & Taviani 2001.

Chapter 11

1 There are shallow marine assemblages from widely distributed areas of the Paleocene of Seymour Island (Stilwell *et al.* 2004), Eocene of Seymour and Cockburn islands, McMurdo Sound and Prydz Bay (Stilwell & Zinsmeister 1992; Quilty *et al.* 1999; Stilwell 2000, 2002, 2005; JDS, unpublished data), Oligocene of King George Island (Birkenmajer & Gaździcki 1986; Gaździcki 1989) and Early–Late Oligocene boundary of McMurdo Sound and Ross Sea and the CRP-3 Drillhole (Dell and Fleming 1975; Beu & Dell 1989; Taviani & Beu 2001), Early Oligocene–Early Miocene? of CRP-1 and CRP-2 drillholes, Cape Roberts (Jonkers & Taviani 1998; Taviani & Zahn 1998; Taviani *et al.* 2000; Taviani & Beu 2003) and macroinvertebrate-rich clasts from the Prince Charles Mountains (Stilwell *et al.* 2002), Late Miocene of James Ross Island (see review by Jonkers 1998), and Pliocene of Cockburn Island and Vestfold Hills (see Jonkers & Taviani 1998, 2003; Quilty *et al.* 2000; Harding 2005; JDS, unpublished data), and Quaternary deposits such as those of the 'Ice Pinnacles' of McMurdo Sound (Hart 1990; Berkman *et al.* 1998; JDS, unpublished data) and CRP-1 drillhole (Taviani and Claps 1998; Taviani *et al.* 1998).
2 Jonkers 1998.
3 Camacho 1992; Griffin & Hünicken 1994; Del Rio *et al.* 2007.
4 Stilwell *et al.* 2004.
5 See papers in Stilwell & Feldmann 2000.
6 Feldmann *et al.* 2003.
7 See papers in Stilwell & Feldmann 2000.
8 See review by Poole & Cantrill (2006); first described by Dusén in 1908 as part of the Swedish South Polar Expedion reports.
9 Stilwell *et al.* 1993; Pole *et al.* 2000; Francis 2000.
10 Currently being described with Sylvia Hope, and new work in 2009 with Julia Clarke.
11 See Case *et al.* 1987; Noriega & Tambussi 1996; Tonni 1980; Tonni & Tambussi 1985.
12 Stilwell *et al.* 1998; Jones 2000.
13 Welton & Zinsmeister 1980, and many subsequent papers by several researchers; Long & Stilwell 2000.
14 Willis & Stilwell 2000.
15 Fuente *et al.* 1995.
16 Köhler 1995.
17 Case *et al.* 1998b.
18 Bond *et al.* 2006; Hooker 1992; Reguero *et al.* 2002.
19 Vizcaíno & Scillato-Yane 1995; Bond *et al.* 2006.
20 Gingerich *et al.* 1994.
21 Heavy large pachyostatic rib; Fordyce 1989 (p. 272) and Locality 443, Unit 5 of Seymour Island; see Stilwell and Zinsmeister, 1992, p. 36, Fig. 48 depicting a vertebra *in situ*.
22 The identification of the large vertebra from Locality 443, presently under study at the University of Otago Geology Department, is uncertain but may either represent a primitive basilosaurine or even be the same species as that found at the top of the sequence, which is a toothed stem-mysticete.
Fossil whale bones have been known from the Antarctic since they were first discovered during the Swedish South Polar Expedition of 1901–03. Carl Wiman described two incomplete vertebrae and assigned them to *Zeuglodon* (= *Basilosaurus*) in 1905.
23 Elliot *et al.* 1975; Mitchell 1989.
24 Based on associated dated sediments by Strontium-Rubidium radiometric dating method, these upper levels of the La Meseta Formation are very close to 34.2 mya. Bits and pieces of this whale were collected over several expeditions, but the bulk of the skeleton was collected by helicopter at the top of the meseta on Seymour Island during the 1986–87 austral summer. Present are the *Basilosaurus*-like cranium (minus the tip of the relatively thin rostrum) and heterodont teeth, cervical vertebrae, manubrium, many ribs, a putative femur and possibly part of the pelvis.

Chapter 12

1 Jonkers & Taviani 1998.
2 Stilwell 2000.
3 Jonkers 2003.
4 Taviani *et al.* 1998.
5 Taviani & Claps 1998.
6 Berkman *et al.* 1998.
7 Berkman *et al.* 1998.

Chapter 13

1 Jeffrey Stilwell remembers seeing the midge and a brightly coloured mite a few times on rocks in East Antarctica, and with a feeling of disbelief as to how they could survive in such extreme conditions.
2 As a scientist participating in five Antarctic expeditions during the 80s and 90s, Jeffrey Stilwell could sense a change in climatic conditions.
3 *Open Access*, February 2009.

GLOSSARY

acanthodian a member of the Acanthodii, an extinct class of primitive bony fishes that lived in Silurian and Permian times; mostly known by isolated fin spines and tiny scales; complete fishes are very rare as fossils.

accretion increase in size by the external addition of new material, such as silty mud balls that often contain fossils as an inner core.

acme an interval of time represented by maximum vigour or numbers of organisms; 'acme zones' used in geology are important to correlate rock bodies and represent a period of time when a particular species was particularly abundant.

Actinopterygii ray-finned fishes; subclass of bony fishes (Osteichthyes) characterised by fins of membranous skin supported by rays or spines; comprises most bony fishes that are live today.

adaptation process of evolutionary modification which ends in improved reproductive efficiency and improved chance of survival; also process of adjustment of an individual organism to environmental conditions.

age in geology, relates to the length of geologic time since the formation of a rock either by solidification from a molten state (igneous) or by sedimentation (sedimentary); interval of time characterised by a specific dominant life form; also, the geologic time unit shorter than an epoch.

agnathan jawless fishes, including the living lampreys and hagfishes, but also many extinct armoured or heavily scaled forms that lived from the Ordovician and Devonian times.

algae predominantly aquatic plants that were some of earth's oldest evolved single-celled organisms.

allosaur any of various carnivorous non-avian theropod dinosaurs of the Late Jurassic and Early Cretaceous periods, exhibiting bird-like features such as fragile, hollow bones; *Allosaurus* is a classic big theropod with a big skull on a short neck, a long tail and reduced forelimbs.

alveoli tiny, delicate air sacs deep within the lungs, where the gas/blood exchange occurs. Oxygen from inhaled air passes through the walls of alveoli and enters the bloodstream while carbon dioxide passes out in the same way when air is exhaled.

ammonite a dominant group of the phylum Mollusca, belonging to the Class Cephalopoda and subclass Ammonoidea, which evolved in the Early Paleozoic Era, but most diverse during the Mesozoic and used widely in biostratigraphy.

amphibian class of quadruped vertebrates (tetrapods) containing frogs, toads, newts, salamanders, etc., including many extinct organism; heyday of fossil amphibians was during the Carboniferous and Permian.

Anatidae a group of modern birds (Aves) including the ducks, geese, swans and many fossil forms.

angiosperm the major group of flower plants, comprising all grasses and deciduous trees which evolved in the Early Cretaceous; in total, about 85 per cent of all land plants are angiosperms.

ankylosaur a group of heavily armoured dinosaurs with a dominant feature of bony plates; heads were small compared to the rest of the body, teeth were reduced or lacking; recorded from the Cretaceous, including Antarctica.

annelid bilaterally symmetrical worms (ringed or segmented), comprising three main classes the Polychaeta, Oligochaeta and Hirudinoidea.

anoxic an environment, whether it be terrestrial or marine, that lacks oxygen; certain intervals of Earth's history have been punctuated by anoxic events, such as the Ocean Anoxic Events during the Cretaceous greenhouse phase.

antiarch widespread, diverse group of Devonian placoderm fishes having pectoral fins enclosed in bone and a headshield with a single opening for eyes and nostrils; *Bothriolepis* was one of the most ubiquitous members of this group, being found on every continent.

aragonite a form of calcium carbonate (CO_3) commonly encountered in the shells of marine organisms, especially invertebrates, and in stalactites in cave systems.

archaeocete ancient group of fossil that originated sometime during the latest Paleocene–Early Eocene (54–53 mya) that comprises the monophyletic group including the common ancestor of *Pakicetus* and all of its descendents including *Ambulocetus,* other archaeocetes, Odontoceti, and Mysticeti.

archaeocyathid shelled, marine invertebrates, related to sponges, that were the dominant reef-formers of the Early Paleozoic and which diversified very quickly in the Cambrian, but then suddenly disappeared in the later part of the period.

Archean the most ancient Epoch; the term has been applied to the oldest rocks of the Precambrian.

arthrodire one of a diverse group of Devonian placoderm fishes having a jointed neck and two pairs of upper jaw tooth plates (supragnathals); includes groenlandaspids and phyllolepids from Antarctica, both of which have specially modified dentitions from the standard arthrodire pattern.

arthropod any of numerous invertebrate animals, including the insects, crustaceans, arachnids and myriapods, that are characterised by a chitinous exoskeleton and a segmented body to which jointed appendages are articulated in pairs.

artiodactyl hoofed ungulates with an even number of toes on each foot; the suborder Artiodactyla includes the living pigs, peccaries, hippopotamuses, camels, etc., and many fossil groups.

asteroid variably small to big stony objects, which are mostly the left-over pieces of planetary formation and which circulate in the Solar System.

astropothere strange group of fossil mammals from the Paleogene and Neogene which combine a perplexing combination of features; for example *Parastrapotherium* was an animal with the 'neck of an elephant and the loins of an Indian rhinoceros'.

austral pertaining to the regions of the southern hemisphere.

Aves formal name for the group of birds with a fossil record extending back into the Late Jurassic; considered today as descendants of the dinosaurs.

bacteria primitive micro-organisms with rigid or semi-rigid external walls; cell division happens by binary fission usually, and not mitosis.

baleen whale the group of cetaceans called formally the Mysticeta.

basalt fine-grained, silica-poor igneous rock with particular composition; associated with tectonic and volcanic activity, often with break-up of the continents during the rifting process, and the birth and opening of new oceans through time.

bedding plane the division surface between individual beds or strata of a stratified (layered) or sedimentary rock.

beech plants belong to the formal group Nothofagaceae, such as *Nothofagus* or southern beech, which has a long fossil record in Antarctica dying out some three million years ago, but still lives today in many austral localities.

belemnite an extinct cephalopod member of Mollusca; mostly bullet-shaped calcitic 'guards', or internal parts only of the cephalopod animal; they died out at the K–T extinction event 65 mya.

Bennettitales an order of the gymnosperm plants that resemble cycads; recorded as fossils in the Mesozoic.

benthic relating to the sea bed, river bed or lacustrine (lake) bottoms, e.g. benthic organism such as a trilobite that lived exclusively on the sea floor, where it was highly mobile, plowing through limey mud in search of food.

biogeography the scientific study of the geographic distribution of fauna, flora and micro-organisms, their life habits and historical factors relating to their present distribution.

biostratigraphy the classification and study of rock strata based on its component organic life forms or fossil content; biostratigaphy is indispensable in organising rock strata and dating geologic units by correlating fossil-bearing sediments.

biota the total fauna and flora of a given area now and in the geologic past.

bioturbation the mixing of sediment by the burrowing, feeding or other activity in terrestrial and marine settings by animals.

bivalve an aquatic mollusc with two shells (valves) often called a 'clam', e.g. cockles, oysters, mussels, many fossil groups; evolved during the Early Cambrian Period and often used in geologic studies to date rocks and interpret palaeoenvironments from terrestrial lacustine/river settings to the abyss of the ocean.

bone bed a specific section of rock strata containing a wealth of vertebrate remains which either lived nearby, or were transported from their initial resting place.

boreal geographically ranging from cool or cold temperature northern hemisphere areas.

brachiopod an invertebrate group of marine invertebrate animals often called a 'lamp shell', comprising two unequal valves that are sessile and attached to hard substrate; brachiopods evolved sometime during the Early Cambrian.

brittlestar a group of Eocene echinoderms called the Ophiuroidea, which abound in the La Meseta Formation of Seymour Island, where they form monotypic concentrations of a high number of individuals.

Bryophyta a division of plants that comprises five classes, including the liverworts, hornworts, true mosses, granite mosses and peat mosses.

Bryozoa a major invertebrate group called 'moss animals', but not related to moss at all; most are marine, and a few are brackish to freshwater; they attach themselves to substrate; abundant fossil forms, extending back to Early Paleozoic, and many groups are extant.

calcareous a descriptive geologic and palaeontologic term applied to a rock or fossil containing calcium carbonate.

Cambrian the first geologic period of the Paleozoic Era, that ranged from 542 to about 490 million years ago, and characterised by the diversification of shell-invertebrates.

carbonate rocks any rock comprising an abundance of carbonate minerals, such as limestone or chalk (calcium carbonate) or dolomite/dolostone (calcium magnesium carbonate); carbonates are chemically precipitated from sea water.

Carboniferous the geological period spanning from about 359 to 299 million years ago, that is subdivided into the lower Mississipian and upper Pennsylvanian epochs, reflecting a stark change in rocks types in the middle of the period about 318 million years ago; the period is characterised by great, coal-rich deposits, especially in the upper part.

cassiduloid comprises the sand dollar and heart-shaped sea urchins in the Phylum Echinodermata, and are predominantly a fossil order.

Cenozoic the most recent Era that has lasted from 65 million years ago to the present, and comprises both Paleogene and Neogene periods; often, the older scheme of Tertiary and Quaternary is still employed.

cephalopod the most advanced class of the Phylum Mollusca, which includes ammonites and belemnites; they may be shelled or lack a shell; shelled forms are often multi-chambered and used for buoyancy; extant forms include the pearly nautilus, argonaut, squid, octopus, cuttlefish, etc.

Cetacea the group of aquatic mammals, including whales, dolphins and porpoises; two extant suborders, Mysticeta (baleen whales) and Odontoceta (toothed whales) occur in the fossil record in Antarctica from Late Eocene to Pliocene.

charadriiform an order of shore birds with a fossil record extending into the Cretaceous.

chiromomid a member of the family of midges that lack piercing mouthparts; as such, they are non-biting insects.

Chondrichthyes the class of jawed fishes with cartilaginous skeletons, the subclass Elasmobranchii includes sharks and rays, and Holocephalii make up the rat fishes or holocephalans.

chordate an animal with a backbone, and member (like us humans) of the major group Chordata.

clastic a type of sedimentary rock (e.g. shales, sandstones, conglomerate, etc.) formed by physical deposition – that is, the breakdown of solid particles that disperse through a water medium or by air; constituent particles of a clastic rock have been broken off other rocks such as igneous granites and others by erosion processes.

Cnidaria (previously coelenterata) an invertebrate animal group of mostly multicellular organisms with a two-fold body plan of ectoderm and endoderm layers, separated by gelatinous connective tissues; two body forms of erect colonial or solitary polyp that is attached to substrate, and other being a medusa, which is either pelagic, solitary or disc-shaped; all cnidarians are carnivorous, with some taking nutrients from algae; comprises extant classes Anthozoa, Cubozoa, Hydrozoa and Scyphyozoa; long fossil record from Late Precambrian throughout Phanerozoic.

coal a sedimentary rock made up of dead, decomposed plant remains; the carbon content of the plants is retained by decomposition in the absence of oxygen; pressure and heat during and after formation increases the fuel content of the coal, and importantly, its commercial value.

cockle the common name for a bivalve (clam) of the order Veneroida.

Conchostraca an order of Early Devonian to Recent, mussel-like crustaceans informally known as 'clam shrimps' of moderate size belonging to the subclass Branchiopoda, entirely enclosed by a bivalve carapace; they live in freshwater environments.

conglomerate a sedimentary rock that is made up of coarse-grained material, such as large boulders and fragments, which are embedded in a fine matrix; usually formed in terrestrial environments, such as rivers and streams, etc., and represent high-energy, turbulent environmental conditions.

conifer group of plants called the Pinatae, largest group of gymnosperms.

conodont a member of an extinct group of small, primitive chordates, preserved primarily in the form of their cone-like teeth; they are the most widespread Paleozoic microfossils and are important for biostratigraphic indexing; recent research questions a relationship of conodonts to vertebrates, which may see a reversal to the early invertebrate allocation.

continental drift the name of the mechanism by which continental landmasses move across the surface of the globe now and in the geologic past.

coracoid pertaining to the bone that in birds, reptiles and monotremes articulates with the scapula and the sternum.

craton the stable ancient centre of a continent, which is normally Precambrian in age, that formed well before the mountain-building episodes of the Paleozoic Era.

Cretaceous the last period of the Mesozoic Era, spanning 145 to 65 million years ago, named after the thick chalk deposits on the northern continents; period also represents the last major greenhouse phase of the Phanerozoic Eon that saw both temperatures and carbon dioxide levels climb to very high levels; at the end of the period was a mass extinction event that exterminated the non-avian dinosaurs (some maybe persisting for a short time in the dawn of the Cenozoic), flying reptiles, marine reptiles, belemnites, etc.

crinoid a major group within the Phylum Echinodermata that were attached to the sea floor by a long stalk, or free-floating and/or quite mobile; often called 'sea lilies', they were quite diverse in the Paleozoic, but were greatly affected in terms of diversity during the Permian–Triassic mass extinction event, and just survived.

crustacea subphylum of Arthropoda that include lots of shrimps, lobsters, crabs, barnacles, water fleas, etc., which live in virtually all environments from terrestrial to abyssal realms.

culmacanthids a group of deep-bodied acanthodian fishes only known from Australia and Antarctica by the genus *Culmacanthus*.

cursorial adapted to a running lifestyle.

cyclorrhaphan non-biting flies with a known fossil record extending back into the Early Cretaceous (c.140 mya) of England in amber deposits.

cynodont ('dog tooth'): a carnivorous therapsid of the late Permian and Triassic periods, whose specialised teeth were well developed; they were one of the most diverse groups of therapsids, with nearly all the characteristics of mammals; they were probably warm-blooded and covered in hair.

decalopod a group of marine arachnids informally known as the 'sea spiders'; they are not swimmers, but crawl slowly over the bottom of the sea or amongst the fronds of seaweed; they occur in polar, temperate and tropical seas.

desiccation the process of drying or removing water.

detritus usually refers to fine pieces of organic matter derived from the breakdown and decomposition of animal and plant remains; some organisms are considered to be detritivores, and feed on particulate organic matter.

Devonian the geologic period extending from 417 to 359 million years ago, commonly known as the Age of Fishes, which saw the establishment of land plants in a big way.

diapsid any of various reptiles having a skull with two pairs of temporal openings and including the lizards, snakes, crocodiles, non-avian dinosaurs and pterosaurs.

diatom microscopic, unicellular chromophycote algae, found in terrestrial and marine environments.

dicynodont ('two dog tooth'); a small herbivorous therapsid from the Late Permian and Early Triassic with reduced dentition and a single pair of tusklike teeth; it was probably warm-blooded.

dinosaur great rulers of the earth for some 160 million years, that comprise two orders of reptiles – the so-called lizard-hipped 'saurischians' and bird-hipped 'ornithischians'. Dinosaurs evolved during the Triassic Period, diversified greatly during the Jurassic, and lasted until the close of the Cretaceous, except for the birds, which are descendants of theropod dinosaurs, and are very diverse today.

Diomedeidae the family of birds that include large oceanic birds such as albatrosses found in the Southern Ocean, specialised for their dynamic soaring flight close to the water surface.

dipnoan the group to which lungfishes belong; osteichthyan fishes characterised by a mosaic of small bones forming the skull roof and snout in early forms, with either powerful crushing toothplates or with rasping denticulate palates.

distal anatomically located far from a point of reference, such as an origin or a point of attachment.

ebridians organic-walled micro-organisms that are unicellular, marine and planktonic with an endoskeleton of silica; their fossil record extends back to the Paleocene, with drastic drop in diversity after the Pliocene.

Echinodermata entirely marine invertebrates with diverse fossil record from late Precambrian to Recent, including Crinozoa (feather stars, sea lilies), Echinozoa (sea urchins, sea cucumbers) and many fossil groups now extinct.

ectoparasite a parasite, such as a flea, that lives on the exterior of another organism.

elasmosaur a very long-necked extinct marine reptile, a type of plesiosaur, which died out at the end of the Cretaceous Period.

elytral of or pertaining to the elytra; either of the leathery or chitinous forewings of a beetle or a related insect, serving to encase the thin, membranous hind wings used in flight.

Eocene geological epoch of the Paleogene Period, extending from 55.5 to 33.7 million years ago; Antarctica has a rich record of Eocene life from Antarctic Peninsula (Seymour and Cockburn islands, in particular) and McMurdo Sound in glacial erratics.

epibyssate of certain bivalves such as marine mussels that secure attachment to firm substrates by byssal threads (byssus – chitinous threads secreted by the foot of the clam).

epifauna a fauna characterised by members whose typical life sites are on the outer surface of their environment, as opposed to within it.

Epoch an interval of geological time, which is a subdivision of a period.

erosion pertaining to the removal or wearing away of the land surface by many processes, including water, ice, wind or other means.

exoskeleton the external skeleton which first developed and evolved at the end of the Precambrian and diversified in the Cambrian Period in certain groups of marine invertebrates, providing not only protection from rapidly evolving predators, but also muscle support and anchorage.

facies a descriptive term that means a general aspect or appearance of a sedimentary rock.

falconid belonging to a family of raptorial birds, including the falcon today, characterised by a short, hooked beak, strong claws and powerful flight.

faunule(s) the localised stratigraphic and geographic distribution of a particular taxon (taxa) of animals, extinct or extant.

femoral of, relating to, or located in the thigh or femur.

femur commonly known as the thigh bone, which is the most proximal (closest to the body) bone of the leg in vertebrates capable of walking or jumping, such as most land mammals.

foraminifera tiny aquatic, solely marine animals with a single or multi-chambered shell, that are particularly useful in biostratigraphic and palaeoenvironmental studies.

Gastropoda a class of snails within the Phylum Mollusca which live in both terrestrial and aquatic environments from mountain tops to deep sea, and have a fossil record extending back into the Cambrian Period.

gaviid the single, monogeneric bird family of the order Gaviiformes, including the loons or divers, which comprise aquatic birds found in many parts of North America and northern Eurasia.

glaciation an event which results from extending climatic cooling which is sufficient to form ice sheets and glaciers; glaciation episodes have occurred throughout Earth's history from the Precambrian to Recent time (now in 'interglacial'), but presence of ice sheets like today is a relatively uncommon occurrence.

glauconite a greenish mineral of the mica group; a hydrous silicate of potassium, iron, aluminium or magnesium, used as a fertiliser and water softener.

Glossopteris a prominent fossil fern of the Late Paleozoic and earliest Mesozoic, recorded from all of the southern continents of the ancient continent of Gondwana, including Antarctica.

Gondwana the great southern land that comprised Antarctica, Australia, South America, Africa, Madagascar, India, New Zealand-Chatham Islands, New Caledonia, parts of southern Asia and many other smaller continental pieces; its existence was hotly disputed until the 1960s, and the evolution of the concept of plate tectonics.

Gondwanathere an extinct order of mammals that lived during the Late Cretaceous through the Eocene in the southern hemisphere, including Antarctica.

granite a coarse-grained igneous rock containing quartz, mica and feldspar, often beautifully coloured pink rock when composed of orthoclase (potassium-rich) feldspars to even rich, deep purples of the granitic rock called charnockite, which is a hypersthene granite, common in the Precambrian shield rocks of India.

graptolite a group of solely Paleozoic marine macroinvertebrates belonging to the group Hemichordata with delicate, branching exoskeletons, that leave impressions in shale; useful in biostratigraphy.

gymnosperms a seed-bearing plant, such as conifers, which have efficient reproductive systems compared to early spore-bearing plants, but dwindled in importance after angiosperms (flowering plants) evolved in the Cretaceous.

hadrosaur any of various amphibious dinosaurs of the genus *Anatosaurus* and related genera that had webbed feet and a duck-like bill; also called the 'duckbill' or 'duck-billed' dinosaurs.

heteromorph in ammonites, one whose shell does not curl up into the tight spiral shape which shells of ammonites from the subclass Ammonoidea typically do.

Hexanchiformes a small order of medium-sized benthic marine squalomorph sharks.

Hiatellidae a family of suspension-feeding marine bivalves that are capable of burrowing deeply in the sea floor, common in the fossil record of Antarctica.

Holasteroida an order of irregular echinoids with an elongate apical system that first appeared in the fossil record in the Mesozoic, Early Cretaceous or earlier.

Holocene the most recent period of geological history, extending from about 10 000 years ago to the present.

Holocephalii a subclass of cartilaginous fishes (Chondrichthyes) made up of a single order called the Chimaeriformes, and with hard bony toothplates.

homosteid a member of the family of arthrodire placoderm or armoured fishes from the Devonian of Europe.

humerus the long bone of the arm or forelimb, extending from the shoulder to the elbow.

Hydrozoa the class of marine and more rarely freshwater cnidarians comprising seven orders in which the life cycle characteristically involves alternation between an asexually reproducing, attached polyp and a sexually reproducing planktonic medusa, sometimes with a calcified exoskeleton.

hypsilophodontid a group of Middle Jurassic to Late Cretaceous, non-avian dinosaurs named after the *Hypsilophus* genus of iguana lizard; they were small dinosaurs, regarded as quick, plant-eating bipeds, approximately one to two metres in length; have been recorded from Asia, Australia, Europe, North America and South America.

Iapetus Ocean the predecessor of the Atlantic Ocean, which was positioned between Laurentia and northern Europe; Iapetus Ocean closed during the Caldedonian and Appalachian orogenies.

Ice Age the common term for an intense period of glaciation or severe cooling episode; peak of last ice age was about 18 000 years ago; now in interglacial period, but still quite a cold spell in terms of Earth's history, and unusual climate, which is set to be more topsy-turvy with global warming.

ichnofossil a trace fossil, such as tracks, footprints, burrows, coprolites, etc., all made by an organism; often the animal making the trace has not been preserved, and is known only by the ichnofossil.

ichthyosaur a marine reptile that existed at the same time of the dinosaurs, of which it was closely allied, but sea-dwelling.

inhomogenous consisting of elements that are not of the same kind or nature.

invertebrates those animals without a backbone.

isopods small to moderately sized crustaceans found in marine, freshwater and terrestrial environments as parasites; common in Antarctic waters, with one of first descriptions by James Eights in 1833, who misidentified these organisms as trilobites (of which they superficially resemble).

Jurassic the middle period of the Mesozoic Era, which lasted from 200 to 145 million years ago, and named after the Jura Mountains; oldest known dinosaurs from Antarctica recorded from the Early Jurassic, such as bizarre, crested, theropod *Cryolophosaurus*.

katabatic a term to describe winds that blow down a slope, such as the frigid winds blowing off the Antarctic ice cap.

lacustrine the formal scientific term for 'lake'.

land bridge an ephemeral link between two landmasses, which is caused by a lowering in sea level; allows the migration of land-based animals between separated continents.

Laurasia a continent encompassing Laurentia (North America and northern Europe) and Angaraland (Siberia and Kazakhstan), which became joined 250 million years ago.

lava molten hot rock (magma) that breaches the surface of Earth and can spill onto land for incredible distances; some lavas erupted explosively, while others flow smoothly from fissures in the crust, and bubble along.

limestone a sedimentary rock, comprised almost entirely of calcium carbonate; limestones mostly arising from the precipitation of invertebrate shells of marine organisms, both big and small (such as forams).

Liptoterna a group of South American and Antarctic Cenozoic-aged mammals with a name meaning 'simple ankles', fossil hoofed mammals (ungulates) that display toe reduction; three-toed, and even a one-toed, horse-like form developed; the Litpopterna occupied ecological roles as browsers and grazers similar to horses and camels in Laurasia.

listroderine a type of weevil with a surprising fossil record in the Neogene in Antarctica, reported from the Meyer Desert Formation at a locality on the Beardmore Glacier in the Transantarctic Mountains about 500 km from the South Pole.

lithification the process of sediment becoming rock which includes compression and hardening; water is squeezed out and chemical changes occur, ending with the formation of rock strata.

lungfish a type of fish that developed lungs alongside gills in the Devonian Period; precursors of amphibians.

lycopsid a group of plants that flourished during the time of swamp forests during the Carboniferous Period.

marsupial group of mammals that give birth to underdeveloped young which are nurtured in pouches in the mother's body; they spread through the continents, including Antarctica in the Eocene but were isolated in Australia and South America as Gondwana broke up.

matrix the hardened sedimentary particles entombing a fossil organism.

megaclasts relatively large pieces of rock of any kind (sedimentary, igneous, metamorphic) immersed in another larger rock.

megalosaur a member of the genus *Megalosaurus*; a huge, flesh-eating theropod dinosaurs of the Jurassic Period.

mesonychids an extinct order of carnivorous hoofed mammals, bearing a strong, albeit superficial, resemblance to wolves.

Mesozoic the 'Middle life' era lasting from about 250 to 65 million years ago, comprising the Triassic, Jurassic and Cretaceous periods.

meteorite extraterrestrial debris that reaches Earth's surface; its composition is similar to terrestrial planets and likely to be a left over piece of planetary material that never accreted onto a planet; provide clues about the early history of the Solar System.

microbotheriid an ancient group of mammals, with a fossil record extending back into the Paleocene with records in the Eocene of Antarctica; new data indicate that microbiotheres are not didelphids at all, but are instead most closely related to the Australasian marsupials; together, the microbiotheres and the Australian orders form the clade Australidelphia.

modiolid a group of medium-sized, marine mussels (bivalves) commonly known as 'horsemussels' that are found in many fossil deposits of Antarctica from Mesozoic to Recent.

Mollusca an animal phylum which includes several invertebrate shelled and 'nude' classes, such as Bivalvia (clams), Gastropoda (snails and slugs), Cephalopoda (ammonites, nautiloids, squid, etc.), Scaphopoda (tusk shells), Amphineura or Polyplacophora (chitons), Rostroconchia (extinct, clam-like molluscs without hinge), Monoplacophora (tiny living fossils, thought to be extinct until re-discovered on Galathea Expedition in 1950s), and other obscure classes including the hyoliths, tentaculitids, etc.; very useful in biostratigraphy and palaeoenvironmental interpretations; evolved sometime during the late Precambrian (Ediacaran) with groups such as *Kimberella*.

monotreme an ancient, egg-laying mammal, of which only the duck-billed platypus and echidna survive to the present day in Australia; fossil record extends back into the Cretaceous (fossils in Australia and southern South America).

mosasaur a member of the family Mosasauridae, very large, extinct, marine, fish-eating lizards of the global Late Cretaceous, including Antarctica, with limbs modified into paddles that are related to the recent monitor lizards.

muricids murex snails or rock snails belonging to a large and varied taxonomic family of small to large predatory sea snails with usually strong sculpture; some 1200 fossil species have been recognised including rare genera and species from Antarctica; many have unusual shells which are considered attractive by shell collectors and by interior designers.

nautiloid a member of a group of mainly extinct marine molluscs (Class Cephalopoda, subclass Nautiloidea) that includes the pearly nautilus today.

nekton collectively, any marine and freshwater organisms that can swim freely and are generally independent of currents, ranging in size from microscopic organisms to whales.

neoselachian a group of elasmobranch fishes, including the sharks and rays.

nodosaurid a family of ankylosaurian, non-avian dinosaurs also known as the 'armoured dinosaurs'; recorded from the Cretaceous Period in what are now North America, Asia, Australia, Antarctica, and Europe.

non-therian a mammal other than Theria, comprising the marsupials and the placental mammals.

non-volant said of an organism incapable of flight.

Oligocene last epoch of the Paleogene Period, extending from 33.7 to 23.8 million years ago.

oligotrophic having low primary productivity; describing substrates or water bodies low in nutrients, or organisms requiring a small nutrient supply.

oncolite a small, variously shaped (often spheroidal), concentrically laminated, calcareous sedimentary structure resembling an oolith; formed by accretion of successive, layered masses of gelatinous sheaths of 'blue-green algae', which are technically not algae, but cyanobacteria.

onychodontid one of a group of Devonian lobe-finned osteichthyans fishes with well-developed tooth whorls on the lower jaw.

Ordovician the geological period that lasted from 490 to 443 million years ago, named after the Ordovices, a Celtic tribe; many of the organisms that evolved and appeared in the Cambrian diversified in the warm shallow seas of the continental shelves of this time.

organic pertaining to life, thus being a living organism, or originating a life form; term also employed to define hydrocarbon chemistry; organic carbons in Earth's oldest rocks may or may not be evidence of living things.

orogenic the process of mountain formation, especially by the folding and faulting of Earth's crust.

ossicle either a small bone, especially one of the three bones of the middle ear; in invertebrates, one of numerous small calcareous structures forming the skeleton of certain echinoderms, as the starfishes; or one of the hard joints of the stem or branches of a crinoid or encrinite.

osteichthyan one of the group of bony fishes comprising the ray-fins (Actinopterygii) and the lobe fins (Sarcopterygii).

osteology the branch of anatomy that deals with the structure and function of bones.

ostracod a member of the Ostracoda, a group of small, freshwater or marine crustaceans, having a bivalve carapace that covers the entire body; they are arguably the most complex organisms studied in micropalaeontology, and are a potent tool in biostratigraphy and palaeoenvironmental interpretations.

otic of or relating to near the ear.

palaeogeography the discipline that studies the ancient geography of the planet by plotting the positions of the continent, mountain chains and oceans at different times through Earth's history.

palaeoniscoid an old but still widely used term for very primitive ray-finned fishes (actinopterygians).

palaeontology the study of ancient life and fossil organic remains.

Paleocene the dawn of the Cenozoic Era and first epoch of the Paleogene Period, extending from 65 to 55.5 million years ago.

Paleogene the first period of the Cenozoic Era extending from 65 to 23.8 million years ago, comprising Paleocene, Eocene and Oligocene epochs.

paleomagnetic the study of the remnant magnetisation in rocks, and study of Earth's magnetic field as it existed in the past.

Paleozoic the geological era spanning the Cambrian to Permian periods, from about 542 to 251 million years ago, ending in the biggest mass extinction event at the close of the Permian.

palynology the scientific study of living or fossil plant spores and pollen.

palynomorph an organic microfossil particle of 5–500 micrometers in diameter found in sedimentary deposits and composed of organic material; such as a spore or pollen that is of interest in palynological studies.

Pangea the supercontinent formed at the close of the Paleozoic Era, comprising all of the major continental blocks; the supercontinent broke up almost as soon as it had formed.

paucispecific said of a biotic assemblage which is low in diversity.

pectens a small scallop inhabiting shallow marine waters; an edible bivalve mollusc with a ribbed, fan-shaped shell.

Pectinidae the family of pecten bivalves.

pectoral fin the front, paired fins of a fish, which later would become the arms of tetrapods.

pelvic fin the rear, paired fins of a fish, which later would become the legs of tetrapods.

Permian the last period of the Paleozoic Era, extending from 299 to 251 million years ago, which ended in Earth's greatest upheaval when more than 90 per cent of all life perished from the globe; named after the region around the town of Perm in Russia.

phocid a hair seal belonging to the cosmopolitan family of marine mammals (Pinnipedia) containing about 20 species and including grey, hooded, harp, monk, crabeater, Ross, leopard, Weddell, and elephant seals, and the sole freshwater Baikal seal; they feed primarily on fishes, squid and crustaceans.

Phoenicopteridae the bird family comprising the flamingos, which make up six living species of large, long-legged wading birds cosmopolitan in tropical freshwater and marine lakes and lagoons; their fossil record extends back into the Eocene.

phorusrhacoid the scientific name for 'terror birds' – large, carnivorous, flightless avians which were top level predators in terrestrial environments.

photosynthesis the chemical process used by plants to manufacture food using carbon dioxide and sunlight; it is believed that the marked expansion in the amount of free oxygen in the early atmosphere reflects the photosynthesis of Earth's oldest known organisms, the cyanobacteria ('blue-green bacteria').

phyllolepid a member of the group of flattended arthrodires (placoderm fishes) with characteristic radiating patterns of ridges on the plates.

placental used to describe mammals that give birth to developed young; placental mammals have superseded marsupials in most part of the world.

placoderm an extinct jawed fish with overlapping bony plates forming a distinct shield over both head and trunk regions, often connected by a ball and socket joint; they were the first jawed vertebrates and the first vertebrates to have internal fertilisation.

plate tectonics the field of geological study that explains the mechanisms that drives the geological processes of Earth; explains how Earth's crust is partitioned into rigid plates, how these plates move around on the surface by mantle currents, and how new plates arise and old ones destroyed.

Pleistocene the period lasting from 1.8 million years ago to 10 000 years ago, which is known for a time of global cooling, leading to a series of ice ages.

plesiosaur a major marine reptile group that lived during the Mesozoic Era, and became extinct at the Cretaceous–Tertiary boundary worldwide.

plioplatecarpines a subfamily of mosasaurs (marine reptiles), Late Cretaceous marine squamates, first appearing in the Turonian stage, becoming extinct in the Maastrichtian at the K–T boundary, having spanned some 24 million years of geologic history.

polydolopid a member of the important group of Paleocene mammals, which were rat- to rabbit-sized animals somewhat similar to rodents in ecology; they declined in diversity during the Eocene (important fossil records on Seymour Island, Antarctic Peninsula) and became extinct in the Oligocene.

Precambrian the interval of time that spans most of Earth's history, from the oldest rocks over four billion years old to about 542 million years ago; the oldest known organisms – about 3.8 billion years old – are from Australia.

prepidolopid a member of the family of fossil marsupials from South America possessing a hypertrophied and procumbent first incisor; they evolved convergently with Australian kangaroos and similar Australasian forms that are grouped in the Order Diprotodontia.

presbyornithid a member of a peculiar family of prehistoric birds that were used to argue for a close relationship between flamingos, waterfowl and waders, but are now known to be unequivocal waterfowl with a peculiarly derived morphology paralleling waders and flamingos.

pronotum a plate-like cover of the first thoracic segment in an insect.

prosauropod a member of the Prosauropoda, mainly herbivorous Triassic dinosaurs that are probably ancestral to sauropods; they possessed blunt teeth, long forelimbs, and extremely large claws on the first finger of the forefoot.

Proterozoic the span of geological time from about 2.5 to 0.5 billion years ago.

proximal nearest; proximate; nearer to a point of reference such as an origin, a point of attachment, or the midline of the body.

pseudodontorn a name used to describe the bony-toothed birds of the family Pelagornithidae, prehistoric family of large seabirds; their fossil remains have been found all over the world in rocks dating between the Late Paleocene and the Late Pliocene, and the Eocene of East and West Antarctica.

pterosaur a flying reptile that was closely related to the dinosaurs, and became extinct at the end of the Cretaceous Period 65 million years ago.

puparium a rigid outer shell formed from the larval skin that covers some insect pupae.

radiometric dating the technique which measures the ratio between different isotopes of a radioactive element and yields an accurate estimate of the age of the formation of the host rock; it assists geologists and palaeontologists in their quest for the ages of rocks and fossils.

ratite a flightless birds having a flat breastbone without the keel-like prominence characteristic of most flying birds; living birds such as the ostrich and emu.

rhizodontid a member of the large to very large (six metres) Devonian–Carboniferous osteichthyans fish with robust lobe fins and well-developed fangs on the lower jaw.

rift valley a valley caused by the separation of tectonic plates; examples are the present-day Great Rift Valley of East Africa, and ancient rifts such as those between Antarctica and Australia, and between Brazil and West Africa.

Rodinia the world supercontinent that formed at the end of the Mesoproterozoic era, then broke into eight continents during the Neoproterozoic; existed between 1300–600 million years ago.

sandstone common sedimentary rock composed of cemented sand grains, formed originally in environments such as beaches, deserts, shallow marine environments, rivers, etc.

sarcopterygian lobe-finned fishes, a group of osteichthyans including the coelacanths, lungfishes, onychdontids and tetrapodomorphans (rhizodontids, tristichopterids, elpistostegalians, etc.).

scaphopod or 'tusk shells'; a member of the class Scaphopoda, marine molluscs distinguished by curved tubular shells open at both ends, resembling a elephant's tusk; mostly small, with some species reaching 15 cm in length, they live in the bottom sediment where they feed on microscopic detritus and organisms such as foraminifera; several hundred known species are found worldwide, and many from the fossil record, including Antarctica.

scapula one of the two large, flat, triangular bones forming the back part of the shoulder; also called the shoulder blade.

Scleractinia an order of the subclass Zoantharia which comprises the true, or stony, corals; these are solitary or colonial anthozoans which attach to a firm substrate.

scrimshaw the art of carving or incising intricate designs on whalebone, whale ivory or walrus tusks, on long oceanic voyages by sailors as a leisure occupation.

sedimentary rock a category of rocks in geology, which includes sandstones, conglomerates, limestones and shales, formed from the deposition of eroded pieces of other rocks and reburied, or from the precipitation of minerals out of sea water or lakes, either by abiotic or biotic processes.

shale a fine-grained sedimentary rock formed in normally quiescent conditions, such as in many terrestrial (swamps, lakes, rivers, etc.), marginal marine (deltas, mudflats, etc.), or offshore marine environments (shallow to abyss).

silicify to convert into or impregnate with silica (SiO_2); specifically, petrify.

silicoflagellates small group of unicellular algae, found in marine environments, making up part of the phytoplankton; In one stage of their life cycle, they produce a siliceous skeleton, composed of a network of bars and spikes arranged to form an internal basket; these form a small component of marine sediments, and are known as microfossils from as far back as the Early Cretaceous.

spatangoid 'heart' or 'bottle' urchins, members of a cosmopolitan order of littoral to abyssal echinoids in the phylum Echinodermata, comprising some 275 living species, and many fossil ones ranging into the Cretaceous; they exhibit pronounced bilateral symmetry with variously specialised spines and tube-feet, and burrow into soft sediments on the sea floor.

spheniscid belonging to the family of 'penguins', flightless marine neonorthine birds widespread in the cold Southern Ocean extending north to the Galapagos Islands; very efficient swimmers using wings modified as flippers; the fossil record extends back into the Paleocene Epoch of Zealandia, where they evolved in Cretaceous–Paleogene greenhouse environments of the southern polar regions.

sternum a long flat bone in most vertebrates that is situated along the ventral midline of the thorax and articulates with the ribs.

striations large grooves and scratches caused by the slow movement of glaciers and ice sheets as they pass over rocks (such as the bedrock); the orientation of striations can indicate the direction of ice flow and provide evidence of former glaciation.

stromatolite an organo-sedimentary structure comprising cyanobacteria ('blue-green bacteria'), which are interlayered with lime 'mats'; stromatolites are the oldest record of organic life on the planet; they form in both terrestrial and marine extreme environments, such as the hypersaline conditions at Shark Bay, Western Australia; they were much more common in the Precambrian with the oldest record of 3.8 billion years in Australia.

stylasterid a member of the family within the Stylasterina or 'hydrocorals', an order of marine invertebrates known from most seas and a range of depths, characterised by a colonial polyp generation which secretes a calcareous skeleton.

subduction a geologic process in which one edge of one crustal plate is forced below the edge of another.

sudamericid a member of an enigmatic group of extinct mammals known as sudamericid gondwanatherians, which lived during the Late Cretaceous–Early Paleogene; they show possible signs of adaptation to a grassy diet; with teeth ideally suited to handling abrasive materials like grass; but because of grass's patchy presence in the fossil record, these features were interpreted as an adaptation to a semi-aquatic, or burrowing, lifestyle, like that of modern beavers.

supercontinent an immense landmass encompassing two or more major continental blocks; examples include Gondwana(land) and Pangea ('all lands').

supragnathal the upper jaw toothplate of a placoderm fish.

symphysis a fibrocartilaginous fusion between two bones, a type of cartilaginous joint, specifically a secondary cartilaginous joint.

synapsid a member of the group vertebrates that developed one hole in their skull behind each eye, during the Late Carboniferous; the living synapsids, known as mammals, are the dominant land animals.

talus a slope formed by an accumulation of rock debris; also the rock debris at the base of a cliff.

tardigrade any of a phylum (Tardigrada) of minute water animals with segmented bodies and four pairs of unsegmented legs, often regarded as primitive arthropods.

tectonic a geologic term related to large-scale forces and movements in the Earth.

teleost belonging or pertaining to the Teleostei, the most diverse and successful group of bony fishes, including most living species; bony fishes have a consolidated internal skeleton, a swim bladder, thin cycloid scales, etc.

temnospondyl any of many primitive amphibians of the Order Temnospondyli; they achieved their highest diversity during the Carboniferous and Permian periods; the last large aquatic form occurred during the Early Jurassic..

Tethys Ocean an ancient sea which separated Gondwana from Laurentia for most of the Mesozoic Era.

tetrapod a four-legged animal, including all groups from the fish-like amphibans in the Devonian Period to all living and extinct amphibians, reptiles, birds, dinosaurs and mammals.

teuthid the formal name for 'squid', belonging to an order of marine carnivorous cephalopod molluscs which are typically streamlined, active swimmers found in shallow and coastal waters today; complete specimens are rare in the fossil record.

thecodont considered to be the most important reptile group of the Triassic period; thecodonts were the ancestors of the dinosaurs, which arose during the mid-Late Triassic.

thelodont one of a group of small jawless fishes covered by elaborately ornamented scales; which are usually all that remains of them as fossils.

therapsid one of a group of synapsids which developed mammal-like features in the Triassic Period, but became extinct in the Mesozoic; believed to be the ancestors of the true mammals.

theropod a large, carnivorous dinosaurs of the suborder Theropoda of the Jurassic and Cretaceous periods, characterised by bipedal locomotion, large jaws and short forelimbs.

tillite a type of glacial material that is formed by an ice sheet; usually chaotic mixtures of mud, sand and boulders; ancient tillites tell geologists of continental movements and glaciation events.

titanichthyid an extinct genus of gigantic arthrodiran fishes, with a very large cranial shield and much reduced edentulous jaw-elements; they were probably filter feeders.

trechine a ground beetles of the tribe Trechini, which includes some of the most spectacular radiations of beetles that live in caves or near the surface of the ground.

Triassic the first period of the Mesozoic Era, following the mass extinction event at the close of the Permian, extending from 251 to 200 million years ago; the few survivors of the Permian–Triassic extinction flourished in the vacant ecospace following the calamity and diversified, with reefs becoming established again about 10 million years into the Triassic; the first dinosaurs and mammals originated in the Triassic.

trilobite a highly mobile, tri-lobed arthropod with a hard exoskeleton that 'bulldozed' through their habitats in ancient shallow seas; trilobites were the most important invertebrate group of the Early Paleozoic Era; they became extinct at the end of the Permian about 251 million years ago; they also assisted with the demise of Cambrian archaeocyathids.

tristichopterid one of a family of moderate to large predatory lobe-finned tetraopdomorph fishes, best typified by the genera *Eusthenopteron* and *Mandageria. Notorhizodon* was a large genus of this group from Antarctica.

tritylodont any member of a genus *Tritylodon*, extinct cynodont therapsids found as fossils in Upper Triassic and Lower Jurassic rocks in southern Africa and North America, and identified possibly from the Lower Jurassic of Antarctica; it is probable that the habits of *Tritylodon* were similar to those of the later rodents and multituberculates.

tylosaur one of a subfamily of mosasaurs, a diverse group of large-bodied marine lizards; found on all continents except Australia and South America; tylosaurs were armed with sturdy, conical teeth and an elongated premaxilla and extensions of the dentaries that do not bear teeth to the very end such as is found in other genera of mosasaurs.

umbonal a knob-like protuberance arising from a surface; in bivalves also known as the 'beak' and is the most protuberant, oldest part of the clam.

ungulate any large, hoof-bearing, grazing mammal; extant species typically belonging to the orders Perissodactyla and Artiodactyla, and many fossil groups.

vascular plant a plants with systems for circulating fluids internally; the first land plants had to have means of distributing nutrients.

ventifact a stone that has been shaped, polished or faceted by wind-driven sand; common in the Dry Valleys, Antarctica, relating to strong katabatic winds from the polar ice cap.

volcaniclastic a rock that is composed of fragmented volcanic products ejected from volcanoes in explosive events.

Weddellian Province a broad, shallow-water biotic province stretching around the southern circum-Pacific during the Late Cretaceous to Early Paleogene; it was proposed by Prof. William Zinsmeister of Purdue University in 1979.

REFERENCES AND FURTHER READING

Aagaard B (1930) *Fangs tog Forskning I Sydishavet.* 2 vols. Gyldendal Norsk, Oslo.

Adamson D and Pickard J (1986) Cainozoic history of the Vestfold Hills. In: *Antarctic Oasis: Terrestrial Environments and History of the Vestfold Hills.* (Ed. J Pickard) pp. 99–139. Academic Press, Sydney.

Adie RJ (1952) Representatives of the Gondwana System in Antarctica. Symposium sur les Series de Gondwana. *19th International Geological Congress, Algires, [1952]*: 393–399.

Adie RJ (1957a) The petrology of Graham Land: III. Metamorphic rocks of the Trinity Peninsula Series. *Science Report Falkland Islands Dependencies Survey* **20**, 1–26.

Adie RJ (1957b) Geological research in Graham Land. *Antarctic Research Advancement of Science* **13**(53), 454–460.

Adie RJ (1958) Geological investigations in the Falkland Island Dependencies since 1940. *Polar Records* **9**(58), 3–17.

Adie RJ (1964) Stratigraphic correlation in West Antarctica. In: *Antarctic Geology.* (Ed. RJ Adie) pp. 307–313. North-Holland Publishing Company, Amsterdam.

Aguirre-Urreta MB, Marenssi S and Santillana S (1995) A new Eocene crab (*Crustacea, Decapoda*) from Seymour Island, Antarctica. *Antarctic Science* **7**, 277–281.

Aguirre-Urreta MB, Olivero EB and Medina FA (1991) A redescription of a Maastrichtian lobster Hoploparia antarctica, Wilckens, 1907 (Crustacea: Decapoda), from Chubut, Argentina. *Journal of Paleontology* **65**, 795.

Aitkenhead N (1975) [1965] The geology of the Duse Bay-Larsen Inlet area north-east Graham Land (with particular reference to the Trinity Peninsula Series). *Scientific Report British Antarctic Survey* **51**, 1–62.

Albright LB, Woodburne MO, Case JA and Chaney DSA (2003) A leatherback sea turtle from the Eocene of Antarctica: implications for antiquity of gigantothermy in Dermochelyidae. *Journal of Vertebrate Paleontology* **23**, 945–949.

Ameghino F (1906) [1905] Enumeración de los impennes fósiles de Patagonia y de la Isla Seymour. *Annales de Museo Nacional de Historia natural de Buenos Aires Series* **3**(6), 97–167.

Andersson JG (1906) On the geology of Graham Land. *Bulletin of the Geological Institute University of Uppsala (Sweden)* **7**, 19–71.

Andersson JG (1907) On the principal geological results of the Swedish Antarctic Expedition. *Congrès Géologique International: compte rendu de la Xème session, Mexico, 1906 / préface José G. Aguilera. – Mexico: Imprenta y Fototipia de la Secretaría de Fomento, 1907,* **2**(6), 725–733.

Andersson JG (1944) *Antarctic.* Saxon & Lindströms Förlag, Stockholm.

Anelli LE, Rocha-Campos AC, dos Santos PR, de Perinotto JAJ and Quaglio F (2006) Early Miocene bivalves from the Cape Melville Formation, King George Island, West Antarctica. *Alcheringa* **30**, 111–132.

Anon. (1936) An Antarctic illusion. The coasts of Grahamland. *The Times*, Saturday 12 December 1936, p. 13.

Araya RA and Hervé FA (1966) Estudio geomorfológico y geológico en las Islas Shetland del Sur, Antártica. Instituto Antártico Chileno, Publicación, Santiago *8*, 1–76.

Archuby F, Reynaldi F, Teodoroff T, Vizcaino S and Reguero M (1995) Analisis de la diversidad de taxones de selaceos (Pisces: Chondrichthyes) del Eoceno de la Isla Seymour, Antartida. In: *Simposio 'Vertebrados Paleogenos de la Antartida'; XI Jornadas Argentinas de Paleontologma de Vertebrados. Resumenes* p. 9. Tucuman, Argentina.

Aronson RB and Blake DB (2001) Global climate change and the origin of modern benthic communities in Antarctica. *American Zoologist* **41**, 27–39. http://icb.oxfordjournals.org/cgi/content/full/41/1/27

Aronson RB, Moody RM, Ivany LC, Blake DB, Werner JE and Glass A (2009) Climate change and trophic response of the Antarctic bottom fauna. *PLoS ONE* **4**(2), e4385.

Arratia G, Scasso RA and Kiessling W (2004) Late Jurassic fishes from Longing Gap, Antarctic Peninsula. *Journal of Vertebrate Paleontology* **24**(1), 41–55.

Ashworth AC and Preece RC (2003) The first freshwater molluscs from Antarctica. *Journal of Molluscan Studies* **69**, 89–92.

Ashworth A and Cantrill DJ (2004) Neogene vegetation of the Meyer Desert Formation (Sirius Group) Transantarctic Mountains, Antarctica. *Palaeogeography, Palaeoclimatology, Palaeoecology* **213**, 65–82.

Askin RA (1988) Campanian to Paleocene palynological succession of Seymour Island and adjacent islands, northeastern Antarctic Peninsula. In: *Geology and Paleontology of Seymour Island, Antarctic Peninsula.* (Eds RM Feldann and MO Woodburne) *Geological Society of America, Memoirs* **169**, 131–153.

Askin RA (1989) Endemism and heterochroneity in the Late Cretaceous (Campanian) to Paleocene palynofloras of Seymour Island, Antarctica: implications for origins, dispersal and palaeoclimates of southern floras. In: *Origins and Evolution of the Antarctic Biota.* (Ed. JA Crame) *Geological Society, London, Special Publications* **47**, 107–119.

Askin RA (1990) Campanian to Paleocene spore and pollen assemblages of Seymour Island, Antarctica. *Review of Palaeobotany and Palynology* **65**, 105–113.

Askin RA, Elliot DH, Stilwell JD and Zinsmeister WJ (1991) Stratigraphy and paleontology of Campanian and Eocene sediments, Cockburn Island, Antarctic Peninsula. *Journal of South American Earth Sciences* **4**(1/2), 99–117.

Askin RA (1992) Late Cretaceous-early Tertiary Antarctic outcrop evidence for past vegetation and climates. In: *The Antarctic Palaeoenvironments: A Perspective on Global Change.* (Eds JP Kennett and DA Warnke) *American Geophysical Union, Antarctic Research Series* **56**, 61–73.

Askin RA (1997) Eocene-earliest Oligocene terrestrial palynology of Seymour Island. In: *The Antarctic Region: Geological Evolution and Processes.* (Ed. CA Ricci) pp. 993–996. *Proceedings of the V11 International Symposium on Antarctic Earth Sciences, Siena, 1995.* Terra Antarctica Publication, Siena.

Askin RA (1997) Permian palynomorphs from southern Victoria Land. *Antarctic Journal of the United States* **30**, 47–48.

Askin RA (1998) Floral trends in the Gondwana high latitudes: palynological evidence from the Transantarctic Mountains. *Journal of African Earth Sciences* **27**(1A), 12–13.

Askin RA (2000) Spores and pollen from the McMurdo Sound erratics, Antarctica. In: *Palaeoenvironments of Eocene Fossiliferous Erratics, McMurdo Sound, East Antarctica.* (Eds JD Stilwell and RM Feldmann) *American Geophysical Union, Antarctic Research Series* **76**, 161–181.

Askin RA, Elliot DH, Stilwell JD and Zinsmeister WJ (1991) Stratigraphy and paleontology of Campanian and Eocene sediments, Cockburn Island, Antarctic Peninsula. *Journal of South American Earth Sciences* **4**(1/2), 99–117.

Askin RA and Raine JI (2000) Oligocene and Early Miocene terrestrial palynology of Cape Roberts Drillhole CRP-2/2A, Victoria Land basin, Antarctica. *Terra Antartica* **7**, 493–501.

Baldoni AM and Medina F (1989) Fauna y microflora del Cretácico en Bahia Brandy, Isla james Ross, Antártida. *Serie Cientifico, Instituto Antártico Chileno* **39**, 45–58.

Balme BE and Playford G (1967) Late Permian plant microfossils from the Prince Charles Mountains, Antarctica. *Revue de Micropalaeontologie* **10**(3), 179–192.

Balushkin AV (1994) *Proeleginops grandeastmanorum* gen. et sp. nov. (Perciformes, Notothenioidei, Eleginopsidae) from the late Eocene of Seymour Island (Antarctica) is a fossil Notothenioid, not a gadiform. *Journal of Ichthyology* **34**, 10–23.

Bargo MS and Reguero MA (1998) Annotated catalogue of the fossil vertebrates from Antarctica housed in the Museo de La Plata: birds and land mammals from La Meseta Formation (Eocene–?early Oligocene) In: *Asociacion Paleontologica Argentina, Publicacion Especial 5, Paleogeno de America del Sur y de la Peninsula Antartica.* (Ed. S Casadio) pp. 211–221.

Baroni C and Ormbelli G (1994a) Abandoned penguin rookeries as Holocene paleoclimatic indicators in Antarctica. *Geology* **22**, 23–26.

Barrera V, Palamarczuk S and Medina F (1999) Palinología de la formación Hidden Lake (Coniaciano-Santoniano), isla James Ross, Antártida. *Revista Espanõla de Micropaleontologia* **31**, 53–72.

Barrett PJ, Baillie RJ and Colbert EH (1968) Triassic amphibian from Antarctica. *Science* **161**, 460–462.

Barton CM (1964) Significance of the Tertiary fossil floras of King George Island, South Shetland Islands. In *Antarctic Geology.* (Ed. RJ Adie) pp. 603–608. North-Holland Publishing Company, Amsterdam.

Barton CM (1965) The geology of the South Shetland Islands: III. The stratigraphy of King George Island. *Scientific Repoert Falkland Island Dependencies Survey* **44**, 1–33.

Barton CE, Hutchinson R, Quilty P, Seers K and Stone T (1987) Quest for the Magnetic Poles: relocation of the South Magnetic Pole at Sea, 1986. *Bureau of Mineral Resources, Geology and Geophysics* Record 1987/**3**, 1–18.

Beatty P (2011) Life in Antarctic lake similar to early fossil records. http://www.lakescientist.com/2011/life-in-antarctic-lake-similar-to-early-fossil-records

Bechervaise J and Quilty PG (1996) Antarctica. In: *The Australian Encyclopaedia.* 6th edn. Vol. 1. (Ed. A Macdougall) pp. 211–220. Australian Geographic P/L, Sydney.

Behrendt JC and Laudon TS (1964) Cretaceous fossils from 'Johnson nunatak', Antarctica. *Science* **143**(3604), 353–354.

Bell CM (1973) The geology of southern Alexander Island. *Bulletin of the British Antarctic Survey* **33 & 34**, 1–16.

Bell PR and Snively E (2008) Polar dinosaurs on parade: a review of dinosaur migration. *Alcheringa* **32**, 271–284.

Berkman PA (1992) Circumpolar distribution of Holocene marine fossils in Antarctic beaches. *Quaternary Research* **37**, 256–260.

Berkman PA (1994a) Geochemical signatures of meltwater in Antarctic coastal marine molluscs during the Holocene. *Memoirs of the National Institute of Polar Research. Special Issue No.* **50**, 49–84.

Berkman PA (1994b) Epizoic zonation on growing scallop shells in McMurdo Sound, Antarctica. *Journal of Experimental Marine Biology and Ecology* **179**, 49–67.

Berkman PA (1997) Ecological variability in Antarctic coastal environments: past and present. In: *Antarctic Communities.* (Eds B Battaglia, J Valencia and DWH Walton) pp. 349–357. Cambridge University Press, Cambridge.

Berkman PA (1998) Indicator species for interpreting environmental variability in the Antarctic coastal zone. *Korean Journal of Polar Research* **8**(1), 69–76.

Berkman PA, Andrews JT, Björck S, Colhoun EA, Emslie SD, Goodwin ID, Hall BL, Hart CP, Hirakawa K, Igsarashi A, Ingólfsson O, López-Martínez J, Lyons WB, Mabin MCG, Quilty PG, Taviani M and Yoshida Y (1998) Circum-Antarctic coastal environmental shifts during the Late Quaternary reflected by emerged marine deposits. *Antarctic Science* **10**(3), 345–362.

Berkman PA and Prentice ML (1996) Pliocene extinction of Antarctic pectinid mollusks. *Science* **21**, 1606–1607.

Berry RJ (2009) Hooker and islands. *Biological Journal of the Linnean Society* **96**, 462–481.

Beu AG and Dell RK (1989) Mollusca. Antarctic Cenozoic history from the CIROS-1 drillhole, McMurdo Sound. (Ed. PJ Barrett) *DSIR Bulletin* **245**, 135–141.

Beu AG (2009) Before the ice: biogeography of Antarctic Paleogene molluscan faunas. *Palaeogeography, Palaeoclimatology, Palaeoecology* **284**, 191–226.

Bibby JS (1966) The stratigraphy of part of north-east Graham Land and the James Ross Island group. *Scientific Reports British Antarctic Survey* **53**, 1–37.

Biernat G, Birkenmajer K and Popiel-Barczyk E (1985) Tertiary Brachiopods from the Moby Dick Group of King George Island (South Shetland Islands, Antarctica). *Studia Geologica Polonica* **81**, 109–141.

Birkenmajer K (1984) Geology of the Cape Melville area, King George Island (South Shetland Island, Antarctica): pre-Pliocene glaciomarine deposits and their substratum. *Studio Geologica Polonica* **79**, 7–36.

Birkenmajer K and Gaździcki A (1986) Oligocene age of the *Pecten* Conglomerate on King George Island, West Antarctica. *Bulletin of the Polish Academy of Sciences* **34**, 219–226.

Birkenmajer K, Gaździcki A and Wrona R (1983) Cretaceous and Tertiary fossils in glacio-marine strata at Cape Melville, Antarctica. *Nature* **303**, 56–59.

Birkenmajer K, Gaździcki A, Krajewski KP, Przybycin A, Solecki A, Tatur A and Yoon HI (2005) First Cenozoic glaciers in West Antarctica. *Polish Polar Research* **26**(1), 3–12.

Birkenmajer K and Zastawniak E (1989) Late Cretaceous-early Tertiary floras of King George Island, West Antarctica: their stratigraphic distribution and paleoclimatic significance. In: Origins and evolution of the Antarctic biota. (Ed. JA Crame) *Geological Society Special Publication* **47**, 227–240.

Blake DB and Aronson RB (1998) Eocene stelleroids (Echinodermata) at Seymour Island, Antarctic Peninsula. *Journal of Paleontology* **72**(2), 339–353.

Blake DB and Zinsmeister WJ (1991) A new marsupiate cidaroid echinoid from the Maastrichtian of Antarctica. *Palaeontology* **34**, 629–635.

Boharty SM, Scherer RP and Harwood DM (1998) Quaternary diatom biostratigraphy and palaeoenvironments of the CPI-1 drillcore, Ross Sea, Antarctica. *Terra Antartica* **5**, 431–453.

Bond M, Cione AL, Goin FJ, Laza JH, Marenssi S, Noriega JI and Reguero MA (1994) Resultados preliminares de la Campana Antartica de Verano 1993/1994. Geologia y Paleontologia de Vertebrados. In: *3ras Jornadas Antarticas, 17–21 octubre 1994, Buenos Aires. Actas.* pp. 47–51. Buenos Aires, Argentina.

Bond M, Pascual R, Reguero MA, Santillana SN and Marenssi SA (1990) Los primeros ungulados extinguidos sudamericanos de la Antartida. In: *VII Jornadas Argentinas de Paleontologia de Vertebrados. Ameghiniana* **26**, 240. Buenos Aires, Argentina.

Bond M, Reguero MA and Vizcaíno SF (1993) Mamiferos continentales de la Formacion La Meseta (Terciario, Antartida): Biocronologia. In: *XIII Congresso Brasileiro de Paleontologia. Boletim de Resumos.* p. 93. Sao Leopoldo, Brazil.

Bond M, Reguero MA, Vizcaíno SF and Marenssi SA (2006) A new 'South American ungulate' (Mammalia: Litopterna) from the Eocene of the Antarctic Peninsula. In: *Cretaceous–Tertiary High-Latitude Palaeoenvironments: James Ross Basin, Antarctica.* (Eds JE Francis, D Pirrie and JA Crame) *Geological Society Special Publication* **258**, 163–176.

Bond M, Vizcaino S and Reguero M (1995) El registro mas moderno de mamiferos fosiles de Antartida. In: *Simposio 'Vertebrados Paleogenos de la Antartida'; XI Jornadas Argentinas de Paleontologia de Vertebrados. Resumenes.* p. 9. Tucuman, Argentina.

Borchgrevink CE (1901) *First on the Antarctic Continent; being an account of the British Antarctic Expedition*, 1898–1900. George Newnes Ltd, London.

Borsuk-Bialnickya E (1988) New remains of Archaeoceti from the Paleogene of Antarctica. *Polish Polar Research* **9**, 437–445.

Bose MN (1966) Fossil plant remains from the Rajmahal and Jabalpur Series in the Upper Gondwanas of India. In: *Symposium on the Floristics and Stratigraphy of Gondwanaland.* Lucknow, Birbal Sahni Institute, pp. 143–154.

Boucot AJ (1965) Lower Devonian fauna of the Horlick Formation, Ohio range, Antarctica. *Antarctic Research Series* **6**, 241–274.

Boucot AJ, Caster KE, Ives D and Talent JA (1963) Relationships of a new Lower Devonian terebratuloid (Brachiopoda) from Antarctica. *Bulletins of American Paleontology* **46**, 81–151.

Boucot AJ, Doumani GA, Johnson JG and Webers GF (1967) Devonian of Antarctica. In: *International Symposium on the Devonian System*, Calgary, Alberta, 1967. (Ed. DH Oswald) *Alberta Society of Petroleum Geologists* **1**, 638–648.

Bowden T (1997) *The Silence Calling. Australians in Antarctica 1947–97.* Allen and Unwin, Sydney.

Bradshaw MA (1981) Paleoenvironmental interpretations and systematics of Devonian trace fossils from the Taylor Group (lower Beacon Supergroup), Antarctica. *New Zealand Journal of Geology and Geophysics* **24**, 615–652.

Bradshaw MA (1984) Permian non-marine bivalves from the Ohio Range, Antarctica. *Alcheringa: An Australasian Journal of Palaeontology* **8**(4), 305–309.

Bradshaw MA (1991) The Devonian Pacific margin of Antarctica. In: *The Geological Evolution of Antarctica.* (Eds MRA Thomson, JA Crame and JW Thomson) pp. 193–197. Cambridge University Press, Cambridge.

Bradshaw MA (2002) A new ichnogenus *Catenarichnus* from the Devonian of the Ohio Range, Antarctica. *Antarctic Science* **14**(4), 422–424.

Bradshaw MA and McCartan C (1991) Palaeoecology and systematics of Lower Devonian bivalves from the Horlick Formation, Ohio Range, Antarctica. *Alcheringa* **15**, 1–42.

Bradshaw MA, Harmsen FJ and Kirkbride MP (1990) Preliminary result of the 1988–89 expedition to the Darwin Glacier area. *New Zealand Antarctric Record* **10**, 28–48.

Bradshaw MA and McCartan L (1983) The depositional environment of the Lower Devonian Horlick Formation, Ohio Range. In: *Antarctic Earth Science.* (Eds RL Oliver, PR James and JB Jago) pp. 238–241. Australian Academy of Science, Canberra, and Cambridge University Press, Cambridge.

Bradshaw MA, Newman J and Aitchison JC (2002) The sedimentary geology, palaeoenvironments and ichnocoenoses of the Lower Devonian Horlick Formation, Ohio Range, Antarctica. *Antarctic Science* **14**, 395–411.

Brazeau MD (2009) The braincase and jaws of a Devonian 'acanthodian' and the origin of modern jawed vertebrates. *Nature* **457**, 305–308.

Brea N (1998) Análisis de los anillos de crecimiento en leños fósiles de coníferas de la Formación La Meseta, Isla Seymour (Marambio), Antártida. In: Paleógeno de América del Sur y de la Península Antártica. (Ed. S Casadío) *Asociación Paleontológica Argentina, Publicación Especial* **5**, 163–175.

Bruce WS (1894) Chapter XIX. In: WG Burn Murdoch's *From Edinburgh to the Antarctic...* pp. 349–364. Longmans, Green and Co., London.

Buatois LA (1995) A new ichnospecies of *Fuersichnus* from the Cretaceous of Antarctica and its palaeoecological and stratigraphic implications. *Ichnos* **3**, 259–263.

Buckeridge JS (2000) A new species of *Austrobalanus* (Cirripedia, Thoracica) from Eocene erratics, Mount Discovery, East Antarctica. In: *Paleobiology and and Paleoenvironments of Eocene Rocks, McMurdo Sound, East Antarctica.* (Eds J Stilwell and R Feldmann), *Antarctic Research Series* **76**, 329–333.

Buckman SS (1910) Antarctic fossil Brachiopoda collected by the Swedish South Polar Expedition, 1901–1903. (With an addendum by J.G. Andersson on the age of the brachiopod yielding beds of Cockburn Island.) *Wissenschaftliche Ergebnisse der Schwedischen Sudpolar-Expedition* **3**(7), 1–43, 3pls.

Buffetaut E (2004) Polar dinosaurs and the question of dinosaur extinction: a brief review. *Palaeogeography, Palaeoclimatology, Palaeoecology* **214**, 225–231.

Buggisch W and Kleinschmidt G (1991) Recovery and recystallization of quartz and 'crystallinity' of illite in the Bowers and Robertson Bay terranes in Victoria Land, Antarctica. In: *Geological Evolution of Antarctica.* (Eds MRA Thomson, JA Crame and W Thomson) pp. 155–159. Cambridge University Press, Cambridge.

Buick DP and Ivany LC (2004) 100 years in the dark: extreme longevity of Eocene bivalves from Antarctica. *Geology* **32**(10), 921–924.

Bull HJ (1896) *The Cruise of the 'Antarctic' to the South Polar Regions.* Edward Arnold, London.

Burn Murdoch WG (1894) *From Edinburgh to the Antarctic. An artist's notes and sketches during the Dundee Antarctic Expedition of 1892–93. With a chapter by WS Bruce Naturalist of the Barque 'Balaena'.* Longmans, Green and Co., London.

Burrett CF and Findlay RH (1984) Cambrian and Ordovician conodonts from the Robertson Bay Group, Antarctica and their tectonic significance. *Nature* **307**, 723–725.

Burrett C, Long J and Stait B (1990) Early-Middle Palaeozoic biogeography of Asian terranes derived from Gondwana. In: *Palaeozoic Palaeogeography and Biogeography.* (Eds WS McKerrow and CR Scotese) pp. 163–174. *Geological Society of London Memoir* **12**.

Burrow CJ, Long JA and Trinajstic K (2008) Disarticulated acanthodian and chondrichthyan remains from the upper Middle Devonian Aztec Siltstone, southern Victoria Land, Antarctica. *Antarctic Science* **21**, 71–88.

Busetti M, Carburlotto A, Armand L, Damiani D, Giorgetti G, Lucchi RG, Quilty PG and Villa G (2003) Plio-Quaternary sedimentation on the Wilkes Land continental rise. Preliminary results. *Deep Sea Research* II, **50**, 1529–1562.

Byrd R (1938) *Alone.* Ace Books Incorporated, New York.

Camacho HH (1957) Observaciones geológicas en Bahia Esperanza. In: Algunas observaciones geológicas y criopedológicas en la Antártida. I. Observaciones geológicas y petrográficas en Bahia Esperanza, Puerto Paraiso e Islas Shetland del Sur y Melchior. (Eds HH Camacho and JV Fabre) *Publicaciones del Instituto Antártida Argentina* **4**, 25–30.

Camacho HH (1959) Antártida a través de sus faunas y floras fósiles. *Contribuciones Instituto Antártida Argentina* **29**, 1–18.

Camacho HH (1992) Algunas consideraciones acerca de la transgression marina Paleocena en la Argentina. *Academia Nacional de Ciencias (Córdoba, Argentina), Miscelanea No.* **85**, 1–41.

Candela AM and Goin FJ (1995) Revision de las especies antárticas de marsupials Polidolopinos (Polydolopimorphia, Polydolopidae). In: *111 Jornadas de Comunicationes sobre Investigationes Antárticas.* pp. 55–56. Instituto Antártico Argentino, Buenos Aires.

Cantrill DJ (1998) Early Cretaceous fern foliage from President Head, Snow Island, Antarctica. *Alcheringa* **22**, 241–258.

Cantrill DJ (2000) A Cretaceous macroflora from a freshwater lake deposit, President Head, Snow Island, Antarctica. *Palaeontolographica Abteilung B* **252**, 153–191.

Cantrill DJ (2001) Early Oligocene *Nothofagus* from CRP-3, Antarctica: implications for the vegetation history. *Terra Antartica* **8**, 401–406.

Cantrill DJ, Drinnan AN and Webb JA (1995) Late Triassic plant fossils from the Prince Charles Mountains, East Antarctica. *Antarctic Science* **7**, 51–62.

Cantrill DJ and Falcon-Lang HJ (2001) Cretaceous (Late Albian) Coniferales of Alexander Island, Antarctica. 2: Leaves, reproductive structures and roots. *Review of Palaeobotany and Palynology* **115**, 119–145.

Cantrill DJ and Nichols GJ (1996) Taxonomy and palaeocology of Early Cretaceous (Late Albian) angiosperm leaves form Alexander Island, Antarctica. *Review of Palaeobotany and Palynology* **92**, 1–28.

Cantrill DJ and Poole I (2002) Cretaceous to Tertiary patterns of floristic change in the Antarctic Peninsula. In: *Palaeobiogeography and Biodiversity Change: the Ordovician and Mesozoic–Cenozoic Radiations.* (Eds JA Crame and AW Owen) *Geological Society, London, Special Publications* **194**, 141–152.

Cantrill DJ and Poole I (2005) Taxonomic turnover in wood floras from the Cretaceous and Tertiary of Antarctica: implications for chages in forest ecology. *Palaeogeography, Palaeoecology, Palaeoclimatology* **215**, 205–219.

Cao L (1992) Late Cretaceous and Eocene palynofloras from Fildes Peninsula, King George Island (South Shetland Islands), Antarctica. In: *Recent Progress in Antarctic Earth Science.* (Ed. Y Yoshida) pp. 363–369. Terra Scientific (Terrapus), Tokyo.

Carlini AA, Pascual R, Reguero MA, Scillato-Yane GJ, Tonni EP and Vizcaino SF (1990) The First Paleogene Land Placental Mammal from Antarctica: its Paleoclimatic and Paleobiogeographical Bearings. In: *IV International Congress of Systematic and Evolutionary Biology. Abstracts.* p. 325. Washington, USA.

Carlquist S (1987) Pliocene *Nothofagus* wood from the Transantarctic Mountains. *Aliso* **11**, 571–583.

Carpenter FM (1969) Fossil insects from Antarctica. *Psyche, Camb., Mass.* **76**(4), 418–425.

Casadio S (Ed.) (1998) Paleógeno de América del sur y de La Península Antártica. *Asociación Paleontológica Argentina Publicación Especial* **5**, 1–224.

Casadio S, Marenssi S and Santillana S (1996) Bioerosiones en Ostrea antarctica y su significado paleoambiental, Eoceno de la isla Marambio (Seymour), Antartida. In: *I Congreso Paleogeno de America del Sur. Reszmenes.* pp. 29–30. Santa Rosa, La Pampa.

Casadio S, Marenssi SA and Santillana SN (2001) Bioerosive endolithic traces attributed to boring bryozoans (Ctenostomata) in the Eocene of Antarctica. *Ameghiniana* **38**, 321–329.

Case JA (1988) Paleogene floras from Seymour Island, Antarctic Peninsula. In: *Geology and Paleontology of Seymour Island, Antarctic Peninsula.* (Eds RM Feldmann and MO Woodburne) *Geological Society of America, Memoir* **169**, 523–530.

Case JA (1989) Antarctica: the effect of high latitude heterochroneity on the origin of the Australian marsupials. In: *Origins and Evolution of the Antarctic Biota.* (Ed. JA Crame) *Geological Society, London, Special Publications* **47**, 217–226.

Case JA (1992) Evidence from fossil vertebrates for a rich Eocene Antarctic marine environment. The Antarctic paleoenvironment: a perspective on global change. *Antarctic Research Series* **56**, 119–130.

Case JA (2006) The late Middle Eocene terrestrial vertebrate fauna from Seymour Island: the tails of the Eocene Patagonian size distribution. In: *Cretaceous–Tertiary High-Latitude Palaeoenvironments: James Ross Basin, Antarctica.* (Eds JE Francis, D Pirrie and JA Crame) *Geological Society Special Publication London* **258**, 177–186.

Case JA, Goin FJ and Woodburne MO (1996) Eocene Antarctic marsupials with tribosphenic molars: implications upon marsupial biogeography. *Journal of Vertebrate Paleontology* **16** (suppl. 3), 26A.

Case J, Martin JE, Chaney DS, Reguero MA and Woodburne MO (1998a) The first hadrosaur from Antarctica. In: *1998 Society of Vertebrate Paleontology Meeting.* Snowbird, Salt Lake City, Utah.

Case JA, Martin JE, Chaney DS, Reguero M and Woodburne MO (1998b) The first hadrosaur from Antarctica. *Journal of Vertebrate Palaontology, Abstracts* **18**, 32A.

Case JA, Martin JE, Chaney DS, Reguero MA, Marenssi SA, Santillana SN and Woodburne MO (2000) The first duck-billed dinosaur (Family Hadrosauridae) from Antarctica. *Journal of Vertebrate Paleontology* **20**, 612–614.

Case JA and Tambussi CP (1999) Maestrichtian record of neornithine birds in Antarctica: comments on a late Cretaceous radiation of modern birds. *Journal of Vertebrate Paleontology* **19** (suppl. 3), 37A.

Case JA, Woodburne MO and Chaney DS (1987) A gigantic Phorhoracoid bird from Antarctica. *Journal of Paleontology* **61**, 1280–1284.

Case JA, Woodburne MO and Chaney DS (1988) A new genus and species of polydolopopid marsupial from the L Meseta Formation, late Eocene, Seymour Island Antarctic Peninsula. In: *Geology and Paleontology of Seymour Island.* (Eds RM Feldmann and MO Woodburne) *Geological Society of America, Memoir* **169**, 505–521.

Casey R (1961) A monograph of the Ammonoidea of the Lower Greensand. Part 2. *Monograph of the Palaeontographical Society, London [1961]*, 45–118.

Césari SN, Marenssi SA and Santillana SN (2001) Conifers from the Upper Cretaceous of Cape Lamb, Vega Island, Antarctica. *Cretaceous Research* **22**, 309–319.

Cesari SN, Remesal M and Parica C (2001) Ferns; a palaeoclimate significant component of the Cretaceous flora of Livingston Island, Antarctica. *VII International Symposium on Mesozoic Terrestrial Ecosystems.* pp. 45–50.

Cesari SN, Parcia CA, Remesal MB and Salani FN (1999) Paleoflora del Cretácico Inferior de la peninsula Byers, islas Shetland del Sur, Antártida. *Ameghiniana* **36**, 3–22.

Chaloner WG and Creber GT (1989) The phenomenon of forest growth in Antarctica: a review. In: *Origins and Evolution of the Antarctic Biota.* (Ed. JA Crame) *Geological Society, London, Special Publications* **47**, 85–88.

Chapman JL and Smellie JL (1992) Cretaceous fossil wood and palynomorphs from Williams Point, Livingston Island, Antarctic Peninsula. *Review of Palaeobotany and Palynology* **74**, 163–192.

Chapman-Smith M (1981) The Taylor Formation (Holocene) and its macrofaunas, Taylor Dry Valley, Antarctica. *Antarctic Research Series* **33**, 365–378.

Chatterjee S (1989) The oldest Antarctic bird. *Journal of Vertebrate Paleontology* **9**, 16A.

Chatterjee S (1997) Multiple impacts at the KT boundary and the death of the dinosaurs. *Proceedings of the 30th International Geological Congress* **26**, 31–54.

Chatterjee S (2002) The morphology and systematics of Polarornis, a Cretaceous loon (Aves: Gaviidae) from Antarctica. In: *Proceedings of the 5th International Meeting of the Society of Avian Paleontology and Evolution.* (Eds Z Zhou and F Zhang) pp. 125–155. Science Press, Beijing.

Chatterjee S and Creisler BS (1994) *Alwalkeria* (Theropoda) and *Morturneria* (Plesiosauria), new names for preoccupied *Walkeria* Chatterjee, 1987 and *Turneria*. *Journal of Vertebrate Paleontology* **14**, 142.

Chatterjee S, Small BJ and Nickell MW (1984) Late Cretaceous marine reptiles from Antarctica. *Antarctic Journal of the United States* 1984 Review, 7–8.

Chatterjee S and Small BJ (1989) New plesiosaurs from the Upper Cretaceous of Antarctica. In: *Origin and Evolution of the Antarctic Biota.* (Ed. JA Crame) *Geological Society of America, Special Publication* **47**, 197–215.

Chatterjee S and Zinsmeister WJ (1982) Late Cretaceous marine vertebrates from Seymour Island, Antarctic Peninsula. *Antarctic Journal of the United States* **17**, 66.

Chiappe LM (1996) Early avian evolution in the southern hemisphere: the fossil record of birds in the Mesozoic of Gondwana. *Memoirs of the Queensland Museum* **39**(3), 533–554.

Choo B, Long JA and Trinajstic K (2009) A new genus and species of basal actinopterygian fish from the Upper Devonian Gogo Formation of Western Australia. *Acta Zoologica* **90** (suppl. 1), 194–210.

Chornogubsky L, Goin F and Reguero M (2009) A reassessment of Antarctic polydolopid marsupials (Middle Eocene, La Meseta Formation). *Antarctic Science* **21**(3), 285–297.

Cichowolski M, Ambrosio A and Concheyro A (2005) Nautilids from the Upper Cretaceous of the James Ross Basin, Antarctic Peninsula. *Antarctic Science* **17**(2), 267–280.

Cione AL, Azpelicueta MM and Bellwood DR (1994) An oplegnathid fish from the Eocene of Antarctica. *Palaeontology* **37**, 931–940.

Cione AL, del Valle RA, Rinaldi CA and Tonni EP (1976) Nota preliminar sobre los pinguinos y tiburones del Terciario inferior de la Isla Vicecomodora Marambio, Antártida. *Contribuciones del Instituto Antártico Argentino* **213**, 3–21.

Cione AL and Medina FA (1987) A record of *Notidanodon pectinatus* (Chondrichthyes, hexanchiformes) in the Upper Cretaceous of the Antarctic Peninsula. *Mesozoic Research* **1**, 79–88.

Cione AL and Reguero MA (1994) New records of the sharks Isurus and Hexanchus from the Eocene of Seymour Island, Antarctica. *Proceedings of the Geologists' Association* **105**, 1–14.

Cione AL and Reguero MA (1998) A middle Eocene basking shark (Lamniformes, Cetorhinidae) from Antarctica. *Antarctic Science* **10**, 83–88.

Cione AL, Reguero MA and Elliot DH (2001) A large osteichthyan vertebra from the Eocene of Antarctica. *Neues Jahrbuch für Geologie und Paläontologie Monatshefte* **9**, 543–552.

Clarke A, Aronson RB, Crame JA, Gili J-M and Blake DB (2004) Evolution and diversity of the benthic fauna of the Southern Ocean continental shelf. *Antarctic Science* **16**, 559–568.

Clarke JA, Ksepka DT, Stucchi M, Urbina M, Giannin N, Bertelli S, Narváez Y and Boyd CA (2007) Paleogene equatorial penguins challenge the proposed relationship between biogeography, diversity, and Cenozoic climate change. *Proceedings of the National Academy of Sciences of the United States, (PNAS)* **104**(28), 11545–11550.

Clarke JA, Tambussi CP, Noriega JI, Erickson GM and Ketcham RA (2005) Definitive fossil evidence for the extant avian radiation in the Cretaceous. *Nature* **433**, 305–308.

Clarke PD, Hughes CP and Thomson MRA (1979) Geological significance of a Middle Cambrian fauna from Antarctica. *Nature* **279**, 791–792.

Clarkson PD, Cooper RA, Hughes CP, Thomson MRA and Webers GF (1981) Ordovician of Antarctica – a review. In: *The Ordovician System in Australia, New Zealand and Antarctica.* (Ed. BD Webby) pp. 46–50. International Union of Geological Sciences Publication.

CNN.com (26 February 2004) 'Two new dinosaur species found in Antarctica'.

Coccozza C and Clarke C (1992) Eocene microplankton from La Meseta Formation. *Antarctic Science* **4**, 355–362.

Cocks LRM (1989) Antarctica's place within Cambrian to Devonian Gondwana. In: *Origins and Evolution of the Antarctic Biota.* (Ed. JA Crame) *The Geological Society, London, UK,* **47**, 9–14.

Cocks LRM and Fortey RA (1988) Lower Palaeozoic facies and faunas around Gondwana. In: *Gondwana and Tethys.* (Eds MG Audley-Charles and A Hallam) *Geological Society Special Publication* **37**, 183–200, London, UK.

Colbert EH (1975) Lystrosaurus from Antarctica. *American Museum Novitates* **2535**, 1–44.

Colbert EH (1982) Triassic vertebrates in the Transantarctic Mountains. In: *Geology of the Central Transantarctic Mountains.* (Eds MD Turner and JF Splettstoesser) *American Geophysical Union, Antarctic Research Series* **36,** 11–35.

Colbert EH (1987) The Triassic reptile *Prolacerta* in Antarctica. *American Museum Novitates* **2882**, 1–19.

Colbert EH (1991) Mesozoic and Cainozoic tetrapod fossils from Antarctica. In: *The Geology of Antarctica.* (Ed. RJ Tingay) pp. 568–587. Clarendon, Oxford. Colbert EH and Cosgriff JW (1974) Labyrinthodont amphibians from Antarctica. *American Museum Novitates* **2552**, 1–30.

Colbert EH and Kitching JW (1975) The Triassic reptile *Procolophon* in Antarctica. *American Museum Novitates* **2566**, 1–23.

Colbert EH and Kitching JW (1977) Triassic cynodont reptiles from Antarctica. *American Museum Novitates* **2611**, 1–30.

Colbert EH and Kitching JW (1981) Scaloposaurian reptiles from the Triassic of Antarctica. *American Museum Novitates* **2709**, 1–22.

Cooper RA, Jago JB, MacKinnon DI, Shergold JH and Vidal G (1982) Late Precambrian and Cambrian fossils from Northern Victoria Land and their stratigraphic implications. In: *Antarctic Geoscience.* (Ed. C Craddock) pp. 629–633. The University of Wisconsin Press, Madison.

Cooper RA and Shergold JH (1989) Palaeozoic invertebrates of Antarctica. In: *The Geology of Antarctica.* (Ed. RJ Tingey) pp. 455–486. Oxford University Press, Oxford, UK.

Cordes AH (2002) A new charadriiform avian specimen from the early Maastrichtian of Cape Lamb, Vega Island, Antarctic Peninsula. *Journal of Vertebrate Paleontology* **22** (suppl. 3), 46A

Cordini IR (1955) Contribución al conocimiento del Sector antártico argentino. *Publicaciones del Instituto Antártida Argentina* **1**, 1–277.

Cordini IR (1959) El concimiento geológico de la Antártida. *Publicaciones del Instituto Antártida Argentina* **6**, 1–159.

Cosgriff JW (1983) Large thecodont reptiles from the Fremouw Formation. *Antarctic Journal of the United States* **18**, 52–55.

Cosgriff JW andHammer WR (1981) New skull of *Lystrosaurus curvatus* from the Fremouw Formation. *Antarctic Journal of the United States* **16**(5), 52–53.

Cosgriff JW and Hammer WR (1984) New material of labyrinthodont amphibians from the Lower Triassic Fremouw Formation of Antarctica. *Journal of Vertebrate Paleontology* **4**, 47–56.

Cosgriff JW, Hammer WR and Ryan WJ (1982) The Pangaean reptile *Lystrosaurus mccaigi* in the Lower Triassic of Antarctica. *Journal of Paleontology* **56**, 371–385.

Cosgriff JW and Zawiski JM (1979) A new species of the Rhytidosteidae from the *Lystrosaurus* zone and a review of the Rhytidosteidae. *Palaeontologia Africana* **22**, 1–27.

Couper RA (1960) Southern Hemisphere Mesozoic and Tertiary Podocarpacaea and Fagaceae and their palaeogeographic significance. In: *A Discussion on the Biology of the Southern Cold Temperate Zone.* (Chairman CFA Pantin) *Proceedings of the Royal Society London. B* **152**, 491–500.

Covacevich V and Hernandez P (1971) Investigaciones paleontólogicas en las isles Shetland del Sur, Antartica 1970–1971. *Boletín Instituto Antártico Chileno* **6**, 3–7.

Covacevich VC and Lamperein CR (1969) Nota sobre el hallazgo de icnitas fosiles de aves en Peninsula Fildes, Isla Rey Jorge, Shetland del Sur, Antartica. *Boletín Instituto Antártico Chileno* **4**, 26–28.

Covacevich V and Lamperein C (1970) Hallazgo de icnitas en Peninsula Fildes Isla Rey Jorge, Archipelago Shetland del Sur, Antartica. *Serie científica – Instituto Antártico Chileno* **1**(1), 55–74.

Covacevich V and Lamperein C (1972) [1971]. Ichnites from Fildes Peninsula, King george Island, South Shetland Islands. In*: Antarctic Geology and Geophysics.* (Ed. RJ Adie) pp. 71–74. *Universitetsforlaget, Oslo.*

Covacevich V and Rich PV (1982) New bird ichnites from Fildes Peninsula, King George Island, West Antarctica. In: *Antarctic Geoscience.* (Ed. C Craddock) pp. 245–254. University of Wisconsin Press, Madison, Wisconsin, USA.

Cox LR (1953) Lower Cretaceous Gastropoda, Lamellibranchia and Annnelida from Alexander I Land (Falkland Islands Dependencies). *Scientific Report Falkland Islands Dependencies Survey* **4**, 1–14.

Craddock C (1969) Geology of the Ellsworth Mountains. (Sheet 4, Ellsworth mountains.) In: *Geologic Maps of Antarctica.* (Eds VC Bushnell and C Craddock) *Antarctic Map Folio Serv., Folio* **12**, Pl. 1V.

Craddock C (1970) Fossil map of Antarctica. In*: Geologic Maps of Antarctica.* (Eds VC Bushnell and C Craddock) *Antarctic Map Folio Ser., Folio* **12**, Pl. XV111.

Craddock C (1982) Antarctica and Gondwanaland. Review paper. In: *Antarctic Geoscience.* (Ed. C Craddock) pp. 3–14. University of Wisconsin Press, Madison, Wisconsin, USA.

Craddock C, Bastien TW, Rutford RH and Anderson J (1965) Glossopteris discovered in West Antarctica. *Science* **148**, 634–637.

Craddock C and Webers GF (1964) Fossils from the Ellsworth Mountains, Antarctica. *Nature* **201**, 174–175.

Craddock C, White CM and Rutford RH (1969) The geology of the Eights Coast. *Antarctic Journal. U.S.* **4**(4), 93–94.

Crame JA (1982) Late Jurassic inoceramid bivalves from the Antarctic Peninsula: their classification and stratigraphic use. *Palaeontology* **25**, 555–603.

Crame JA (1983) Cretaceous inoceramid bivalves from Antarctica. In: *Antarctic Earth Science.* (Eds RL Oliver, PR James and JB Jago) pp. 298–302. Australian Academy of Science, Canberra, and Cambridge University Press, Cambridge.

Crame JA (1985) Lower Cretaceous inoceramid bivalves from the Antarctic Peninsula region. *Palaeontology* **28**, 475–525.

Crame JA (1986) Polar origins of marine invertebrate faunas. *Palaios* **1**, 616–617.

Crame JA (Ed.) (1989a) *Origins and Evolution of the Antarctic Biota.* Geological Society Special Publications. The Geological Society, London, UK.

Crame JA (1989b) Origins and evolution of the Antarctic biota: an introduction. In: *Origins and Evolution of the Antarctic Biota.* (Ed. JA Crame) pp. 1–8. The Geological Society, London, UK.

Crame JA (1994) Evolutionary history of Antarctica. In: *Antacrtic Science.* (Ed. G Hemple) pp. 188–214. Springer-Verlag, Berlin.

Crame JA (1995) Occurrence of the bivalve genus Manticula in the Early Cretaceous of Antarctica. *Palaeontology* **38**, 299–312.

Crame JA (1996) A new oxytomid bivalve from the Upper Jurassic–Lower Cretaceous of Antarctica. *Palaeontology* **39**, 615–628.

Crame JA, Lomas SA, Pirrie D and Luther A (1996) Late Cretaceous extinction patterns in Antarctica. *Journal of the Geological Society, London* **153**, 503–506.

Cranwell LM (1959) Fossil pollen from Seymour Island, Antarctica. *Nature* **184**(4701), 1782–1785.

Cranwell LM (1969a) Antarctic and circum-Antarctic paly–nological contributions. *Antarctic Journal U.S.* **4**(5), 197–198.

Cranwell LM (1969b) Palynological intimations of some pre-Oligocene Antarctic climates. In*: Palaeoecology of Africa and of the Surrounding Islands and Antarctica.* (Ed. EM van Zinderen Bakker) **V**, 1–19, A.A. Balkema, Cape Town.

Creber GT and Chaloner WG (1984) Climatic indications from growth in fossil woods. In: *Fossils and Climate.* (Ed. JP Brenchley) pp. 49–74. Wiley Publishing, Chichester, UK.

Cridland AA (1963) A Glossopteris flora from the Ohio Range, Antarctica. *American Journal of Botany* **5**, 186–195.

Cunningham WL (1997) The use of modern and fossil diatom assemblages as climate proxies in the Central and Western Ross Sea, Antarctica. Masters thesis, University of Colorado.

Dalziel IWD (1991) Pacific margins of Laurentia and East Antarctica–Australia as a conjugate rift pair: evidence and implications for an Eocambrian supercontinent. *Geology* **19**, 598–601.

Dalziel IWD (2007) The Ellsworth Mountains: critical and enduringly enigmatic. In: *Antarctica: A Keystone in a Changing World. Online Proceedings of the 10th ISAES.* (Eds AK Cooper, CR Raymond *et al.*), USGS Open-File-Report 2007–1047, Short Research Paper 004, doi:10.3133/of2007–1047.srp004.

Daniels J (1996) Systematics of Pliocene dolphins (Odontoceti: Delphinida) from marine Plain, Antarctica. Unpublished Msc thesis, University of Otago, Dunedin, New Zealand.

Debrenne F and Gravestock DI (1990) Archaeocyatha from the Sellick Hill Formation and Fork Tree Limestone on Fleurieu Peninsula, South Australia. In: *The Evolution of a Late Precambrian–Early Palaeozoic Rift Complex: The Adelaide Geosyncline.* (Eds JB Jago and PS Moore) *Geological Society of Australia. Special Publication* **16**, 290–309.

Debrenne F and Kruse PD (1986) Shackleton Limestone archaeocyaths. *Alcheringa* **10**, 235–278.

Debrenne F and Kruse PD (1989) Cambrian Antarctic archaeocyaths. In: *Origins and Evolution of the Antarctic Biota.* (Ed. JA Crame) *The Geological Society, London, Special Publications* **47**, 15–28.

Debrenne F, Rozanov AY and Webers GF (1984) Upper Cambrian Archaeocyatha from Antarctica. *Geological Magazine* **121**, 291–299.

Defauw SL (1989) Patterns of evolution in the Dicynodontia with special reference to austral taxa. In: *Origins and Evolution of the Antarctic Biota.* (Ed. JA Crame) *The Geological Society, London, Special Publications* **47**, 63–84.

de la Fuente M, Santillana S and Marenssi S (1995) An Eocene Leatherback turtle (Cryptodira: Dermochelyidae) from Seymour Island, Antarctica. *Studia Geologica Salamanticensia* **31**, 17–30.

Dell RK (1990) Antarctic Mollusca with special reference to the fauna of the Ross Sea. *Royal Society of New Zealand* **27**, 1–311.

Dell RK and Fleming CA (1975) Oligocene–Miocene bivalve Mollusca and other macrofossils from Sites 270 and 272 (Ross Sea), DSDP, Leg 28. *Ubutuak Reports of the Deep Sea Drilling Project* **28**, 693–703.

Del Rio CJ, Martínez S, Stilwell JD and Concheyro A (2007) Paleontology of the Cerros Bayos section, Roca Formation (Danian), La Pampa Province, Argentina. *Alcheringa* **31**, 241–269.

Del Valle R, Fourcade NH and Rinaldi CA (1976) Nota sobre el hallazgo de una nueva especie de nautiloideo del género *Eutrephoceras* Owen, en la Isla Vicecomodoro Marambio, Antártida. *Dirección Nacional del Antártico – Instituto Antártico Argentino, Contribución no.* **208**, 1–10.

Del Valle R, Medina F and De Brandoni Z (1977) Nova preliminary sobre el hallazgo de reptiles fósiles marinos del suborden Plesiosauria en las isles James Ross y Vega, Antártida. *Contribución Instituto Antártico Argentino* **212**, 1–13.

Dettmann ME (1989) Antarctica: Cretaceous cradle of austral temperate rainforests? In: *Origins and Evolution of Antarctic Biota.* (Ed. JA Crame) *Geological Society, London, Special Publications* **47**, 89–105.

Dettmann ME and Thomson MRA (1987) Cretaceous palynomorphs from the James Ross Island area Antarctica – a pilot study. *British Antarctic Survey Bulletin* **77**, 13–59.

Doktor M, Gaździcki A, Jerzmańska A, Porębski SJ and Zastawniak E (1996) A plant-and-fish assemblage from the Eocene La Meseta Formation of Seymour Island (Antarctic Peninsula) and its environmental implications. In: *Palaeontological Results of the Polish Antaractic Expeditions.* Part 11. (Ed. A Gaździcki) *Palaeontologia Polonica* **55**, 127–147.

Dolding PJD (1992) Palynology of the Marambio Group (Upper Cretaceous) of northern Humps Island. *Antarctic Science* **4**, 311–326.

Domack E, O'Brien P, Harris P, Taylor F, Quilty PG, DeSantis L and Raker B (1998) Late Quaternary sediment faces in Prydz Bay, East Antarctica and their relationship to glacial advance onto the continental shelf. *Antarctic Science* **10**, 236–246.

Donoghue PCJ, Forey PL and Aldridge RJ (2000) Conodont affinity and chordate phylogeny. *Biological Reviews* **75**, 191–251.

Doumani GA, Boardman RS, Rowell AJ, Boucot AJ, Johnson JG, McAlester AL, Saul J, Fisher DW and Miles RS (1965) Lower Devonian fauna of the Horlick Formation, Ohio Range, Antarctica. *Antarctic Research Series* **6**, 241–281.

Doyle P (1991) Teuthid cephalopods from the Upper Jurassic of Antarctica. *Palaeontology* **34**, 169–178.

Drinnan AN and Crane PR (1990) Cretaceous Paleobotany and its bearing on the biogeography of austral Angiosperms. In: *Antarctic Paleobiology: Its Role in the Reconstruction of Gondwana.* (Eds TN Taylor and EL Taylor) pp. 192–219. Springer, New York.

Duane AM (1996) Palynology of the Byers Group (Late Jurassic–Early Cretaceous) of Livingston and Snow islands, Antarctic Peninsula: its biostratigraphical and palaeoenvironmental significance. *Review of Palaeobotany and Palynology* **91**, 241–281.

Dunbar MJ (1968) *Ecological Development in Polar Regions: A Study in Evolution.* Prentice Hall, Englewood Cliffs, USA.

Dusén P (1908) Über die tertiäre Flora der Seymour Insel. *Wissenschaftliche Ergebnisse der Schwedischen Südpolar-Expedition 1901–1903* **3**(3), 1–27.

Dutra TL and Batten DJ (2000) Upper Cretaceous floras of King George Island, West Antarctica, and their palaeoenvironmental and phytogeographic implications. *Cretaceous Research* **21**, 181–209.

Dutton AL, Lohmann KC and Zinsmeister WJ (2002) Stable isotope and minor element proxies for Eocene climate of Seymour Island. *Paleoceanography* **17**, 6–13.

Dzik J and Gaździcki A (2001) The Eocene expansion of nautilids to high latitudes. *Palaeogeography Palaeoclimatology Palaeoecology* **17**, 279–312.

Edwards CW (1980) Early Mesozoic marine fossils from central Alexander Island. *British Antarctic Survey Bulletin* **49**[1979], 33–58.

Edwards D (1990) Silurian-Devonian paleobotany: problems, progress and potential. In: *Antarctic Paleobiology.* (Eds TN Taylor and EL Taylor) pp. 89–101. Springer, New York.

Edwards WN (1928) The occurrence of Glossopteris in the Beacon Sandstone of Ferrar Glacier, South Victoria Land. *Geological Magazine* **65**, 323–327.

Eights J (1833) Description of a new crustaceous animal found on the shores of the South Shetland Islands, with remarks on their natural history. In *Transactions of the Albany Institute, New York* **2**(1), 53–69.

Eklund H (2003) First Cretaceous flowers from Antarctica. *Review of Palaeobotany and Palynology* **256**, 1–31.

Eklund H, Cantrill DJ and Francis JE (2004) Late Cretaceous plant mesofossils from Table Nanatak, Antarctica. *Cretaceous Research* **25**, 211–228.

Elliot DH (1966) Geology of Nordenskjöld Coast and a comparison with north-west Trinity Peninsula, Graham Land. *Bulletin British Antarctic Survey* **10**, 1–43.

Elliot DH (1967) The Geology of Joinville Island. *Bulletin British Antarctic Survey* **12**, 23–40.

Elliot DH (1974) Stratigaraphy and sedimentary petrology of the Ablation Point area Alexander Island. *Bulletin British Antarctic Survey* **39**, 87–113.

Elliot DH and Tasch P (1967) Lioestheriid Conchostracans: a new Jurassic locality and regional and Gondwana correlations. Journal of Paleontology **41**(6), 1561–1563.

Elliot DH, Colbert EH, Breed WJ, Jensen JA and Powell JS (1970) Triassic tetrapods from Antarctica: evidence for continental drift. *Science* **169**, 1197–1201.

Elliot DH, Rinaldi C, Zinsmeister WJ, Trautman TA, Bryant WA and Del Valle R (1975) Geological investigations on Seymour Island, Antarctic Peninsula. *Antarctic Journal U.S.* **10** (4), 182–186.

Evans KR (1992) Marocella: Antarctic specimens of an enigmatic Cambrian animal. *Journal of Paleontology* **66**, 558–562.

Evans KR and Rowell AJ (1990) Small shelly fossils from Antarctica: an Early Cambrian faunal connection with Australia. *Journal of Paleontology* **64**(5), 692–700.

Falcon-Lang HJ (2011) Secrets of Antarctica's fossilized forests. http://www.bbc.co.uk/news/science-environment-12378934

Falcon-Lang HJ and Cantrill DJ (2000a) Terrestrial paleoecology of the Cretaceous (early Aptian) Cerro Negro Formation, South Shetlands Islands, Antarctica: a record of polar vegetation in a volcanic arc environment. *Palaios* **17**, 491–506.

Falcon-Lang HJ and Cantrill DJ (2000b) Cretaceous (Late Albian) Coniferales of Alexander Island, Antarctica. 1: Wood taxonomy: a quantitative approach. *Review of Palaeobotany and Palynology* **111**, 1–17.

Falcon-Lang HJ and Cantrill DJ (2001a) Gymnosperm woods from the Cretaceous (Mid-Aptian) Cerro Negro Formation, Byers Peninsula, Livingston Island, Antarctica: the aborescent vegetation of a volcanic arc. *Cretaceous Research* **22**, 227–293.

Falcon-Lang HJ and Cantrill DJ (2001b) Leaf phrenology of some mid-Cretaceous polar forests, Alexander Island, Antarctica. *Geological Magazine* **138**, 39–52.

Falcon-Lang HJ, Cantrill DJ and Nichols GJ (2002) Biodiversity and terrestrial ecology of a mid-Cretaceous, high latitude flood plain, Alexander Island, Antarctica. *Journal of the Geological Society, London* **158**, 709–725.

Farabee MJ, Taylor EL and Taylor TN (1991) Late Permian palynomorphs from the buckley Formation, central Transantarctic Mountains, Antarctica. *Review of Palaeobotany and Palynology* **69**, 353–368.

Farquharson GW (1983) The Nordenskjöd Formation of the northern Antarctic Penindsula: an Upper Jurassic radiolarian mudstone and tuff sequence. *British Antarctic Survey Bulletin* **60**, 1–22.

Feldmann RM (1989) *Metanephrops jenkinsi* n. sp. (Decapoda: Nephropidae) from the Cretaceous and Paleocene of Seymour Island, Antarctica. *Journal of Paleontology* **63**, 64–69.

Feldmann RM (1993) Fossil Mole Crabs from Late Cretaceous rocks on James Ross Island, Antarctic Peninsula. *Antarctic Journal of the United States* **28**, 46.

Feldmann RM (1994) *Antarctomithrax thomsoni*, a new genus and species of crab (Brachyura; Majidae) from the La Meseta formation (Eocene) of Seymour Island, Antarctica. *Journal of Paleontology* **68**, 174–176.

Feldmann RM and Gaździcki A (1997) A new species of *Glyphea* (Decapoda: Palinura) from the La Meseta Formation, Seymour Island, Antarctic Peninsula. *Acta Palaeontologica Polonica* **42**, 437–445.

Feldmann RM and Quilty PG (1997) First Pliocene decapod crustacean (Malacostraca: Palinuridae) from Antarctica. *Antarctic Science* **9**(1), 56–60.

Feldmann RM, Tschudy DM and Thomson MRA (1993) Late Cretaceous and Paleocene decapod crustaceans from the James Ross Basinm Antarctic Peninsula. *The Palaeontological Society Memoirs* **28**, 1–41.

Feldmann RM, Schweitzer CE and Marenssi SA (2003) Decapod crustaceans from the Eocene La Meseta Formation, Seymour Island, Antarctica: a model for preservation of decapods. *Journal of the Geological Society, London* **160**, 151–160.

Feldmann RM and Schweitzer CE (2006) Paleobiogeography of Southern Hemisphere decapod Crustacea. *Journal of Paleontology* **80**, 83–103.

Feldmann RM and Woodburne MO (Eds) (1988) Geology and paleontology of Seymour Island, Antarctic Peninsula. *Geological Society of America, Memoir* **169**, 1–566.

Felix J (1909) Uber die fossilen Korallen der Snow Hill-Insel und Seymour-Insel. *Wissenschaftliche Ergebnisse der Schwedischen Sudpolar-Expedition* **3**(5), 1–15.

Filkorn HF (1990) Scleractinia from the López de Bertodano and Sobral formations (Late Cretaceous – Early Tertiary), Seymour and Snow Hill islands, Antarctic Peninsula, Antarctica. *Geological Society of America, Abstracts with Programs* **22**(5), 10.

Filkorn HF (1994) Fossil Scleractinian corals from James Ross Basin, Antarctica. *Antarctic Research Series* **65**, 1–96. American Geophysical Union, Washington DC.

Filkorn HF and Feldmann RM (1993) Fossil corals from Seymour Island, Antarctica. *Antarctic Journal of the United States* **27**(5), 20–21.

Findlay RH and Jordan H (1984) The volcanic rocks of Mt. Black Prince and Lawrence Peaks, North Victoria Land, Antarctica, pp. 143–151. In: *German Antarctic North Victoria Land Expedition 1982/83, GANOVEX III, Vol. 1.* (Ed. NW Roland) *Geologisches Jahrbuch* **B60,** Hannover.

Fitzgerald EMG (2006) A bizarre new toothed mysticete (Cetacea) from Australia and the early evolution of baleen whales. *Proceedings of the Royal Society B* **273**, 2955–2963.

Fitzgerald PG and Barrett PJ (1986) *Skolithos* in a Permian braided rivers deposit, Southern Victoria Land, Antarctica. *Palaeogeography Palaeoclimatology Palaeoecology* **52**(3/4), 237–247.

Flannery TF (2005) *The Weather Makers. The Future and History Impact of Climate Change.* The Text Publishing Company, Melbourne.

Fleet M (1966) Occurrence of fossiliferous Upper Cretaceous sediments at Cape Marsh, Robertson Island. *Bulletin British Antarctic Survey* **8**, 89–91.

Fleming CA (1963) Paleontology and southern biogeography. In: *Pacific Basin Biogeography.* (Ed. JL Gressitt) pp. 369–385. Bishop Museum Press, Honolulu, USA.

Fleming CA (1975) The geological history of New Zealand and its biota. In: *Biogeography and Ecology in New Zealand.* (Ed. G Kuschel) pp. 1–86. Dr. W. Junk Publishers, The Hague, The Netherlands.

Fleming WLS (1938) Geology and glaciology. In: *Notes on the Scientific Work of the British Graham Land Expedition, 1934–37.* (WLS Fleming, A Stephenson, BB Roberts and GCL Bertram) *Geographical Journal* **91**(6), 508–512.

Florin R (1940a) Die heutige und frühere Verbreitung der Koniferengattung. *Acmopyle Pilger. Svensk Botanisk Tidskrift* **34**(2), 117–140.

Florin R (1940b) The Tertiary fossil conifers of south Chile and their phytogeographical significance, with a review of the fossil conifers of southern lands. *Kongliga Svenska Vetenskaps-Akademiens Handlingar. Ny Följd, Stockholm* **2**, 1–107.

Fordyce RE (1989) Origins and evolution of Antarctic marine mammals. In: *Origins and Evolution of the Antarctic Biota.* (Ed. HA Crame) *Special Publications of the Geological Society of London* **47**, 269–281.

Fordyce RE (1994) The evolutionary history of whales and dolphins. *Annual Reviews Earth Planet Science* **22**, 419–455.

Fordyce RE and Quilty PG (1993) Pliocene whales and dolphins (Cetacea) from the Vestfold Hills, Antarctica. *Caveps '93. Conference on Australasian Vertebrate Evolution, Palaeontology and Systematics, Abstracts.* [unpaginated]

Fordyce RE and Quilty PG (1994) Pliocene whales and dolphins (Cetacea) from the Vestfold Hills, Antarctica. *Records of the South Australian Museum* **27**, 219.

Fordyce RE and Quilty P (2008) New Pliocene *Australodelphis* (Delphinidae) and right whales (Balaenidae) from Vestfold Hills, Antarctica, imply taxonomic and ecological turnover in the Southern Ocean since 4 Ma. *Journal of Vertebrate Paleontology* **28** (suppl. 3), 79a.

Fordyce E, Quilty PG and Daniels J (2002) *Australodelphis mirus*, a bizarre new toothless ziphiid-like fossil dolphin (Cetacea: Delphinidae) from the Pliocene of Vestfold Hills, East Antarctica. *Antarctic Science* **14**, 37–54.

Fostowicz-Frelik L and Gaździcki A (2000) Anatomy and histology of plesiosaur bones from the Late Cretaceous of Seymour Island, Antarctic Paninsula. *Palaeontological Results of the Polish Antarctic Expeditions. Part III. Palaeontologica Polonica* **60**, 7–32.

Fourcade NH (1960) Estudio geológico-petrográfico de Caleta Potter, Isla 25 de Mayo, Islas Shetland del Sur. *Publicaciones del Instituto Antártida Argentina* **8**, 1–121.

Francis JE (1986) Growth rings in Cretaceous and Tertiary wood from Antarctica and their palaeoclimatic implications. *Palaeontology* **29**, 665–684.

Francis JE (1994) Permian climates of the southern margins of Pangea: evidence from fossil wood in Antarctica. *Memoirs of the Canadian Society of Petroleum Geologists* **17**, 275–282.

Francis JE (1999) Evidence from fossil plants for Antarctic palaeoclimates over the past 100 million years. *Terra Antartica Reports* **3**, 43–52.

Francis JE (2000) Fossil wood from Eocene high latitude forests. In: *Paleobiology and Paleoenvironments of Eocene Rocks, McMurdo Sound, East Antarctica.* (Eds JD Stilwell and RM Feldmann) *Antarctic Research Series* **76**, 253–260.

Francis JE and Hill RS (1996) Fossil plants from the Pliocene Sirius Group, Transantarctic Mountains: evidence for climate from growth rings and fossil leaves. *Palaios* **11**, 389–396.

Francis JE and Poole I (2002) Cretaceous and early Tertiary climates of Antarctica: evidence from fossil wood. *Palaeogeography, Palaeoclimatology, Palaeoecology* **182**, 47–64.

Francis J, Tosolina AM and Cantrill DJ (2003) Biodiversity and climatic change in Antarctic Palaeogene floras. In: *Antarctic Contributions to Global Earth Sciences. 9th International Symposium on Antarctic Earth Sciences (ISAES IX), Potsdam, Germany.* Alfred Wegener Institute for Polar and Marine Research, Berlin, p. 107.

Frey FA, Coffin MF, Wallace PJ and Quilty PG (Eds) (2003) *Proceedings of the Ocean Drilling Program, Scientific Results,* **183**. Available from: Ocean Drilling Program, Texas A&M University, College Station, TX 7845-9547, USA.

Fricker CA (1999) Composition and paleobiology of a Late Cretaceous (Maastrichtian) gastropod faunule from the Lopez de Bertodano Formation of Seymour Island, Antarctic Peninsula. Unpublished MSc thesis, Purdue University.

Fuente MS, de la Santillana SN and Marenssi S (1995) An Eocene leatherback turtle (Cryptodira; Dermochelyidae) from Seymour Island, Anarctica. *Studia. Geologica. Salmanticensia.* **31**, 21–34.

Fuenzalida H, Araya R and Hervé F (1972) Middle Jurassic flora from north-eastern Snow Island, South Shetland Islands. In: *Antarctic Geology and Geophysics.* (Ed. RJ Adie) pp. 93–97. Universitetsforlaget, Oslo.

Fulford-Smith SP and Sikes EL (1996) The evolution of Ace Lake, Antarctica, determined from sedimentary diatom assemblages. *Palaeogeography, Palaeoclimatology, Palaeoecology* **124**, 73–86.

Gabites HI (1985) Triassic paleoecology of the Lashly Formation, Transantarcic Mountains, Antarctica. Unpublished Msc thesis, Victoria University, Wellington.

Gandolfo MA, Hoc P, Santillana SN and Marenssi SA (1998) Una flór fosil con afinidades a la Familia Grossulariaceae (Orden Rosales) colectada en la Formacion La Meseta (Eoceno), isla Marambio, Antartida. In: *Paleógeno de América del Sur y de la Península Antártica.* (Ed. S Casadio) *Asociación Paleontológica Argentina, Publicación Especial* **5**, 147–153.

Gandolfo MA, Marenssi SA and Santillana SN (1998) Flora y Paleoclima de la Formación La Meseta (Eoceno-Oligoceno inferior?), isla Marambio (Seymour), Antártida. In: *Paleógeno de América del Sur y de la Peninsula Antártica.* (Ed. S Casadio) *Asociación Paleontológica Argentina, Publicacion Especial* **5**, 155–162.

Gasparini Z (1988) Un dinosaurio antártico. *Antártida* **17**, 6–8.

Gasparini Z (1996) New data on the ankylosaurian dinosaur from the Late Cretaceous of the Antarctic Peninsula. *Memoirs of the Queensland Museum* **39**, 583–594.

Gasparini Z, Bardet N, Martin JE and Fernandez M (2003) The Elasmosaurid Plesiosaur *Aristonectes cabrera* from the latest Cretaceous of South America and Antarctica. *Journal of Vertebrate Paleontology* **23**, 104–115.

Gasparini Z and del Valle RA (1981) Mosasaurios: primer hallazgo en el continente Antarctico. *Antartida* **11**, 16–20.

Gasparini Z and del Valle R (1984) Mosasaurios (Reptilia, Sauria) Cretácicos, en el continente Antártico. *IX Congreso Geológico Argentino, Bariloche* **4**, 423–431.

Gasparini Z, del Valle R and Goñi R (1984) Un elasmosáurido (Reptilia, Plesiosauria) del Cretácico Superior de la Antártida. *Instituito Antártico Argentino, Contribución* **305**, 1–24.

Gasparini Z, Fernandez M, Reguero M and Paulina A (1997) Reptiles cretacicos de la Penmnsula Antartica. In: *Reunion de Comunicaciones Cientmficas de la Facultad de Ciencias Naturales y Museo, UNLP, 19–25 de agosto de 1997.*

Gasparini Z and Goñi R (1985) Los plesiosaurus Cretácicos de América del Sur y del continente Antártico. *Actas, VII Congresso Brasileiro de Paleontologica, Coletanea de Trabalhos Paleontológicos. Série Geologica* **27** (2), 55–63.

Gasparini Z and Olivero EB (1989) El dinosaurio antártico. *Ciencia Hoy* **1**(4), 9–10.

Gasparini Z, Olivero E, Scasso R and Rinaldi C (1987) Un ankylosaurio (Reptilia, Ornithischia) Campaniano en el continente Antarcttico. *Anais do X Congresso Brasileiro de Paleontologia* **13**, 131–141.

Gasparini Z, Pereda-Suberbiola X and Molnar RE (1996a) New data on the ankylosaur dinosaur from the Late Cretaceous of the Antarctic Peninsula. *Memoirs of the Queensland Museum* **39**, 583–594.

Gasparini Z, Pereda-Suberbiola X and Molnar RE (1996b) New data on the ankylosaur dinosaur fron the Late Cretaceous of the Antarctic Peninsula. In: *First Gondwanan Dinosaur Symposium*. Trelew, Argentina.

Gaździcki A (1989) Planktonic foraminifera from the Oligocene. Polonez Cove Formation of King George Island, West Antarctica. *Polish Polar Research* **10**, 47–56.

Gaździcki A (2004) Eocene-Pliocene ecosystems of South Shetlands and Antarctic Peninsula. *Bollettino di Geofisica Teorica ed Applicata* **45**, 258–260.

Gaździcki A, Gruszczyński M, Hoffman A, Małkowski K, Marenssi SA, Hałas S and Tatur A (1992) Stable carbon and oxygen isotope record in the Paleogene la meseta Formation, Seymour Island, Antarctica. *Antarctic Science* **4**, 461–468.

Gaździcki A and Pugaczewska H (1984) Biota of the 'Pecten conglomerate' (Polonez Cove Formation, Pliocene) of King george Island (DSouth Shetland Islands, Antarctica). *Studia Geologica Polonica* **79**, 59–120.

Gaździcki A and Wrona R (1982) Late Cretaceous and Tertiary fossils from glaciomarine sediments of Melville Peninsula, King George Island, West Antarctica. Summary. *Przeglad Geologiczny* **8**(353), 399–404.

Gaździcki A and Wrona R (1986) Polskie Badania paleontologiczne w Antarktyce zachodniej. *Przeglad Geologiczny* **11**, 609–617.

Geikie A (1898) Notes on some specimens of rocks from the Antarctic regions, with petrographical notes by J.J.H. Teall. *Proceedings of the Royal Society Edinburgh* **22**(1), 66–70.

Gibson J, Quilty PG, Swadling KM, Newman L and Paterson KS. (2008) *Paratrochammina minutissima* n. sp. in brackish, marine-derived lakes of the Vestfold Hills, East Antarctica. *Journal of Foraminiferal Research* **38**, 292–297.

Gingerich PD, Raza SM, Arif M, Annar M and Zhou X (1994) New whale from the Eocene of Pakistan and the origin of cetacean swimming. *Nature (London)* **368**, 844–847.

Goin FJ and Carlini AA (1995) An Early Tertiary Microbiotheriid marsupial from Antarctica. *Journal of Vertebrate Paleontology* **15**, 205.

Goin FJ, Case JA, Woodburne MO, Vizcaino SF and Reguero MA (1999) New discoveries of opossum-like marsupials from Antarctica (Seymour Island, Medial Eocene). *Journal of Mammalian Evolution* **6**, 335–365.

Goin FJ, Reguero MA, Pascual R, Koenigswald von W, Woodburne Mo, Case JA, Marenssi SA, Vieytes C and Vizcaíno SF (2006) First gondwanatherian mammal from Antarctica. In: *Cretaceous-Tertiary High-Latitude Palaeoenvironments: James Ross Basin, Antarctica.* (Eds JE Francis, D Pirrie and JA Crame) *Geological Society Special Publication London* **258**, 135–144.

Goin FJ, Reguero MA and Vizcaino SF (1994) Novedosos hallazgos de 'comadrejas' (Marsupialia) del Eoceno medio de Antartida. In: III *Jornadas de Comunicaciones sobre Investigaciones Antarticas. Abstracts,* pp. 59–61. Dirección Nacional del Antártico-Instituto Antártico Argentino, Buenos Aires.

Goin FJ, Vizcaino S and Reguero M (1995) Las 'comadrejas' (Mammalia, Marsupialia) del Eoceno de Antartida. In: *Simposio 'Vertebrados Paleogenos de la Antartida'; XI Jornadas Argentinas de Paleontologia de Vertebrados. Resumenes.* p. 11. Tucuman, Argentina.

González-Ferran O, Katsui Y and Tavera JJ (1970) Contribución al conocimiento geológico de la Peninsula Byers de la Isla Livingston; Islas Shetland del Sur, Antártica. *Publicaciones INACH Serie Cientifica* **1**(1), 41–54.

Gordon WT (1920) Scottish National Antarctic Expedition 1902–1904: Cambrian organic remains from a dredging in the Weddell Sea. *Transactions of the Royal Society of Edinburgh* **52**, 681–714.

Gordon WT (1930) A note on *Dadoxylon (Araucarioxylon)* from the Bay of Isles. In: *Report on the geological collections made during the voyage of the 'Quest' on the Shackleton-Rowett Expedition to the South Atlantic and Weddell Sea in 1921–22.* pp. 28–54. Trustees of the British Museum, London.

Gothan W (1908) Die fossilen Hölzer von der Seymour- und Snow Hill-Insel. *Wissenschaftliche Ergebnisse der Schwedischen Südpolar Expedition 1901–1903* **3**(8), 1–33.

Gould LM (1931) *Cold. The Record of an Antarctic Sledge Journey.* Brewer, Warren & Putnam, New York.

Gow CE (1992) An enigmatic new reptile from the lower Fremouw Formation of Antarctica. *Palaeontologia Africana* **29**, 21–33.

Grande L and Chatterjee S (1987) New Cretaceous fish fossils from Seymour Island, Antarctic Peninsula. *Paleontology* **30**(4), 829–837.

Grande L and Eastman JT (1986) A review of the Antarctic ichthyofauna in the light of new fossil discoveries. *Palaeontology* **29**, 113–137.

Gregory JW (1915) The geological relations and some fossils of South Georgia. *Transactions of the Royal Society of Edinburgh* **50**(Pt. 4)(24), 817–822, pls. XCII–XCIII.

Griffin M and Hünicken MA (1994) Late Cretaceous-early Tertiary gastropods from south-western Patagonia, Argentina. *Journal of Paleontology* **68**, 257–274.

Grikurov GE and Dibner AF (1968) Novyye dannyye o serii Triniti (C_{1-3}) v Zapanoy Antarktide. [More information on the Trinity Series (C_{1-3}) of Western Antarctica.] *Akademiya Nauk SSSR (Doklady Earth Science Sections)* **179**(2), 410–412. [English translation in *Doklady Earth Science Sections, American Geological Institute* **179**, 39–41.]

Grindley GW, Mildenhall DC and Schopf JM (1980) A mid-Late Devonian flora from the Ruppert Coast, Marie Byrd Land, Antarctica. *Journal of the Royal Society of New Zealand* **10**(3), 271–285.

Grindley GW and Mildenhall DC (1981) Geological background to a Devonian plant fossil discovery, Ruppert Coast, Marie Byrd Land, West Antarctica. In: *Gondwana Five.* (Eds MM Cresswell and P Vella) pp. 23–30. A. A. Balkema, Rotterdam.

Grindley GW, Mildenhall DC and Schopf JM (1987) A mid-late Devonian flora from the Ruppert Coast, Marie Byrd Land, West Antarctica. *Royal Society of New Zealand Journal* **10**, 271–285.

Hain S (1990) Die beschalten benthischen Mollusken (Gastropoda und Bivalvia) des Weddellmeeres, Antarktis. The benthic seashells (Gastropoda and Bivalvia) of the Weddell Sea, Antarctica. *Reports on Polar Research* **70**, 1–180.

Halle TG (1911) On the geological structure and history of the Falkland Islands. Almqvist and Wiksells Boktryckeri-A.-B, Uppsala. [Reprinted from the *Bulletin of the Geological Insitute, University of Uppsala* **11**, 115–229.]

Halle TG (1913) The Mesozoic flora of Graham Land. *Wissenschaftliche Ergebnisse der Schwedischen Südpolar Expedition* **3**(14), 1–123.

Halpern M (1964) Cretaceous sedimentation in the 'General Bernardo O'Hifins' area of north-west Antarctic Peninsula. In: *Antarctic Geology.* (Ed. JB Hadley) pp. 334–347. North-Holland Publishing Company, Amsterdam.

Halpern M (1965) The geology of the General Bernardo O'Higgins area, northwest Antarctic Peninsula. In: *Geology and Paleontology of the Antarctic.* (Ed. JB Hadley) Washington, American Geophysical Union. *Antarctic Research Series* **6**, 177–209.

Hammer WR (1987) Takrouna Formation fossils of Northern Victoria Land. *Antarctic Research Series* **46**, 243–247.

Hammer WR (1988) The *Cynognathus* zone (late early Triassic) vertebrate fauna from Antarctica. *Antarctic Journal of the United States* **23**, 10–11.

Hammer WR (1990) Triassic terrestrial vertebrate faunas of Antarctica. In: *Antarctic Paleobiology: Its Role in the Reconstruction of Gondwana.* (Eds TN Taylor and EL Taylor) pp. 42–50. Springer Verlag, New York.

Hammer WR (1995) New Therapsids from the Upper Fremouw Formation (Triassic) of Antarctica. *Journal of Vertebrate Paleontology* **15**, 105–112.

Hammer WR (1997) Jurassic dinosaurs from Antarctica. In: *Dinofest International.* (Eds DL Wolberg, E Stump and GD Rosenberg) pp. 249–251. Academy of Natural Sciences, Philadelphia.

Hammer WR, Collinson JW and Ryan WJ (1990) A new Triassic vertebrate fauna from Antarctica and its depositional setting. *Antarctic Science* **2**, 163–167.

Hammer WR and Cosgriff JW (1981) *Myosaurus gracilis*; an anomodont reptile from the Lower Triassic of Antarctica and South Africa. *Journal of Paleontology* **55**, 410–424.

Hammer WR and Hickerson WJ (1993) A new Jurassic dinosaur fauna from Antarctica. *Journal of Vertebrate Palaeontology* **13**, 40A

Hammer WR and Hickerson WJ (1994) A crested theropod dinosaur from Antarctica. *Science* **264**, 828–830.

Hammer WR and Hickerson WJ (1996) Implications of an Early Jurassic vertebrate fauna from Antarctica. In: *The Continental Jurassic.* (Ed. M Morales) *Museum of Northern Arizona Bulletin* **60**, 215–218.

Hammer WR and Hickerson WJ (1998) Antarctic dinosaurs: a glimpse into the early Jurassic of Gondwana. *Journal of Vertebrate Paleontology, Abstracts* **18**, 48A.

Hammer WR, Hickerson WJ and Slaughter R (1994) A dinosaur assemblage from the Transantarctic Mountains. *Antarctic Journal of the United States* **29**(5), 31–32.

Hammer WR and Smith ND (2008) A tritylodont postcanine from the Hanson Formation of Antarctica. *Journal of Vertebrate Paleontology* **28**, 269–273.

Hammer WR and Zawiskie JM (1982) Beacon fossils from northern Victoria Land. *Antarcic Journal of the United States* **17**(5), 13–14.

Hampe O and Long JA (1999) The histology of Middle-Late Devonian chondrichthyan teeth from South Victoria Land, Antarctica. *Records of the Western Australian Museum Supplement* **57**, 23–36.

Harasewych MG, Oleinik A and Zinsmeister WJ (2009) The Cretaceous and Paleogene pleurotomariid (Gastropoda: Vetigastropoda) fauna of Seymour Island, Antarctica. *Journal of Paleontology* **83**(5), 750–766.

Harding LA (2005) Early Pliocene Mollusca from Marine Plain, East Antarctica. Unpublished MSc thesis, University of Melbourne.

Harrington HJ (1969) Fossiliferous rocks in moraines at Minna Bluff, McMurdo Sound. *Antarctic Journal of the United States* **4**(4), 134–135.

Harris PT, O'Brien PE, Quilty PG, Taylor F, Domack E, DeSantis L and Raker B (1997) Vincennes Bay, Prydz Bay and Mac. Robertson Shelf: AGSO Cruise 186, ANARE Voyage 5, 1996/97 (BRAD) *AGSO Record* 1997/**51.**

Harris P, O'Brien PE, Quilty P, McMinn A, Holdway D, Exon NF, Hill PJ and Wilson CW (1999) Sedimentation and continental slope processes in the vicinity of an ocean waste disposal site, southeastern Tasmania. *Australian Journal of Earth Sciences* **46**, 577–591.

Harris PT, Taylor F, Domack E, DeSantis L, Goodwin I, Quilty PG and O'Brien PE (1998) Glacimarine Siliciclastic Muds from Vincennes Bay, East Antarctica; Preliminary Results of an Exploratory Cruise in 1997. *Terra Antartica* **4**, 11–20.

Hart MB (1990) Cretaceous sea level changes and global eustatic curves: evidence from SW England. *Proceedings of the Ussher Society* **7**, 268–272.

Harwood DM (1986) Cretaceous and Cenozoic siliceous microfossil biostratigraphy and Antarctic glacial and marine geologic history. Unpublished PhD thesis, Ohio State University, Columbus.

Harwood D (1988) Upper Cretaceous and Lower Paleocene diatom and silicoflagelate biostratigraphy of Seymour Island, eastern Antartic Peninsula. Seymour Island. *Geological Society of America, Memoir* **169**, 55–129.

Harwood DM and Levy RH (2000) Introduction and overview. In: *Paleobiology and Paleoenvironments of Eocene Rocks, McMurdo Sound, East Antarctica.* (Eds JD Stilwell and RM Feldmann) *American Geophysical Union Antarctic Research Series* **67**, 1–18.

Harwood DM, McMinn A and Quilty PG (2000) Diatom stratigraphy and age of the Pliocene Sørsdal Formation, Vestfold Hills, East Antarctica. *Antarctic Science* **12**, 443–462.

Harwood DM and Webb PN (1998) Glacial transport of diatoms in the Antarctic Sirius Group: Pliocene refrigerator. *GSA Today* **8**(4), 1–8.

Hathway B, Duane AM, Cantrill DJ and Kelley SP (1999) 40Ar/39Ar geochronology and palynology of the Cerro Negro formation, South Shetland Islands, Antarctica: a new radiometric tie for Cretaceous terrestrial biostratigraphy in the southern Hemisphere. *Australian Journal of Earth sciences* **46**, 593–606.

Hayes P (1996) The Coniacian-Santonian (Late Cretaceous) flora of the Hidden Lake Formation, James Ross Basin, Antarctic Peninsula. In: *Fifth Quadrennial Conference of the International Organisation of Paleobotany Abstract Volume 41.* International Organisation of Palaeobotany and Department of Geological Sciences, University of California, Santa Barbara, USA.

Hayes P (1999) Cretaceous angiosperm leaf floras from Antarctica. PhD thesis, University of Leeds.

Hayes P, Francis JE, Cantrill DJ and Crame JA (2006) Palaeoclimate analysis of Late Cretaceous angiosperm leaf floras, James Ross Island, Antarctica. In: *Cretaceous-Tertiary High-latitude Palaeoenvironments, James Ross Basin, Antarctica.* (Eds JE Francis, D Pirrie and JA Crame) *Geological Society, London, Special Publications* **258**, 49–62.

Heim F (1912) Geologische Beobachtungen über Süd-Georgien. *Gesellschaft Erdkunde Berlin Zeitschrift* **6**, 451–456.

Helby RJ and McElroy CT (1969) Microfloras from the Devonian and Triassic of the Beacon Group, Antarctica. *New Zealand Journal of Geology and Geophysics* **12**, 376–382.

Hennig A (1911) Le conglomerate pleistocène à Pecten de l'île Cockburn. *Wissenschaftliche Ergebnosse der Schwedischen Südpolar-Expedition* **3**(10), 1–72.

Hernandez PJ and Azcarate V (1971) Estudio paleobotanico preliminary sobre restos de una tafoflora de la Peninsula Byers (Cerro Negro), Isla Livingston; Islas Shetland del Sur, Antártica. *Serie Cientifica, Instituto Antárctico Chileno* **2**(1), 15–50.

Hikuroa D (2004) Late Jurassic microinvertebrates of the Antarctic Peninsula. PhD thesis, Auckland University, New Zealand.

Hill D (1964a) Archaeocyatha from loose material at Plunket Point at the head of Beardmore Glacier. In: *Antarctic Geology.* (Ed. RJ Adie) pp. 609–622. North Holland Publishing, Amsterdam, The Netherlands.

Hill D (1964b) Archaeocyatha from the Shackleton Limestone of the Ross System, Nimrod Glacier area, Antarctica. *Transactions of the Royal Society of New Zealand, Geology* **2**, 137–146.

Hill D (1965) Archaeocyatha from Antarctica and a review of the phylum. *Scientific Reports, Transantarctic Expedition 1955–58* **10**, Geology 3, 1–150.

Hill RS (1993) Nothofagus fossils in the Sirius Group, Transantarctic Mountains: leaves and pollen and their climatic implications. *American Geophysical Union, Antarctic Research Series* **60**, 67–74.

Hill RS, Harwood DM and Webb PN (1996) *Nothofagus beardmorensis* (Nothofagaceae), a new species based on leaves from the Pliocene Sirius Group, Transantarctic Mountains, Antarctica. *Review of Palaeobotany Palynology* **94**, 11–24.

Hill RS and Jordan GJ (1996) Macrofossils as indicators of Plio-Pleistocene climates in Tasmania and Antarctica. *Papers and Proceedings of the Royal Society of Tasmania* **130**, 9–15.

Hill RS and Truswell EM (1993) *Nothofagus* fossils in the Sirius Group Transantarctic Mountains. In: *The Antarctic Paleoenvironment: A Perspective on Global Change.* (Eds JP Kennet and D Warnke) *American Geophysical Union, Antarctic Research Series* **60**, 67–73.

Hirvas H, Nenonen K and Quilty PG (1993) Till stratigraphy and glacial history of the Vestfold Hills area, East Antarctica. *Quaternary International* **18**, 81–95.

Hoffman P (1991) Did the breakout of Laurentia turn Gondwanaland inside out? *Science* **252**, 1409–1412.

Holland R (1910) The fossil Foraminifera. *Wissenschaftliche Ergebnisse der Schwedischen Südpolar-Expedition* **3**(9), 1–12.

Hooker JD (1844–1860) *The botany of the Antarctic voyage of H.M. Discovery Ships Erebus and Terror in the years 1839–1843. Part 1. Flora Antarctica, in two parts (1844–47); Part 2. Flora Novae-Zelandiae, volume 1 (1853–55), volume 2 (1855); Part 3. Flora Tasmaniae, volume 1 (1855–60), volume 2 (1860).* Reeve Brothers, London.

Hooker JJ (1992) An additional record of a placental mammal (Order Astrapotheria) from the Eocene of Western Antarctica. *Antarctic Science* **4**, 107–108.

Hooker JJ, Milner AC and Sequiera SEK (1991) An ornithopod dinosaur from the Late Cretaceous of West Antarctica. *Antarctic Science* **3**, 331–332.

Hooker WJ (1843) Notes on the botany of the Antarctic Voyage conducted by Captain James Cook Ross in Her Majesty's Discovesry Ships Erebus and Terror; with observations on the tussac grass of the Falkland Islands. *London Journal of Botany* **2**, 247–329.

Horne RR (1969) Sedimentology and palaeongeography of the Lower Cretaceous depositional trough of south-eastern Alexander Island. *Bulletin of the British Antarctic Survey* **22**, 61–76.

Howarth MK (1958) Upper Jurassic and Cretaceous ammonite faunas of Alexander Land and Graham Land. *Scientific Report of the Falkland Islands Dependencies Survey* **21**, 1–16.

Howarth MK (1966) Ammonites from the Upper Cretaceous of the James Ross Island group. *Bulletin of the British Antarctic Survey* **10**, 55–69.

Howe J and Cantrill DJ (2001) Palaeoecology and taxonomy of pentoxylales from the Albian of Antarctica. *Cretaceous Research* **23**, 779–793.

Howe J and Francis JE (2005) Metamorphosed palaeosoils associated with Cretacaeous fossil forest, Alexander Island, Antarctica. *Journal of the Geological Society of London* **162**, 951–957.

Howlett PJ (1989) Late Jurassic – Early Cretaceous cephalopods of eastern Alexander Island, Antarctica. *Special Papers in Palaeontology* **41**, 1–72.

Huber BT (1988) Upper Campanian – Paleocene Foraminifera from James Ross Island region, Antarctic Peninsula. *Geological Society of America, Memoir* **169**, 163–252.

Huber BT (1998) Tropical paradise at the Cretaceous poles? *Science* **282**, 2199–2200.

Huber BT and Watkins DK (1992) Biogeography of Campanian-Maastrichtian calcareous plankton in the region of the Southern Ocean: paleogeographic and paleoclimatic implications. In: *A Perspective on Global Change.* (Eds JP Kennett and DA Warnke) *American Geophysical Union Antarctic Research Series* **56**, 31–60.

Hunt RH (2001) Biodiversity and palaeocology of Tertiary fossil floras in Antarctica. PhD thesis, University of Leeds.

Hunt RJ and Poole I (2003) Revising Palaeogene West Antarctic climate and vegetation history in light of new data from King George Island. In: *Causes and Consequences of Globally Warm Climates in the Early Paleogene.* (Eds B Schmitz and E Thomas) *Geological Society of America Special Papers* **369**, 395–412.

Igarashi A, Harada N and Moriwaki K (1995) Marine fossils of 30–40 ka in raised beach deposits and late Pleistocene glacial history around Lützow-Holm Bay, East Antarctica. *Proceedings NIPR Symposium on Antarctic Geoscience* **8**, 219–229.

Iltchenko LN (1972) [1971]. Late Precambrian acritarchs of Antarctica. In: *Antarctic Geology and Geophysics.* (Ed. RJ Adie) pp. 599–602. Universitetsforlaget, Oslo.

Isbell JL, Taylor TN and Taylor EL (1990) Depositional setting of Permian and Triassic fossil plants in the Allan Hills, southern Victoria Land. *Antarctic Journal of the United States* **25**, 22–27.

Ivany LC, Blake DB, Lohmann KC and Aronson RB (2004) Eocene cooling recorded in the chemistry of La Meseta Formation mollusks, Seymour Island, Antarctic Peninsula. *Bollettino di Geofisica Teorica ed Applicata* **45**, 242–245.

Ivany LC, Lohmann KC, Hasiuk F, Blake DB, Glass A, Aronson RB and Moody RM (2008) Eocene climate record of a high southern latitude continental shelf: Seymour Island, Antarctica. *Geological Society of America Bulletin* **120**(5–6), 659–678.

Jadwiszczak P (2001) Body size of Eocene Antarctic penguins. *Polish Polar Research* **22**, 147–158.

Jadwiszczak P (2003) Eocene penguins of Seymour Island: systematics, evolution and paleoecology. In: *Terra Nostra 2003/4: 9th International Symposium on Antarctic Earth Sciences (ISAES IX), 8–12 September 2003, Potsdam, Germany,* p. 169.

Jadwiszczak P (2008) Short Note: An intriguing penguin bone from the Late Eocene of Seymour Island, Antarctic Peninsula. *Antarctic Science* **20**, 589–590.

Jago JB and Oliver RL (1985) An alleged archaeocyath from Cape Denison, Antarctica. *Transactions of the Royal Society of South Australia* **109**, 183–185.

Jefferson TH (1980) Angiosperm fossils in supposed Jurassic volcanogenic shales, Antarctica. *Nature* **285**, 157–158.

Jefferson TH (1981) Palaeobotanical contributions to geology of Alexander Island, Antarctica. PhD thesis, University of Cambridge.

Jefferson TH (1982) Fossil forest from the Lower Cretaceous of Alexander Island, Antarctica. *Palaeontology* **25**, 681–708.

Jefferson TJ (1983) Palaeoclimatic significance of some Mesozoic Antarctic fossil floras. In: *Antarctic Earth Science.* (Eds RL Oliver, PR James and JB Jago) pp. 593–598. Australian Academy of Science, Canberra.

Jefferson TJ and Taylor TN (1983) Permian and Triassic woods from the Transantarctic Mountains: paleoenvironmental indicators. *Antarctic Journal of the United States* **18**(5), 14–16.

Jell JS (1981) Silurian tabulate corals from the Cloudmaker Moraine, Beardmore Glacier, Antarctica. *Alcheringa* **5**, 311–316.

Jerzmanska A and Swidnicki J (1992) Gadiform remains from the La Meseta Formation (Eocene) of Seymour Island, West Antarctica. *Polish Polar Research* **13**, 241–253.

Johanson Z (2004) Late Devonian sarcopterygian fishes from Eastern Gondwana (Australia and Antarctica) and their importance in phylogeny and biogeography. In: *Recent Advances in the Origin and Early Radiation of Vertebrates.* (Eds G Arratia, MVH Wilson and R Cloutier) pp. 287–308. Verlag Dr. F. Pfeil, München.

Johanson Z and Ahlberg P (2001) Devonian rhizodontids (Sarcopterygii; Tetrapodomorpha) from East Gondwana. *Transactions of the Royal Society of Edinburgh: Earth Sciences* **92**, 43–74.

Jones CM (2000) The first record of a fossil bird from East Antarctica. In: *Paleobiology and Paleoenvironments of Eocene Rocks, McMurdo Sound, East Antarctica. Antarctic Research Series* **76**, 359–364.

Jonkers HA (1998) Stratigraphy of Antarctic late Cenozoic pectinid-bearing deposits. *Antarctic Science* **10**(2), 161–170.

Jonkers HA (2003) Late Cenozoic-Recent Pectinidae (Mollusca: Bivalvia) of the Southern Ocean and neighbouring regions. *Monographs of Marine Mollusca* **5**, 1–125.

Jonkers HA and Taviani M (1998) Lower Miocene macrofossils from the CRP-1 Drillhole, Cape Roberts (Victoria Land Basin, Antarctica). *Terra Antartica* **5**, 493–498.

Kandefer HM (1994) Róznorodnosc fauny pingwinów kopalnych antarktycznej Wyspy Seymour w oparciu o analize humeri z kolekcji Instytutu Biologii Filii Uniwersytetu Warszawskiego w Bialymstoku. Unpublished dissertation, Warszawski University, Poland.

Keating JM (1992) Palynology of the Lachman Crags Member, Santa Marta Formation (Upper Cretaceous) of north-west James Ross Island. *Antarctic Science* **4**, 293–304.

Keating JM, Spenceer-Jones M and Newham S (1992) The stratigraphical palynology of the Kotick Point and Whisky Bay formations, Gustav Group (Cretaceous), James Ross Island. *Antarctic Science* **4**, 279–292.

Keller G, Adatte T, Stinnesbeck W, Rebolledo-Vleyra M, Fucugauchi JU, Kramar U and Stüben D (2004) Chixulub impact predates the K-T boundary mass extinction. *PNAS* **101**(11), 3753–3758.

Keller G, Adatte T, Berner Z, Harting M, Baum G, Prauss M, Tantawny A and Stüben D (2007) Chicxulub impact predates the K-T boundary: new evidence from Brazos, Texas. *Earth and Planetary Letters* **255**, 339–356.

Kelly SRA (1991) Antarctic stromatolites from the Late Jurassic of Alexander Island (abstract). *Palaeontological Association Newsletter* **12**, 12.

Kelly SRA (1993) Biofacies and biostratigraphic constraints on regression in the uppermost Fossil Bluff Group (Aptian-Albian), Alexander Island, Antarctica. In: Gondwana 8. (Ed. RH Findlay) Balkema, Rotterdam.

Kelly SRA (1995a) New trigonioid bivalves from the Albian (Early Cretaceous) of Alexander Island, Antarctic Peninsula: systematics, paleoecology, and austral Cretaceous paleobiogeography. *Journal of Paleontology* **69**, 264–279.

Kelly SRA (1995b) New trigonioid bivalves from the Early Jurassic to earliest Cretaceous of the Antarctic Peninsula region: systematics and austral paleobiogeography. *Journal of Paleontology* **69**, 66–84.

Kelly SRA (1996) The austral palaeobiogeography of an Early Cretaceous (Albian) trigoniid bivalve assemblage from the upper part of the Fossil Bluff Group, Alexander Islands, Antarctica (International Cretaceous Symposium, 4th, 1992). *Mitteilungen – Geologisch Palaontologischen Institut der Universitat Hamburg* **77**, 129–139.

Kelly SRA and Doyle P (1991) The bivalve *Aulacomyella* from the Early Tithonian (Late Jurassic) of Antarctica. *Antarctic Science* **3**, 97–107.

Kelly SRA and Montcrieff (1992) Marine molluscan constraints on the age of Cretaceous fossil forests of Alexander Island, Antarctica. *Geological Magazine* **129**(6), 771–778.

Kemp EM (1972) Lower Devonian palynomorphs from the Horlick Formation, Ohio Range, Antarctica. *Palaeontographica B* **139**, 105–124.

Kemp EM (1973) Permian flora from the Beaver Lake area, Prince Charles Mountains, Antarctica. I. Palynological examination of samples. *Palaeontological Papers 1969, Bulletin of the Burearu of Minerals and Resources Geology and Geophysics Australia* **126**, 7–12.

Kemp EM and Barrett PJ (1975) Antarctic glaciation and early Tertiary vegetation. *Nature* **258**, 507–508.

Kemp AES, Pearce RB, Grigorov I, Rance J, Lange CB, Quilty PG and Salter I (2006), Production of giant marine diatoms and their export at oceanic frontal zones: Implications for Si and C flux from stratified oceans. *Global Biogeochemical Cycles* **20**, GB4S04, doi:10.1029/2006GB002698.

Kennett JP and Warnke DA (Eds) (1992) The Antarctic paleoenvironment: a perspective on global change. Part One. *Antarctic Research Series* **56**, 1–385.

Kennett JP and Warnke DA (Eds) (1993) The Antarctic paleoenvironment: a perspective on global change. Part Two. *Antarctic Research Series* **60**, 1–273.

Kilian W (1906) Sur une faune d'ammonites néocrétacée recueillie par l'expédition antarctique suédoise. *Comptes rendus hebdomadaires des Séances de l'Académie des Sciences Paris* **142**(5), 306–308.

Kilian W and Reboul P (1909) Les Céphalopodes néocrétacés des isles Seymour et Snow Hill. *Wissenschaftliche Ergebnisse der Schwedischen Südpolar-Expedition* **3**(6), 1–75.

Kirwan LP (1962) *A History of Polar Exploration.* Pelican, Harmondsworth.

Kitching JW, Collinson JW, Elliot DH and Colbert EH (1972) *Lystrosaurus* Sone (Triassic) fauna from Antarctica. *Science* **175**, 524–527.

Köhler R (1995) A new species of the fossil turtle *Psephophorus* (Order Testudines) from the Eocene of the South Island, New Zealand. *Journal of the Royal Society of New Zealand* **25**, 371–384.

Konyushkov KN and Shulyatin OG (1980) Ob arkheotsiatakh Antarktidy i ikh sopostavlenii s arkheotsiatami Sibiri (On the archaeocyaths of Antarctica and their comparison with the archaeocyaths of Siberia). In: *Kembriy Altae-Sayanskoy skladchatoy oblasty (Cambrian of the Altay Sayan Fold Belt).* (Ed. IT Zhuralevla) pp. 143–150. Nauka Publishers, Moscow.

Krause DW (1997) Cosmopolitanism among Gondwanan Late Cretaceous mammals. *Nature* **390**, 504–507.

Kriwet J (2003) First record of an Early Cretaceous shark (Chondrichthyes, Neoselachii) from Antarctica. *Antarctic Science* **15**(4), 507–511.

Kriwet J (2005) Additions to the Eocene Selachian fauna of Antarctica with comments on Antarctic Selachian diversity. *Journal of Vertebrate Paleontology* **25**(1), 1–7.

Kriwet J and Gaździcki A (2003) New Eocene Antarctic chimeroid fish (Holocephali, Chimaeriformes). *Polish Polar Research* **24**(1), 29–51.

Kriwet J, Lirio JM, Nuñez HJ, Puceat E and Lécuyer C (2006) Late Cretaceous Antarctic fish diversity. In: *Cretaceous-Tertiary High-Latitude Palaeoenvironments: James Ross Basin, Antarctica.* (Eds JE Francis, D Pirrie and JA Crame) *Geological Society Special Publication* **258**, 83–100.

Kriwet J, Nuñez HJ and Llirio JM (2002) Late Cretaceous fish faunas from the Antarctic Peninsula. *Journal of Vertebrate Paleontology* **23** (suppl. 3), 76A.

Ksepka DT and Clarke JA (in press) The basal penguin (Aves: Sphenisciformes) *Perudyptes devriesi* and a phylogenetic evaluation of the penguin fossil record. *Bulletin of the American Museum of Natural History.*

Kyle RA (1974) *Plumsteadia ovata* n.sp., a Glossopterid fructification from south Victoria Land. *New Zealand Journal of Geology and Geophysics* **17**, 719–720.

Kyle RA (1977a) Devonian palynomorphs from the basal Beacon supergroup of south Victoria Land, Antarctica. *New Zealand Journal of Geology and Geophysics* **20**, 1147–1150.

Kyle RA (1977b) Palynostratigraphy of the Victoria Group of South Victoria Land, Antarctica. *New Zealand Journal of Geology and Geophysics* **20**, 1081–1102.

Kyle RA and Fascola A (1978) Triassic palynology of the Beardmore Glacier area of Antarctica. *Palinologia* **1**, 313–319.

Kyle RA and Schopf JM (1982) Permian and Triassic palynology of the Victoria group, Transantarctic Mountains. In: *Antarctic Geosciences.* (Ed. C Craddock) pp. 649–659. University of Wisconsin Press, Madison.

Lacey WS and Lucas RC (1981a) A lower Permian flora from the Theron Mountains, Coats Land. *British Antarctic Survey Bulletin* **53**, 153–156.

Lacey WS and Lucas RC (1981b) Triassic flora of Livingston Island, South Shetland Islands. *British Antarctic Survey Bulletin* **53**, 157–173.

Laird MG and Waterhouse JG (1962) Archaeocyathine limestones of Antarctica. *Nature* **194**, 861.

Laird MG, Andrews PB, Kyle PR and Jennings P (1972) Late Cambrian fossils and the age of the Ross Orogeny, Antarctica. *Nature* **238**, 34–36.

Laird MG (1991) The Late Proterozoic-Middle Paleozoic rocks of Antarctica. In: *The Geology of Antarctica.* (Ed. RJ Tingey) pp. 74–119. Oxford University Press, Oxford.

Laird MG (1991) Lower mid-Paleozoic sedimentation and tectonic patterns on the paleo-Pacific margin of Antarctica. In: *Geological Evolution of Antarctica.* (Eds MRA Thomson, JA Crame and JW Thomson) pp. 177–185. Cambridge University Press, Cambridge.

Lamanna MC (2011) Paleobiogeographic significance of Cretaceous continental vertebrates from Antarctic and Australia. Northeastern (46th Annual) and North-Central (45th Annual) Joint Meeting (20–22 March 2011), Paper No. 40-9. Geological Society of America *Abstracts with Programs* **43**(1), 120.

Lambert J (1910) Les echinides des îles Snow-Hill et Seymour. *Wissenschaftliche Ergebnisse der Schwedischen Südpolar-Expedition* **3**(11), 1–15.

Lamabrecht L, Lacey WS and Smith CS (1972) Observations on the Permian flora of the Law Glacier area, central Transantarctic Mountains. *Bulletin de la Societé Belge de Géologie, de Paléontologie et d'Hydrologie* **81**, 169–179.

Larwood GP (Ed.) (1973) *Living and Fossil Bryozoa: Recent Advances in Research.* Academic Press, London.

Laudon TS, Lackey LL and Quilty PG (1966) Geology of eastern Ellsworth Land Antarctica. *Proceedings of the 11th Pacific Science Congress* **4**(1), 29.

Laudon TS, Lackey LL, Quilty PG and Otway PM (1969) Geology of eastern Ellsworth Land. (Sheet 3, eastern Ellsworth Land). In: *Geologic Maps of Antarctica.* (Eds VC Bushnell and C Craddock) *Antarctic Map Folio Series, Folio* **12**, Pl. 111.

Lazarus D and Caulet JP (1993) Ceozoic Southern Ocean reconstructions from sedimentologic, radiolarian, and other microfossil data. In: *The Antarctic Paleonenvironment: A Perspective on Global Change Part Two.* (Eds JP Kennett and Warnke DA) *Antarctic Research Series* **60**, 145–174.

Lee DE and Stilwell JD (2000) Rynchonellide brachiopods from Late Eocene erratics in the McMurdo Sound region, Antarctica. In: *Paleobiology and Paleoenvironments of Eocene Rocks, McMurdo Sound, East Antarctica.* (Eds JD Stilwell and RM Feldmann) *American Geophysical Union Antarctic Research Series Volume* **67**, 325–327.

Lewis AR, Marchant DR, Ashworth AC, Hedenäs L, Hemming SR, Johnson JV, Long MJ, Machlus ML, Newton AE, Raine JI, Willenbring JK, Williams M, Wolfe AP (2008) Mid-Miocene cooling and the extinction of tundra in continental Antarctica. *Proceedings of the National Academy of Sciences* **105**(31), 10676–10680.

Li H (1994) Tertiary Fossil Hill flora from Fildes Peninsula of King George Island, Antarctica. In: *Stratigraphy and Palaeontology of Fildes Peninsula, King George Island, Antarctica.* (Ed. Y Shen) *State Antarctic Committee Monograph* **3**, 133–171.

Li H and Shen Y (1990) A primary study of Fossil Hill floras from Fildes Peninsula of King George Island, Antarctica. *Acta Palaeontologica Sinica* **29**, 147–153.

Li ZX (1997a) Did Rodinia form by the closure of a Mesoproterozoic Palaeo-Pacific Ocean? *International Conference on Terrane Geology, Christchurch,* pp. 99–102.

Li ZX (1997b) When was Rodinia born? *Terra Nova 9, European Union of Geosciences, EUG9, Strasbourg, France,* p. 162.

Lindstrom S (1994) Late Palaeozoic palynology of western Dronning Maud Land, Antarctica. *Lund Publications in Geology* **121**, 1–33.

Lipps JH and Hickman CS (1984) Origin, age, and evolution of Antarctic and deep-sea faunas. In: *The Environment of the Deep Sea.* (Eds WG Ernst and JG Morin) pp. 325–356. Rubey Volume 11. Prentice-Hall Inc., New Jersey.

Long AG (1966) Mesozoic plants and the problem of angiosperm ancestry. *Lethaia* **6**, 163–178.

Long DJ (1991) Fossil Cutlassfish (Perciformes: Trichiuridae) teeth from the La Meseta Formation (Eocene), Seymour Island, Antarctic Peninsula. *PaleoBios* **13**, 3–6.

Long DJ (1992) Sharks from the La Meseta Formation (Eocene), Seymour Island, Antarctic Peninsula. *Journal of Vertebrate Paleontology* **12**, 11–32.

Long DJ and Stilwell JD (2000) Fish remains from the Eocene of Mount Discovery, East Antarctica. In: *Paleobiology and Paleoenvironments of Eocene Rocks, McMurdo Sound, East Antarctica.* (Eds JD Stilwell and RM Feldmann) *American Geophysical Union Antarctic Research Series* **76**, 349–353.

Long J (2000) *Mountains of Madness – A Scientist's Odyssey through Antarctica.* Joseph Henry Press, USA (National Academies Press edition).

Long JA (1983a) New bothriolepid fishes from the Late Devonian of Victoria, Australia. *Palaeontology* **26**, 295–320.

Long JA (1983b) A new diplacanthoid acanthodian from the Late Devonian of Victoria, Australia. *Memoirs of the Association of Australasian Palaeontologists* **1**, 51–65.

Long JA (1984) New phyllolepids from Victoria and the relationships of the group. *Proceedings of the Linnean Society of New South Wales* **107**, 263–308.

Long JA (1989) Biogeographic and biostratigraphic significance of the Devonian fish fauna from the Aztec Siltstone, Antarctica. *New Zealand Antarctic Record* **9**(1), 45–51.

Long JA (1995) A new groenlandaspidid arthrodire (Pisces; Placodermi) from the Middle Devonian Aztec Siltstone, South Victoria Land, Antarctica. *Records of the Western Australian Museum* **17**, 35–43.

Long JA (2001) On the relationships of *Onychodus* and *Psarolepis. Journal of Vertebrate Paleontology* **21**, 815–820.

Long JA (2010) *The Rise of Fishes: 500 Million Years of Evolution.* University of New South Wales Press, Sydney.

Long JA, Choo B and Young GC (2008) A new basal actinopterygian fish from the Middle Devonian Aztec Siltstone of Antarctica. *Antarctic Science* **20**, 393–412.

Long JA and Molnar RE (1998) A new Jurassic theropod dinosaur from Western Australia. *Records of the Western Australian Museum* **19**, 221–229.

Long JA and Young GC (1995) Sharks from the Middle-Late Devonian Aztec Siltstone, southern Victoria Land, Antarctica. *Records of the Western Australian Museum* **17**, 287–308.

Long JA, Young GC, Holland T, Senden T and Fitzgerald MEH (2006) An exceptional Devonian fish from Australia sheds light on tetrapod origins. *Nature* **444**, 199–202.

Long WE (1962) Sedimentary rocks of the Buckeye Range, Horlick Mountains, Antarctica. *Science* **136**, 319–321.

Long WE (1965) Stratigraphy of the Ohio Range, Antarctica. In: *Geology and Paleontology of the Antarctic.* (Ed. JB Hadley) *American Geophysical Union Antarctic Research Series* **6**, 71–116.

Lusky JC, Reguero MA and Vizcaíno SF (1994) Geographical position applying Global Position System (GPS) in the Eocene land-vertebrate bearing localities from Seymour (Marambio) Island, Antarctic Peninsula. In: *III Jornadas de Comunicaciones sobre Investigaciones Antárticas. Dirección Nacional del Antártico-Instituto Antártico Argentino, Buenos Aires, Abstracts,* 53–54.

Macdonald DIM, Isabell JL and Hammer WR (1991) Vertebrate trackways from the Triassic Fremouw formation, Queen Alexandra Range, Antarctica. *Antarctic Journal of the United States* **26**(5), 20–22.

Macellari CE (1984) Late Cretaceous stratigraphy, sedimentology, and macropaleontology of Seymour Island, Antarctic Peninsula. Unpublished PhD dissertation, The Ohio State University, USA.

Macellari CE (1985) Paleobiogeografia y edad de la fauna de *Maorites-Gunnarites* (Ammonoidea) del Cretácico Superiór de la Antártida y Patagonia. *Ameghiana* **21**, 223–242.

Macellari CE (1986) Late Campanian-Maastrichtian ammonite fauna from Seymour Island (Antarctic Peninsula) *Paleontological Society Memoir* **18**, 1–55.

Macellari CE (1988) Stratigraphy, sedimentology and paleoecology of Upper Cretaceous/Paleocene shelf-deltaic sediments of Seymour Island. *Geological Society of America Memoir* **169**, 25–53.

Macellari CE and Zinsmeister WJ (1983) Sedimentology and macropaleontology of the Upper Cretaceous to Paleocene sequence of Seymour Island. *Antarctic Journal of the United States of America* **18**, 69–71.

Macfadyen WA (1966) Foraminifera from the Upper Cretaceous of James Ross Island. *Bulletin of the British Antarctic Survey* **8**, 75–87.

Macphail MK and Truswell EM (2004) Palynology of Neogene slope and rise deposits from ODP sites 1165 and 1167, East Antarctica. *Proceedings ODP Scientific Research* **188**, 1–20.

Maheshwari HK (1982) Permian wood from Antarctica and revision of some lower Gondwana wood taxa. *Palaeontographica B* **138**, 1–43.

Marchant HJ, Lugg DJ and Quilty PG (Eds) (2003) *Australian Antarctic Science: The First 50 Years of ANARE.* Artemis Press: Hobart.

Marenssi SA, Reguero MA, Santillana SN and Vizcaino SF (1989) Analisis geologico y paleontologico de un nivel de costa del eoceno tardio (Fn La Meseta) de la Isla Marambio (Seymour Island, Antartida). In: *VI Jornadas Argentinas de Paleontologia de Vertebrados. Resumenes.* pp. 45–46. San Juan, Argentina.

Marenssi SN, Reguero MA, Santillana SN and Vizcaino SF (1994) Eocene land mammals from Seymour Island, Antarctica: paleobiogeographical implications. *Antarctic Science* **6**, 3–15.

Marenssi S, Feldmann RM, and Schweitzer CE (2001) Distribution of fossil decapod crustaceans from the La Meseta Formation (Eocene) on Seymour Island, Antarctica: Changing times, changing patterns. *Geological Society of America, Abstracts with Programs* **33**(6), A-378.

Marples BJ (1953) Fossil penguins from the mid Tertiary of Seymour Island. *Scientific Reports Falklands Islands Dependencies Survey* **5**, 1–15.

Marshall CR (1995) Distinguishing between sudden and gradual extinctions in the fossil record: predicting the position of the Cretaceous-Tertiary iridium anomaly using the ammonite fossil record on Seymour Island, Antarctica. *Geology* **23**, 731–734.

Martin JE (2006) Biostratigraphy of the Mosasauridae (Reptilia) from the Cretaceous of Antarctica. In: *Cretaceous-Tertiary High-Latitude Palaeoenvironments: James Ross Basin, Antarctica. Geological Society Special Publication London* **258**, 101–108.

Martin JE, Bell G, Case J, Chaney DS, Fernández MS, Gasparini Z, Reguero M and Woodburne MO (2002) Mosasaurs (Reptilia) from the Late Cretaceous of Antarctic Peninsula. In: *Antarctica at the Close of a Millennium.* (Eds JA Gamble, DNB Skinner and S Henrys) *Eighth International Symposium on Antarctic Earth Sciences, Bulletin of the Royal Society of New Zealand* **35**, 293–299.

Martin JE, Bell GL, Fernandez MS, Case JA and Woodburne MO (1999) Mosasaurs from Antarctica, and their bearing on global distributions. *European Union of Geosciences, Journal of Conference Abstracts* **4**, 734.

Martin JE, Bell G, Fernández M, Reguero M, Case J, Chaney DS and Woodburne MO (1999) Mosasaurs from the Late Cretaceous of Antarctica and their bearing on marine paleobiogeography. In: *8th International Symposium on Antarctic Earth Sciences, Wellington, New Zealand, Programme and Abstracts,* p. 201.

Martin JE and Crame JA (2006) Palaeobiological significance of high-latitude late Cretaceous vertebrate fossils from the James Ross Basin, Antarctica. In: *Cretaceous-Tertiary High-Latitude Palaeoenvironments: James Ross Basin, Antarctica.* (Eds JE Francis, D Pirrie and JA Crame) *Geological Society Special Publication* **258**, 109–124.

Martin JE and Fernandez M (2005) The synonymy of the Late Cretacaeous mosasaur (Reptilia) genus *Lakumasaurus* from Antarctica with *Taniwhasaurus* from New Zealand and its bearing upon faunal similarity. In: *Gondwana 12: Geological and Biological Heritage of Gondwana.* (Eds RJ Pankhurst and GD Veiga) Academia Nacional de Ciencias, Cordobo, Argentina, p. 244.

Matthews DH (1959) Aspects of the geology of the Scotia arc. *Geological Magazine* **95**(6), 425–441.

Mawson D (Sir) (no date; c1950s) South Australian participation in Antarctic exploration. *Proceedings of the Royal Geographical Society of Australasia, S.A. Branch.* Centenary History of South Australia.

McCormick R (1884) *Voyages of Discovery in the Arctic and Antarctic Seas and Round the World, 2 volumes.* Sampson Low, Marston, Searle & Rivington, London.

McGhee GR Jr (2001) The 'multiple impact' hypothesis for mass extinction: a comparison of the Late Devonian and the late Eocene. *Palaeogeography, Palaeoclimatology, Palaeoecology* **176**, 47–58.

McHenry CR, Cook AG and Wroe S (2005) Bottom-feeding plesiosaurs. *Science* **310**, 75.

McKelvey BC, Webb P-N, Gorton MP and Kohn BP (1972) Stratigraphy of the Beacon Supergroup between the Olympus and Boomerang Ranges, Victoria Land. In: *Antarctic Geology and Geophysics.* (Ed. RJ Adie) pp. 345–352. Universitetsforlaget, Oslo, Norway.

McKinney ML, McNamara KJ and Wiedman LA (1988) Echinoids from the La Meseta Formation (Eocene), Seymour Island, Antarctica. In: *Geology and Paleontology of Seymour Island, Antarctic Peninsula.* (Eds RM Feldmann and MO Woodburne) *Geological Society of America Memoir* **169**, 499–504.

McLoughlin S and Drinnan AN (1996) Anatomically preserved Permian *Noeggerathiopsis* leaves from East Antarctica. *Review of Palaeobotany and Palynology* **92**, 207–227.

McLoughlin S and Drinnan AN (1997) Revised stratigraphy of the Permian Bainmedart Coal Measures, northern Prince Charles Mountains, East Antarctica. *Geological Magazine* **134**, 335–353.

McLoughlin S, Lindström S and Drinnan AM (1997) Gondwanan floristic and sedimentological trends during the Permian-Triassic transition: new evidence from the Amery Group, northern Prince Charles Mountains, East Antarctica. *Antarctic Science* **9**, 281–293.

McLoughlin S and Long JA (1994) New records of Devonian plants from southern Victoria Land, Antarctica. *Geological Magazine* **131**, 81–90.

McManus HA, Taylor EL, Taylor TN and Collinson JW (2002) A petrified *Glossopteris* flora from Collinson Ridge, central Transantarctic Mountains: Late Permian or Early Triassic? *Review of Palaeobotany and Palynology* **120**, 233–246.

Medina FA, Fourcade NH and del Valle RA (1983) La fauna del Jurasico superior del Refugio Ameghino y Cerro el Manco, Peninsula Antarctica. *Contribuciones Instituto Antarctica Argentino* **293**, 118.

Medina FA, Riccardi AC and Ghiglione M (2003) The Albian ammonite *Aioloceras* Whitehouse in Antarctica. *Alcheringa* **27**(2), 85–91.

Mela P and Solinus J (1519) *Itinerarivm Antonini Avg. Vibivs Seqvester.* P. Victor de regionibus urbis Romæ. Dionysius Afer de Situ orbis Prisciano Interprete. Impressum Florentiæ per hæredes Philippi Juntæ. Anno.D. XIX.à Christiana salu te supra mille mense Aprili. Leone.X. Pontisice.

Meyer-Berthaud B, Taylor TN and Taylor EL (1993) Petrified stems bearing *Dicroidium* leaves from the Triassic of Antarctica. *Palaeontology* **36**, 337–356.

Meyer DL and Oji T (1993) Eocene crinoids from Seymour Island, Antarctic Peninsula: paleobiogeographic and paleoecologic implications. *Journal of Paleontology* **67**, 250–257.

Miller D, Summers J and Silber S (2004) Environmental versus genetic sex determination: a possible factor in dinosaur extinction? *Fertility and Sterility* **81**, 954–964.

Miller MF (1998) Benthic aquatic ecosystems across the Permian-Triassic transition: record from biogenic structures in fluvial sandstones, central Transantarctic Mountains. *Journal of African Earth Sciences* **31**, 157–164.

Miller MF and Collinson JW (1994) Trace fossils from Permian and Triassic sandy braided stream deposits, central Transantarctic Mountains. *Palaios* **9**, 605–610.

Miller MF, Hasiotis ST, Babacock LE, Isbell JL and Collinson JW (2001) Tetrapod and large burrows of uncertain origin in Triassic high paleolatitude flood plain deposits, Antarctica. *Palaios* **16**, 218–232.

Milner AC, Hooker JJ and Sequeira S (1992) An ornithopod dinosaur from the Upper Cretaceous of the Antarctic Peninsula. *Journal of Vertebrate Paleontology* **12**, 44A (abstract).

Mitchell ED (1989) A new cetacean from the Late Eocene La Meseta Formation, Seymour Island, Antarctic Peninsula. *Canadian Journal of Fisheries and Aquatic Science* **46**, 2219–2235.

Mohr BAR (2001) The development of Antarctic fern floras during the Tertiary, and palaeoclimatic and palaeobiogeographic implications. *Palaeontographica Abteilung B,* **259**, 167–208.

Molnar RE (1989) Terrestrial tetrapods in Cretaceous Antarctica. In: *Origins and Evolution of the Antarctic Biota.* (Ed. JA Crame) *Geological Society Special Publication* **47**, 131–140.

Molnar RE (1996) An Antarctic Cretaceous theropod. *Memoirs of the Queensland Museum* **39**, 669–674.

Molnar RE, Angriman AL and Gasparini Z (1996a) An Antarctic Cretaceous dinosaur. *Memoirs of the Queensland Museum* **39**, 669–674.

Molnar RE, Lopez Angriman A and Gasparini Z (1996b) An Antarctic Cretaceous dinosaur. In: *First Gondwanan Dinosaur Symposium.* Trelew, Argentina.

Molnar RE and Wiffen J (1994) Polar dinosaur faunas from New Zealand. *Cretaceous Research* **15**, 689–706.

Moore EM (1991) Southwest US East Antarctic (SWEAT) connection: a hypothesis. *Geology* **5**, 425–428.

Moores EM and Dalziel IWD (1992) Southwest U.S.-east Antarctic (SWEAT) connection: a hypothesis and Pacific margins of Laurentia and East Antarctica-Australia as a conjugate rift pair: evidence and implications for an Eocambrian supercontinent (reply to comment by A. J. Boucot). *Geology* **20**, 87–88.

Morycowa E, Rubinowsli Z and Tokarski AK (1982) Archeaocyathids from a moraine at Three Sisters Point, King George Island (South Shetlands Island, Antarctica). *Studia Geologica Polonica* **74**, 73–80.

Myrcha A, Jadzwiszcak P, Tambussi CP, Noriega JI, Ga dzicki A, Tatur A and Del Valle RA (2002) Taxonomic revision of Eocene Antarctic penguins based on tarsometatarsal morphology. *Polish Polar Research* **23**, 5–46.

Myrcha A, Tatur A and del Valle R (1990) A new species of fossil penguin from Seymour Island, West Antarctica. *Alcheringa* **14**, 195–205.

Nathorst AG (1904) Sur la flore fossil des regions antarctiques. *Comptes rendus hebdomadaires des Séances de l'Académie des Sciences. Paris* **138**(23), 1447–1450.

Nathorst AG (1907), On the Upper Jurassic flora of Hope Bay, Graham Land. *Comptes rendus 10th International Geological Congress, Mexico, 1906,* **10**(2), 1269–1270.

Nelson PHH (1975) [1966] The James Ross Island Volcanic Group of north-east Graham Land. *Scientific Reports of the British Antarctic Survey* **54**, 1–62.

Néraudeau D and Mathey B (2000) Biogeography and diversity of South Atlantic Cretaceous echinoids: implications for circulation patterns. *Palaeogeography, Palaeoclimatology, Palaeoecology* **156**, 71–88.

Néraudeau D, Crame JA and Kooser M (2000) Upper Cretaceous echinoids from James Ross Basin, Antarctica. *Geobios* **33**, 455–466.

New Scientist (6 March 2004) 'Bizarre dinos of the deep freeze'. p. 16.

Nichols RL (1955) 'Bedrock geology of Marguirite Bay area, Palmer Peninsula, Antarctica'. Ronne Antactic Expedition, Technical Report 13. Department of the Navy, Office of Naval Research, Washington DC.

Nordenskjöld O, Andersson JG, Larsen CA and Skottsberg C (1904) *Antarctic Två År Bland Sydpolens Isar.* Vols 1 & 2. Albert Bonniers Förlag, Stockholm.

Noriega JI and Tambussi CP (1995) A late Cretaceous Presbyornithidae (Aves: Anseriformes) from Vega Island, Antarctic Peninsula: Paleobiogeographic implications. *Ameghiniana* **32**(1), 57–61.

Noriega JI and Tambussi CP (1996) The non penguin avifauna from the Eocene (early Oligocene?) of Seymour Island, Antarctic Peninsula. *Fourth International Meeting of the Society of Avian Paleontology and Evolution, Abstracts 13.* Washington.

Norris G (1965) Triassic and Jurassic miospores and acritarchs from the Breacon and Ferrar Groups, Victoria Land, Antarctica. *New Zealand Journal of Geology and Geophysics* **8**, 236–277.

Novas FE, Fernández M, Gasparini Z, Lirio JM, Nuñez HJ and Puerta P (2002) *Lakumasaurus antarcticus*, n. gen. et sp., a new mosasaur (Reptilia, Squamata) from the Upper Cretaceous of Antarctica. *Ameghiniana* **39**(2), 245–249.

O'Brien P, Borisova I, Quilty P, Bardsley T, Constable A and Harris P (2003) History of the Southern Ocean and Antarctic continental shelf – the ANARE contribution. In: *Australian Antarctic Science: The First 50 years of ANARE.* (Eds HJ Marchant, DJ Lugg and PG Quilty) pp. 139–163. *Proceedings of the ANARE Jubilee meeting.* Artemis Press, USA.

O'Brien PE, De Santis L, Harris PT, Domack E and Quilty PG (1999) Ice shelf grounding zone features of western Prydz bay, Antarctica: sedimentary processes from seismic and sidescan images. *Antarctic Science* **11**, 78–91.

Ociepa AM and Barbacka M (2011) *Spesia antarctica* gen. et sp. nov. – a new fertile fern spike from the Jurassic of Antarctica. *Polish Polar Research* **32**(1), 59–66.

O'Keefe FR (2001) Ecomorphology of plesiosaur flipper geometry. *Journal of Evolutionay Biology* **14**, 987–991.

Oleinik AE and Zinsmeister WJ (1996) Paleocene diversification of bucciniform gastropods on Seymour Island, Antarctica. *Journal of Paleontology* **70**, 923–934.

Olivero EB (1998) Early Campanian heteromorph ammonites from James Ross Island, Antarctica. *National Geographic Research* **4**, 259–271.

Olivero E (1992) Asociaciones de ammonites de la Formación Santa Marta (Cretácico tardío), Isla James Ross. In: *Geología de la Isla James Ross.* (Ed. C Rinaldi) pp. 47–75. Instituto Aantártico Argentino, Buenos Aires.

Olivero EB and Aguirre-Urreta MB (1994) A new tube-builder hydractinian, symbiotic with hermit crabs, from the Cretaceous of Antarctica. *Journal of Paleontology* **68**, 1169–1182.

Olivero EB, Gasparini Z, Rinaldi CA and Scasso R (1991) First record of dinosaurs in Antarctica (Upper Cretaceous, James Ross Island): palaeogeographic implications. In: *Geological Evolution of Antarctica.* (Eds MRA Thomson, JA Crame and JW Thomson) pp. 617–622. Cambridge University Press, Cambridge, UK.

Olivero EB and Medina FA (2000) Patterns of late Cretaceous ammonite biogeography in southern high latitudes: the family Kossmaticeratidae in Antaractica. *Cretaceous Research* **21**, 269–279.

Olivero EB and Palamarczuk S (1987) Ammonites y dinoflagelados Cenomanianos de la Isla James Ross, Antártida. *Ameghiniana* **24**, 35–49.

Olivero EB and Zinsmeister WJ (1989) Large heteromorph ammonites from the Upper Cretaceous of Seymour Island, Antarctica. *Journal of Paleontology* **63**, 626–636.

Olson SL (1985) The fossil record of birds. In: *Avian Biology.* (Eds DS Franer, JR King and KC Parkes) pp. 80–238. Academic Press, New York.

Olson SL (1992) *Neogaeornis wetzeli* Lambrecht, a Cretaceous loon from Chile (Aves: Gaviidae). *Journal of Vertebrate Paleontology* **12**(1), 122–124.

Orlando HA (1963) La flora fósil en las inmediaciones de la Peninsula Ardley, Isla 25 de Mayo, isles Shetland del Sur. *Contribuciones Instituto Antártida Argentina* **79**, 1–14.

Orlando HA (1964) The fossil flora of the surroundings of Ardley Peninsula (Ardley Island), 25 de Mayo Island (King George Island), South Shetland Islands. In *Antarctic Geology.* (Ed. RJ Adie) pp. 629–636. North-Holland Publishing Company, Amsterdam.

Orlando HA (1967) Primera flórula Triásica de la Antártida occidental. *Contribuciones Instituto Antártida Argentina* **118**, 1–16.

Orlando HA (1968) A new Triassic flora from Livingston Island, South Shetland Islands. *Britsih Antarctic Survey Bulletin* **16**, 1–13.

Orlando HA (1971) Las floras fósiles de Antártida occidental y sus relaciones estratigráficas. *Contribuciones Instituto Antártida Argentina* **140**, 1–12.

Osborn JM, Phipps CJ, Taylor TN and Taylor EL (2000) Structurally preserved sphenophytes from the Triassic of Antarctica: reproductive remains of *Spaciinodum. Review of Palaeobotany and Palynology* **111**, 225–235.

Ottone EG and Medina FA (1998) A wood from the Early Cretaceous of James Ross Island. *Ameghiniana* **35**, 291–298.

Palmer AR and Gatehouse CG (1972) Early and Middle Cambrian trilobites from Antarctica. *Geological Survey Professional Paper* **456-D**, 1–37.

Palmer AR and Rowell AJ (1995) Early Cambrian trilobites from the Shackleton Limestone of the Central Transantarctic Mountains. *Paleontological Society Memoir* **45** (*Journal of Paleontology* **69** (suppl. 6)), 1–28.

Pascual RF (1997) Fossil land mammals and the geobiotic history of southern South America. In: *Symposium I 'Past and Present Biota from Southern South America: A Model of Historical Biogeography' (R.Pascual & J.Crisci, Coords.) Southern Temperate Biota and Ecosystems. 'Past, Present, and Future'. II Southern Connection Congress. Programs and Abstracts.* p. 59. Valdivia, Chile.

Pavlov VV (1958) Results of a palynological analysis of samples from the sedimentary-volcanic Beacon Series (Antarctica, King george V Coast, Horn Bluff). *Po Paleontologii i Biostratigrafii, Naucho-Issledovatelyskii Institut Geologii Arktiki, Leningrad,* **12**, 77–79.

Peck LS and Brey T (1996) Bomb signals in old Antarctic brachiopods. *Nature* **380**, 207–208.

Perovich NE and Taylor EL (1989) Structurally preserved fossil plants from Antarctica. IV Triassic ovules. *American Journal of Botany* **76**, 992–999.

Pettigrew TH and Willey LE (1975) Belemnite fragments from Annenkov Island. *British Antarctic Survey Bulletin* **40**, 33–36.

Pewe TL, Rivard NR and Llano GA (1959) Mummified seal carcasses in the McMurdo Sound region, Antarctica. *Science* **130**, 716.

Philippe M, Barale G, Torres T and Covacevich V (1993) First study of *in situ* fossil woods from the Upper Cretaceous of Livingston Island, South Shetland Islands, Antarctica: paleoecological implications. *Comptes rendus de l'Académie des sciences. Série 2, Mécanique, Physique, Chimie, Sciences de l'univers, Sciences de la Terre* **317**, 103–108.

Pickard J (1985) The Holocene fossil marine macrofauna of the Vestfold Hills, East Antarctica. *Boreas* **14**, 189–202.

Pickard J (Ed.) (1986) *Antarctic Oases: Terrestrial Environments and History of Antarctica.* Academic Press, Sydney.

Pickard J, Adamson DA, Harwood DM, Miller GH, Quilty PG and Dell RK (1986) Early Pliocene marine sediments in the Vestfold Hills, East Antarctica: implications for coastline, ice sheet and climate. *South African Journal of Science* **82**, 520–521.

Pickard J, Adamson DA, Harwood DM, Miller GH, Quilty PG and Dell RK (1988) Early Pliocene marine sediments, coastline, and climate of East Antarctica. *Geolog* **16**, 158–161.

Pickard J and Seppelt RD (1984) Holocene occurrence of the moss *Bryum algens* Card. in the Vestfold Hills, Antarctica. *Journal of Bryology* **13**, 209–217.

Pigg KB and Taylor TN (1990) Permineralized *Glossopteris* and *Dicroidium* from Antarctica. In: *Antarctic Paleobiology.* (Eds TN Taylor and EL Taylor) pp. 164–172. Springer, New York.

Pirie JHH (1905) On the graptolite-bearing rocks of the South Orkneys, with a note by Dr Peach on specimens from the South Orkneys. *Proceedings of the Royal Society of Edinburgh* **25**(1), 463–470.

Pirrie D, Crame JA and Riding JB (1991) Late Cretaceous stratigraphy and sedimentology of Cape Lamb, Vega Island, Antarctica. *Cretaceous Research* **12**, 227–258.

Pirrie D, Duane AM and Riding JB (1992) Jurassic to Tertiary stratigraphy and palynology of the James Ross Basin; review and introduction. *Antarctic Science* **4**, 259–266.

Pirrie D, Feldmann RM and Buatois LA (2004) A new decapod trackway from the Upper Cretaceous, James Ross Island, Antarctica. *Palaeontology* **47**(1), 1–12.

Playford G (1990) Antarctic Palaeobiology: its role in the reconstruction of Gondwana. In: *Proterozoic and Paleozoic Palynology of Antarctica: A Review.* (Eds TN Taylor and EL Taylor) pp. 51–70. Springer-Verlag, New York.

Plimer IR (2001) *A Short History of Planet Earth.* ABC Books, Sydney.

Plimer IR (2003) The past is the key to the present. *IPA Review* **55**, 9–13.

Plumstead EP (1962) Fossil floras of Antarctica. *Trans-Antarctic Expedition 1955–58, Scientific Reports. Geology* **9**, 1–154.

Plumstead EP (1964) Palaeobotany of Antarctica. In: *Antarctic Geology.* (Ed. RJ Adie) pp. 637–654. North-Holland Publishing Company, Amsterdam.

Plumstead EP (1975) A new assemblage of plant fossils from Milorgfjella, Dronning Maud Land. *British Antarctic Survey Scientific Report* **83**, 1–30.

Pole M, Hill RS and Harwood D (2000) Eocene plant macrofossils from erratics, McMurdo Sound, Antarctica. In: *Palaeobiology and Palaeoenvironments of Eocene Fossiliferous Erratics, McMurdo Sound, East Antarctica.* (Eds JD Stilwell and RM Feldmann) *American Geophysical Union, Antarctic Research Series* **76**, 243–251.

Poole I and Cantrill DJ (2001) Fossil woods from Williams Point Beds, Livingston Island, Antarctica: a late Cretaceous southern high latitude flora. *Palaeontology* **44**, 1081–1112.

Poole I and Cantrill DJ (2006) Cretaceous and Cenozoic vegetation of Antarctica integrating the fossil wood record. In: *Cretaceous-Tertiary High-Latitude Palaeoenvironments: James Ross Basin, Antarctica.* (Eds JE Francis, D Poirrie and JA Crame) *Geological Society Special Publication, London* **258**, 63–82.

Poole I, Cantrill DJ, Hayes P and Francis JE (2000) The fossil record of Cunoniaceae: new evidence from late Cretaceous wood of Antarctica. *Review of Palaeobotany and Palynology* **111**, 127–144.

Poole I and Francis JE (1999) The first record of fossil atherospermataceous wood from the upper Cretaceous of Antarctica. *Review of Palaeobotany and Palynology* **107**, 97–107.

Poole L and Francis JE (2000) The first record of fossil wood of Winteraceae from the Upper Cretaceous of Antarctica. *Annals of Botany* **85**, 307–315.

Poole I and Gottwald H (2001) Monimiaceae *sensu lato*, an element of Gondwanan polar forests: evidence from the late Cretaceous-early Tertiary wood flora of Antactica. *Australian systematic Botany* **14**, 207–230.

Poole I, Gottwald H and Francis JE (2000) Illiciaceae, an element of Gondwanan polar forests? Late Cretaceous and Early Tertiary woods of Antarctica. *Annals of Botany* **86**, 421–432.

Poole I, Hunt RJ and Cantrill DJ (2001) A fossil wood flora from King George Island: ecological implications for an Antarctic Eocene vegetation. *Annals of Botany* **88**, 33–54.

Poole I, Mennega AMW and Cantrill DJ (2003) Valdivian ecosystems in the late Cretaceous and early Tetiary of Antarctica as evidenced from fossil wood. *Review of Palaeobotany and Palynology* **124**, 9–27.

Poole I, Richter HG and Francis JE (2000) Evidence for Gondwanan origins for *Sassafras* (Lauraceae)? Late Cretaceous fossil wood of Antarctica. *International Association of Wood Anatomists Journal* **21**, 463–475.

Putz MK and Taylor EL (1996) Wound response in fossil trees from Antarctica and its potential as a Paleoenvironmental indicator. *International Association of Wood Anatomists Journal* **17**, 77–88.

Quilty, PG (1970) Jurassic ammonites from Ellsworth land, Antarctica. *Journal of Paleontology* **44**(1), 110–116.

Quilty PG (1972) Middle Jurassic brachiopods from Ellsworth Land, Antarctica. *New Zealand Journal of Geology and Geophysics* **15**(1), 140–147.

Quilty PG (1972a) Pentacrinites and (?) apiocrinus from the Jurassic of Ellsworth Land, Antarctica. *Neues Jahrbuch für Mineralogie, Geologie, und Paläontologie Monatshefte* **8**, 484–489.

Quilty PG (1972b) Middle Jurassic brachiopods from Ellsworth Land, Antarctica. *New Zealand Journal Geology & Geophysics* **15**(1), 140–147.

Quilty PG (1973) Cenomanian-Turonian and Neogene sediments from northeast of Kerguelen Ridge, Indian Ocean. *Journal of the Geological Society of Australia* **20**(3), 361–368.

Quilty PG (1977) Late Jurassic bivalves from Ellsworth Land, Antarctica; their systematics and paleogeographic implications. *New Zealand Journal of Geology and Geophysics* **20**, 1033–1080.

Quilty PG (1979) Foraminiferida and Diatomaceae from sediment samples near the Balleny Islands. In: *Voyage to the Ice.* (Ed. D Lewis) pp. 122–127. ABC, Sydney.

Quilty PG (1982) Tectonic and other implications of Middle-Upper Jurassic rocks and marine faunas from Ellsworth Land, Antarctica. In *Antarctic Geoscience.* (Ed. C Craddock) pp. 669–678. University of Wisconsin Press, Madison, Wisconsin, USA.

Quilty PG (1983) Bajocian bivalves from Ellsworth Land, Antarctica. *New Zealand Journal of Geology and Geophysics* **26**, 395–418.

Quilty PG (1984) *Psammosphaera longsightensis* Quilty, new name for *P. parva* Crespin 1953 (not *P. parva* Flint, 1899). *Alcheringa* **8**, 242.

Quilty PG (1985a) Mineral resources of the Australian Antarctic Territory. In: *Australia's Antarctic Policy Options.* (Ed. S Harris) *Centre for Resource and Environmental Studies, Australian National University Monograph* **11**, 165–203. (Reproduced with permission as ANARE Research Note **27**.)

Quilty PG (1985b) Distribution of foraminiferids in sediments of Prydz Bay, Antarctica. N.H. Ludbrook Memorial Volume. pp. 329–340. Department of Mines and Energy, South Australia.

Quilty PG (1986) Australia's Antarctic Basins – identification and evolution. *Petroleum Exploration Society of Australia (PESA) Journal* **8**, 20–46.

Quilty PG (1988a) Foraminiferida from Quaternary sediments, Vestfold Hills, Antarctica. *Hydrobiologia* **165**, 213–220.

Quilty PG (1988b) *Cycleryon* (Crustacea, Decapoda) from the Jurassic of Ellsworth Land Antarctica. *Journal of Paleontology* **62**, 619–622.

Quilty PG (1988c) Book review. Walton, D.W.H., Antarctic Science. *Historical Records of Australian Science* **7**, 211, 212.

Quilty PG (1989a) Landslides: extent and economic significance in Antarctica. In: *Landslides: Extent and Economic Significance in the World.* (Eds EE Brabb and BL Harrod) pp. 127–132. A.A. Balkema, Rotterdam.Quilty PG (1989b) Changes in the Antarctic environment over the last five million years. In: *Antarctic Science: Geology and Biology.* (Eds HT Huh, B-K Park and S-H Lee) pp. 11–16. Korea Ocean Research & Development Institute, Seoul, Seo Deung Co. Ltd.

Quilty PG (1990a) New evidence for changes in the Antarctic marine environment over the last five million years. Text of keynote address. In: *Antarctic Ecosystems.* (Eds KR Kerry and G Hempel) pp. 3–8. Proceeding 5th SCAR Symposium on Antarctic Biology, Hobart, September, 1988.

Quilty PG (1990b) Antarctica as a continent for Science. In: *Antarctica's Future: Continuity or Change.* (Eds RA Herr, HR Hall, and MG Haward) pp. 29–37. Australian Institute for International Affairs Conference, Hobart.

Quilty PG (1991a) The Geology of Marine Plain Vestfold Hills, East Antarctica. In: *Geological Evolution of Antarctica.* (Eds MRA Thomson, JA Crame and JW Thomson) pp. 683–687. Cambridge University Press, Cambridge.

Quilty PG (1991b) Implications of Late Neogene sediments, East Antarctica. In *Proceedings First Latinamerican Conference on Geophysics, Geodesy and Space Sciences, Buenos Aires, July-August, 1990.* (Ed. O Schneider) pp. 19–24.

Quilty PG (1991c) Sources of data on the Quaternary of East Antarctica. In: *Quaternary Research in Australian Antarctica: Future Directions.* Department of Geography and Oceanography, University College, Australian Defence Force Academy, Canberra. (Eds D Gillieson and S Fitzsimons) *Special Publication* **3**, 107–114.

Quilty PG (1992a) Upper Cretaceous planktonic foraminifers and biostratigraphy, Southern Kerguelen Plateau, Indian Ocean. *Proceedings ODP Sci, Results* **120**: College Station TX (Ocean Drilling Program), pp. 371–392.

Quilty PG (1992b) Upper Cretaceous benthic foraminifers and paleoenvironments, Southern Kerguelen Plateau, Indian Ocean. *Proceedings ODP Sci, Results* **120**: College Station TX (Ocean Drilling Program), pp. 393–443.

Quilty PG (1992c) Unsuccessful search for fish otoliths, Cretaceous and younger, ODP Leg 120. Southern Indian Ocean. *Proceedings ODP Sci. Results* **120**: College Station TX (Ocean Drilling Program), pp. 1065–1066.

Quilty PG (1992d) Cretaceous – Tertiary invertebrate macrofaunas, ODP Leg 120, Southern Kerguelen Plateau, Indian Ocean. *Proceedings ODP Sci. Results* **120**: College Station TX (Ocean Drilling Program), pp. 445–448.

Quilty PG (1992e) Evolution of the modern glaciation of Antarctica. In *The Impact of Global Change on Antarctica.* (PG Quilty convener/editor) Department of Arts, Sport, the Environment, Tourism and Territories. Workshop, Hobart, May, 1991, pp.10–16.

Quilty PG (convener/editor) (1992f) Impact of Global Change on the Antarctic. Proceedings of workshop convened under the auspices of the Department of the Arts, Sport, the Environment, Tourism, and Territories, 1–3 May, 1991. pp. 1–78.

Quilty PG (1993a) Late Neogene sediments of coastal East Antarctica – an overview. In: *Recent Progress in Antarctic Earth Science.* (Eds Y Yoshida, K Kaminuma and K Shiraishi) pp. 699–705. Proceedings of the Sixth International symposium on Antarctic Earth Sciences. Tokyo, September 1991. Terra Scientific Publishing Company, Tokyo.

Quilty PG (1993b) Coastal East Antarctic Neogene sections and their contribution to the Icecap Evolution Debate. In: *The Antarctic Paleoenvironment: A Perspective on Global Change.* (Eds JP Kennett and DA Warnke) *American Geophysical Union, Antarctic Research Series* **60**, 251–264.

Quilty PG (1994a) Preliminary results of 1993/1994 fieldwork, Marine Plain, Vestfold Hills, East Antarctica. *Terra Antartica* **1**(2), 409–410.

Quilty PG (1994b) *Antarctica.* In *Cambridge Encyclopaedia of Australia.* (Ed. S Bambrick) pp. 285–287. Cambridge University Press, Cambridge.

Quilty PG (1995a) Heard Island; Offshore McDonald Islands. *Bulletin of Volcanic Eruptions* **32**, 32–33.

Quilty PG (1995b) Heard Island; Offshore McDonald Islands. *Bulletin of Volcanic Eruptions* **32**, 127–128.

Quilty PG (compiler) (1995b) Climate succession and glacial history of the Southern Hemisphere over the last five million years. *ANARE Research Notes* **94**, 39 (volume of conference abstracts)

Quilty PG (1996) The Pliocene environment of Antarctica. *Papers Proceedings Royal Society Tasmania* **130**(2), 1–8.

Quilty PG (1997a) Tasmania and Antarctica – a long association. *The Australian Gemmologist* **19**, 368–374.

Quilty PG (1997b) The potential for better Antarctic Neogene biostratigraphy. In: *Geology and Seismic Stratigraphy of the Antarctic Margin.* (Eds P Barker and A Cooper) *American Geophysical Union, Antarctic Research Series* **71**, 153–167.

Quilty PG (1998a) Future directions for the Australian Antarctic research program and implications for Tasmania. In: *Tasmania, the Antarctic and the Sub-Antarctic. Proceedings of the Forum held by the Governor of Tasmania,* 17–18 September 1996, pp. 80–86.

Quilty PG (1998) Book review: *The Antarctic Region: Geological Evolution and Processes.* (Ed. CA Ricci) Terra Antartica Publication, Siena, 1997. *EOS Transactions of the American Geophysical Union* **79**.

Quilty PG (1998) Surveyors and their products: mapping in the Antarctic. *Proceedings of the 39th Australian Surveyors Congress,* Launceston, November 1998, pp. 1–11.

Quilty PG (1999) Geophysics as centrepiece: its role in the Australian Antarctic Territory. *Exploration Geophysics* **30**, 68–75.

Quilty PG (2001) Reworked Paleocene and Eocene Foraminifera, Mac.Robertson Shelf, East Antarctica: paleoenvironmental implications. *Journal of Foraminiferal Research* **31**, 369–384.

Quilty PG (2002) The Antarctic past and you. In: *The Antarctic: Past, Present and Future.* (Eds J Jabour-Green and M Haward) *Antarctic CRC Research Report* **28**, 35–39.

Quilty PG (2002) Late Cretaceous foraminifera from a dredge and from ODP Site 1138, Kerguelen Plateau, southern Indian Ocean. Now available on http://www-odp.tamu.edu/publications/183_SR/003/003.htm.

Quilty PG (2003a) Solid Earth Geophysics: its role in the Australian National Antarctic Research Expeditions (ANARE). In: *Australian Antarctic Science: The First 50 Years of ANARE.* (Eds HJ Marchant, DJ Lugg and PG Quilty) pp. 203–223. *Proceedings of the ANARE Jubilee meeting.* Artemis Press, USA.

Quilty PG (2003b) Opening address, 20 July 1997. Antarctic Science – the ANARE perspective. In: *Australian Antarctic Science: The First 50 Years of ANARE.* (Eds HJ Marchant, DJ Lugg, and PG Quilty) pp. 21–26. *Proceedings of the ANARE Jubilee meeting.* Artemis Press, USA.

Quilty PG (2003c) Jubilee closing speech, Tuesday 22 July 1997. Influences on the future directions of Australian Antarctic Research. In: *Australian Antarctic Science: The First 50 Years of ANARE.* (Eds HJ Marchant, DJ Lugg and PG Quilty) pp. 579–588. *Proceedings of the ANARE Jubilee meeting.* Artemis Press, USA.

Quilty PG (2003) Neogene foraminifers and accessories from ODP Leg 188, Sites 1165, 1166 and 1167, Prydz Bay, Antarctica. *Proceedings ODP, Scientific Results* (also formally published 2004).

Quilty PG (2006) Landform evolution in the Marine Plain region, Vestfold Hills, East Antarctica. *Antarctic Science* **18**, 239–259.

Quilty PG (2006) Heard Island and the McDonald Islands: built on the Kerguelen Plateau foundation. In *Heard Island; Sentinel of the Southern Ocean.* (Ed. K Green) pp. 1–9. Surrey Beattie, Sydney.

Quilty PG (2007) Origin and evolution of the sub-Antarctic Islands: the foundation. In: *The sub-Antarctic.* (Eds P Selkirk, PG Quilty and M Davies) *Papers and Proceedings of the Royal Society of Tasmania* **141**(1), 35–58.

Quilty PG (in press) Identification and evolution of Antarctic Sedimentary Basins. In: *Circum-Pacific.* (Ed. MK Horn) *Council Transaction Series* **4**.

Quilty PG and Franklin D (1997) Geology of possible runway sites in the Davis region, Vestfold Hills, East Antarctica. *ANARE Research Notes* **98**, 1–57.

Quilty PG and Goddard P (2004) Diary of H.V. Goddard, crew member of S.Y. *Aurora* on Antarctic voyage to retrieve Dr Douglas Mawson, 19 November 1913 – 19 March 1914. *Polar Record* **40**, 193–203.

Quilty PG, Gillieson D, Burgess J, Gardiner G, Spate A and Pidgeon D (1990) *Ammoelphidiella* and associated benthic foraminifera, Arsemann Hills, East Antarctica. *Journal of Foraminiferal Research* **20**, 1–7.

Quilty PG, Kerry KR and Marchant HJ (1985) A seasonally recurrent patch of Antarctic planktonic diatoms. *Search* **16**, 48.

Quilty PG, Lirio JM and Jillet D (1994) Revision de la Geologia en Marine Plain, Vestfold Hills, Antartida Oriental. *Terceras Jornadas de Comunicaciones sobre Investigaciones Antarticas*, pp. 229–236.

Quilty PG, Lirio JM and Jillett D (2000) Stratigraphy of the Pliocene Sorsdal Formation, marine Plain, Vestfold Hills, East Antarctica. *Antarctic Science* **12**(2), 205–216.

Quilty PG, Murray-Wallace C and Whitehead JM (2004) *Austrochlamys heardensis* (Fleming, 1957) from Central Kerguelen Plateau, Indian Ocean: palaeontology and possible tectonic significance. *Antarctic Science* **16**, 329–338.

Quilty PG, O'Brien PE, Taylor F and Truswell EM (1995) The Palaeogene environment of the MacRobertson Shelf region, East Antarctica – preliminary results. *Abstracts, 7th International Symposium on Antarctic Earth Sciences, Siena*, p. 318.

Quilty PG, Rubenach M and Wilcoxon JA (1973) Miocene ooze from Macquarie Island. *Search.* **4**(5), 163–164.

Quilty PG, Shafik S, McMinn A, Brady H and Clarke I (1983) Microfossil evidence for the age and environment of deposition of sediments of Heard and McDonald Islands. In *Antarctic Earth Science.* (Eds R Oliver *et al.*) pp. 636–639. Australian Academy of Science, Canberra.

Quilty PG, Stagg HMJ, Thwaites R, Burbury R and Krummell R (1984) Does Gribb Bank exist? *Polar Record* **22**, 320–325.

Quilty PG, Truswell EM, O'Brien PE and Taylor F (1999) Paleocene-Eocene biostratigraphy and palaeoenvironment of East Antarctica: new data from the Mac. Robertson Shelf and western parts of Prydz Bay. *AGSO Journal of Geology and Geophysics* **17**, 133–143.

Quilty PG and Wheller GE (2000) Heard Island and the McDonald Islands: a window on the Kerguelen Plateau. *Papers and Proceedings of the Royal Society of Tasmania* **133**(2), 1–12.

Rabot C (1904) Découvertes de paléontology végétale faites par l'expédition antarctique suédoise. *La Géographie* **10**(1), 50–51.

Raine JI (1998) Terrestrial palynomorphs from Cape Roberts Project drillhole CRP-1, Ross Sea, Antarctica. *Terra Antartica* **5**, 539–548.

Raine JI and Askin RA (2001) Terrestrial palynology of Cape Roberts Project drillhole CRP-3, Victoria Land Basin, Antarctica. *Terra Antartica* **8**, 389–400.

Rao AR (1953) Some observations on the Rajmahal flora. *Palaeobotanist* **2**, 25–28.

Rao CP (1999) Cold water polar argonitic bivalve elemental composition, East Antarctica. *Carbonate Evaporites* **14**(1), 56–63.

Rees PM (1993) Dipterid ferns from the Mesozoic of Antarctica and New Zealand and their stratigraphical significance. *Palaeontology* **36**, 637.

Rees PM and Smellie JL (1989) Cretaceous angiosperms from an allegedly Triassic flora at Williams Point, Livingston Island, South Shetland Islands. *Antarctic Science* **1**, 239–248.

Rees MN, Pratt BR and Rowell AJ (1989) Early Cambrian reefs, reef complexes and associated lithofacies of the Shackleton Limestone, Transantarctic Mountains. *Sedimentology* **36**, 341–361.

Reguero MA, Marenssi SA and Santillana SN (2002) Antarctic Peninsula and South America (Patagonia) Paleogene terrestrial faunas and environments: biogeographic relationships. *Paleogeography Palaeoclimatology Palaeoecology* **179**, 189–210.

Reguero MA, Santillana S, Marenssi S, Vizcaino S, Elliot D, Lusky J and Moly J (1995) Nuevas localidades con vertebrados fosiles de la isla Marambio (Seymour), Antartida. In: *Simposio 'Vertebrados Paleogenos de la Antartida'; XI Jornadas Argentinas de Paleontologma de Vertebrados. Resumenes.* p. 13. Tucuman, Argentina.

Reguero MA, Vizcaíno SF, Goin FJ, Marenssi SA and Santillana SN (1998) Eocene high-latitude terrestrial vertebrates from Antarctica as biogeographic evidence. In: *Paleógeno de América del Sur y de la Peninsula Antártica.* (Ed. S Casadio) *Asociación Paleontológica Argentina, Publicación Especial* **5**, 185–198.

Repetski JE (1977) A fish from the Upper Cambrian of North America. *Science* **200**(4341), 529–531.

Retallack GJ (1997) Permian and Triassic driftwood from the Allan Hills, Antarctica. *Antarctic Journal of the United States* **30**(5), 37–39.

Retallack GJ and Hammer WR (1998) Paleoenvironment of the Triassic therapsid *Lystrosaurus* in the central Transantarctic Mountains, Antarctica. *Antarctic Journal of the United States* **31**(5), 33–35.

Retallack GJ, Krull ES and Robinson SE (1997) Permian and Triassic palaeosols and palaeoenvironments of southern Victoria Land, Antarctica. *Antarctic Journal of the United States* **30**(5), 33–35.

Retallack GJ, Seyedolali A, Krull ES, Holser WT, Ambers CP and Kyte FT (1998) Search for evidence of impact at the Permian-Triassic boundary in Antarctica and Australia. *Geology* **26**, 979–982.

Retallack GJ, Jahren AH, Sheldon ND, Chakrabarti R, Metzger CA and Smith RMH (2005) The Permian-Triassic boundary in Antarctica. *Antarctic Science* **17**(2), 241–258.

Retallack GJ and Krull ES (1999a) Landscape ecological shift at the Permian-Triassic boundary in Antarctica. *Australian Journal of Earth Sciences* **46**, 785–812.

Retallack GJ and Krull ES (1999b) Permian coprolites from Graphite Peak, Antarctica. *Antarctic Journal of the United States* **32**(5), 7–9.

Riccardi AC (1991) Jurassic and Cretaceous marine connections between the southeast Pacific and Tethys. *Palaeogeography, Palaeoclimatology, Palaeoecology* **87**, 155–189.

Rich TH (1996) Significance of polar dinosaurs in Gondwana. *Memoirs of the Queensland Museum* **39**(3), 711–717.

Rich TH, Gangloff RA and Hammer WR (1997) Polar dinosaurs. In: *Encyclopedia of Dinosaurs.* (Eds PJ Currie and K Padian) pp. 562–573. Academic Press, San Diego.

Rich THV and Rich PV (1989) Polar dinosaur and biotas of the Early Cretaceous of southeastern Australia. *National Geographic Research* **5**, 15–53.

Rich TH and Vickers-Rich P (2000, revised 2001) *Dinosaurs of Darkness.* Allen & Unwin, Sydney.

Rich TH, Vickers-Rich P, Fernandez M and Santillana S (1999) A probable hadrosaur from Seymour Island, Antarctic Peninsula. In: *Proceedings of the Second Gondwanan Dinosaur Symposium.* Tokyo, Japan. *National Science Museum Monographs* **15**, 219–222.

Richter M and Thomson MRA (1989) First Aspidorhynchidae (Pisces: Teleostei) from Antarctica. *Antarctic Science* **1**, 57–64.

Richter M and Ward DJ (1990) Fish remains from the Santa Marta Formation (Late Cretaceous) of James Ross Island, Antarctica. *Antarctic Science* **2**, 67–76.

Riding JB (1999) A compilation and interpretation of palynological data on the Hidden Lake Formation (Gustav Group) from James Ross Island, Antarctic Peninsula. *British Geological Survey, Technical Report* **WH/99/115R**, 1–21.

Riding JB and Crame JA (2002) Aptian to Coniacian (Early-late Cretaceous) palynostratigraphy of the Gustav Group, James Ross Basin, Antarctica. *Cretaceous Research* **23**, 739–760.

Riding JB, Keating JM, Snape MG, Newham S and Pirrie D (1992) Preliminary Jurassic and Cretaceous dinoflagellate cyst stratigraphy of the James Ross Island area, Antarctic Peninsula. *Newsletters on Stratigraphy* **26**, 19–39.

Rigby JF (1969) Permian sphenopsids from Antarctica. *Professional papers U.S. Geologocal Survey* **613**, F1–F13.

Rigby JF and Schopf JM (1969) Stratigraphic implications of Antarctic paleobotanical studies. In: *Earth Sciences. Gondwana Stratigrahy. IUGS Symposium, Buenos Aires, 1967.* Vol. 2. pp. 91–106. United Nations Educational, Scientific and Cultural Organization, Paris.

Ritchie A (1971a) Ancient animals of Antarctica – part 2. *Hemisphere* **15**, 12–17.

Ritchie A (1971b) Fossil fish discoveries in Antarctica. *Australian Natural History* **17**, 65–71.

Ritchie A (1972) Appendix: Devonian fish. In: *Stratigraphy of the Beacon Supergroup between the Olympus and Boomerang Ranges, Victoria Land.* (Eds RJ Adie, BC McKelvey, PN Webb, MP Gorton and BP Kohn) pp. 345–352. Universitetsforlaget, Oslo, Norway.

Ritchie A (1975) *Groenlandaspis* in Antarctica, Australia and Europe. *Nature* **254**, 569–573.

Ritchie A (1984) A new Placoderm, *Placolepis* gen. nov. (Phyllolepidae), from the Late Devonian of New South Wales, Australia. *Proceedings of the Linnean Society of New South Wales* **107**, 321–353.

Ritchie A (2004) A new genus and two new species of groenlandaspidid arthrodire (Pisces: Placodermi) from the Early-Middle Devonian Mulga Downs Group of western New South Wales, Australia. *Fossils and Strata* **50**, 1–26.

Rode AL, Lieberman BS and Rowell AJ (2003) A new Early Cambrian bradoriid (Arthropoda) from East Antarctica. *Journal of Paleontology* **77**(4), 691–697.

Roniewicz E and Morycowa E (1985) Fossil *Flabellum* (Scleractinia) of Antarctica. *Acta Palaeontologica Polonica* **30**, 99–106.

Roniewicz E and Morycowa E (1987) Development and variability of Tertiary *Flabellum rariseptatum* (Scleractinia), King George Island, West Antarctica. *Palaeontologica Polonica* **43**, 83–103.

Ross Captain Sir James Clark R.N. (1847a) *A Voyage or Discovery and Research in the Southern and Antarctic Regions, during the years 1839–43. With Plates, Maps and Woodcuts. Volume 1.* John Murray, Albemarle Street, London.

Ross Captain Sir James Clark R.N. (1847b) *A Voyage or Discovery and Research in the Southern and Antarctic Regions, during the years 1839–43. With Plates, Maps and Woodcuts.* Volume 2. John Murray, Albemarle Street, London.

Ross MJ (1982) *Ross in the Antarctic.* Whitby, Caedmon.

Rowell AJ, Evans KR and Rees MN (1988) Diversity and significance of the Early Cambrian Shackleton Limestone fauna, central Transantarctic Mountains. In: *Origin and Evolution of the Antarctic Biota, Abstracts Volume,* 37.

Rowell AJ, Rees MN and Braddock P (1987) Silurian marine fauna not confirmed from Antarctica. *Alcheringa* **11**, 137.

Rowell AJ, Rees MN and Braddock P (1988) Pre-Devonian Paleozoic rocks of the central Transantarctic Mountains. *Antarctic Journal of the United States* **21**, 48–50.

Rowell AJ, Rees MN and Evans KR (1992) Depositional setting of the Lower and Middle Cambrian in the Pensacola Mountains. *Antarctic Journal of the United States* **25**, 40–42.

Sandy MR (1991) Cretaceous brachiopods from James Ross Island, Antarctic Peninsula, and their paleobiogeographic affinities. *Journal of Paleontology* **65**, 396–411.

Scasso RA, Olivero EB and Buatois L (1991) Lithofacies, biofacies and ichnoassemblage evolution of a shallow submarine volcaniclastic fan-shelf depositional system (Upper Cretaceous-James Ross Island, Antarctica). *South American Journal of Earth Sciences* **4**, 239–260.

Schaeffer B (1972) A Jurassic fish from Antarctica. *American Museum Novitates* **2495**, 1–17.

Schlich R, Wise SW and shipboard party (1988) Ocean drilling program: that sinking feeling. *Nature* **334**, 385–386.

Schlich R, Munschy M, Whitechurch H, Aubry MP, Wise SW, Palmer A, Berggren WA, Bitschene PR, Blackburn NA, Breza J, Coffin M, Harwood DM, Heider F, Holmes MA, Howard WR, Inokuchi H, Kelts KR, Lazarus D, Mackensen A, Maruyama T, Pratson E, Quilty PG, Rack F, Salters VJM, Sevigny JH, Storey M, Takemura A, Watkins D and Zachos JC (1989) Resultats preliminaires de la campagne de forages du JOIDES RESOLUTION sur le plateau de Kerguelen (ODP Leg 120) *Comptes Rendus de l'Académie des Sciences, Paris* **308**(2), 1597–1603.

Schmidt-Tome M and Wolfart R (1984) Tremadocian faunas (trilobites, brachiopods) from Reilly Ridge, North Victoria Land, Antarctica. *Newsletters on Stratigraphy* **13**, 88–93.

Schopf JM (1967) Antarctic fossil plant collected during the 1966–1967 season. *Antarctic Journal of the United States* **2**(4), 114–116.

Schopf JM (1968) Studies in Antarctic paleobotany. *Antarctic Journal of the United States* **3**(5), 176–177.

Schopf JM (1973) Plant material from the Miers Bluff Formation of the South Shetland Islands. *Report of the Institute of Polar Studies, Ohio State University* **45**, 1–45.

Schweitzer CE, Feldmann RM, Marenssi S and Waugh DA (2005) Remarkably preserved annelid worms from the La Meseta Formation (Eocene), Seymour Island, Antarctica. *Palaeontology* **48**, 1–13.

Schweitzer CE and Feldmann RM (2005) Decapod crustaceans, the K/P event, and Palaeocene recovery. In: *Crustacea and Arthropod Relationships.* (Eds S Koenemann and RA Jenner) pp. 17–53. Taylor & Francis, Boca Raton, Florida.

Schwendemann AB, Decombeix AL, Taylor EL and Taylor TN (2010) *Collinsonites schopfii* gen. et sp. nov., a herbaceous lycopsid from the Upper Permian of Antarctica. *Review of Palaeobotany and Palynology* **158**(3–4), 291–297.

Seitz J (2011) 'Scientific expedition in Antarctica discovers treasure trove of fossils, microorganisms'. *The Santiago Times*, 27 March 2011. http://www.santiagotimes.cl/index.php?option=com_content&view=article&id=21067:scientific-expedition-in-antarctica-discovers-treasure-trove-of-fossils-new-microorganisms&catid=45:health-and-science&Itemid=41.

Selkirk P, Quilty PG and Davies M (Eds) The sub-Antarctic. *Papers and Proceedings of the Royal Society of Tasmania* **141**(1), 1–181.

Seward AC (1914) Antarctic fossil plants. In: *British Antarctic ('Terra Nova') Expedition, 1910. Natural History Report. Geology* **1**, 1–149. British Museum (Natural History), London, UK.

Sharman G and Newton ET (1894) Note on some fossils from Seymour Island, in the Antarctic regions, obtained by Dr Donald. *Transactions of the Royal Society of Edinburgh* **37**(3), 707–709.

Sharman G and Newton ET (1898) Notes on some additional fossils collected at Seymour Island, Graham's Land, by Dr Donald and Captain Larsen. *Proceedings of the Royal Society of Edinburgh* **22**(1), 58–61.

Sharpe RB (1902) Aves. Report on the Collections of Natural History made in the Antarctic regions during the voyage of the 'Southern Cross'. British Museum (Natural History), London **IV**, 106–173.

Shergold JH, Cooper RA, MacKinnon DI and Yochelson EL (1976) Late Cambrian Brachiopoda, Mollusca and Trilobita from Northern Victoria Land, Antarctica. *Palaeontology* **19**, 247–291.

Shipboard Scientific Party (2001) *Proceedings of the Ocean Drilling Program, Initial Report,* **188**. College Station TX (Ocean Drilling Program).

Shu D-G, Luo H-L, Morris SC, Zhang X-L, Hu S-X, Chen L, Han J, Zhu M, Li Y and Chen L-Z (1999) Lower Cambrian vertebrates from South China. *Nature* **402**, 42–46.

Shu D-G, Conway Morris S, Han J, Zhang Z-F, Yasui K, Janvier P, Chen L, Zhang X-L, Liu J-N, Li Y and Liu H-Q (2003) Head and backbone of the Early Cambrian vertebrate *Haikouichthys. Nature* **421**, 526–529.

Sidor CA, Damiani R and Hammer WR (2008) A new Triassic Temnospondyl from Antarctica and a review of Fremouw Formation Biostratigraphy. *Journal of Vertebrate Paleontology* **28**(3), 656–663.

Sidor CA, Miller MF and Isbell JL (2008) Tetrapod burrows from the Triassic of Antarctica. *Journal of Vertebrate Paleontology* **28**, 277–284.

Sidor CA, Steyer J-S and Damiani R (2007) *Paratosuchus* (Temnospondyli: Mastodonsauridae) from the Triassic of Antaractica. *Journal of Vertebrate Paleontology* **27**, 232–235.

Simpson GG (1946) Fossil penguins. *Bulletin of the American Museum of Natural History* **87** (Art. 1), 1–100.

Simpson GG (1971) Review of fossil penguins from Seymour Island. *Proceedings of the Royal Society of London* **B178**, 357–387.

Simpson GG (1972) Conspectus of Patagonian fossil penguins. *American Museum Novitates* **2488**, 1–37.

Skidmore MJ (1972) The Geology of South Georgia: 111. Prince Olav Harbour and Stromness Bay areas. *Scientific Reports of the British Antarctic Survey* **73**, 1–50.

Slack KE, Jones CM, Ando T, Harrison GL, Fordyce RE and Penny D (2006) Early penguin fossils, plus mitochondrial genomes, calibrate avian evolution. *Molecular Biology and Evolution* **23**, 1144–1155.

Slaughter RW, Hickerson WJ and Hammer WR (1994) Analysis of Antarctic theropod teeth based on serration densities and patterns. *Abstracts with Programs. Geological Society of America* **26**, 61.

Smith EA (1902) Mollusca. Report on the Collections of natural History, made in the Antarctic Regions during the voyage of the 'Southern Cross'. British Museum (Natural History), London **V11**, 201–213.

Smith ND, Hammer WR and Currie PJ (2005) Osteology and phylogenetic relationships of the Early Jurassic theropod *Cryolophosaurus ellioti* (Dinosauria: Theropoda): implications for basal theropod evolution. *Journal of Vertebrate Paleontology* **25**(supp to #3), 116A.

Smith ND, Makovicky PJ, Pol D, Hammer WR and Currie PJ (2007) The dinosaurs of the Early Jurassic Hanson Formation of the Central Transantarctic Mountains: phylogenic review and synthesis, in Antarctica. A keystone in a changing world. *Online Proceedings of the 10th ISAES.* (Eds AK Cooper and CR Raymond *et al.*) USGS Open-File Report 2007-1047. *U.S. Geological Survey and The National Acadamies. Short Research Paper* 003, <doi:10.3133/of2007-1047.srp003>.

Smith ND, Makovicky PJ, Hammer WR and Currie PJ (2008) Osteology of *Cryolophosaurus ellioti* (Dinosauria: Theropoda) from the Early Jurassic of Antarctica and implications of early theropod evolution. *Zoological Journal of the Linnean Society* **151**(2), 377–421.

Smith ND and Pol D (2007) Anatomy of a basal sauropodomorph dinosaur from the Early Jurassic Hanson Formation of Antarctica. *Acta Palaeoantological Polonica* **52**(4), 657–674.

Smith AB and Tranter TH (1985) Protremaster, a new Lower Jurassic genus of asteroid from Antarctica. *Geological Magazine* **122**, 351–359.

Smith Woodward A (1908) On fossil fish remains from Snow Hill and Seymour Islands. *Wissenschaftliche Ergebnisse der Schwedischen Südpolar-expedition 1901–1903* **3**(4), 1–4.

Smith Woodward A (1921) Fish-remains from the Upper Old Red Sandstone of Granite Harbour, Antarctica. *British Antarctic Terra Nova Expedition, 1910, Natural History Report, Geology* **1**(2), 51–62.

Smoot EL and Taylor TN (1986) Evidence of simple polyembryony in Permian seeds from Antarctica. *American Journal of Botany* **73**, 1079–1081.

Smoot EL, Taylor TN and Delevoryas T (1985) Structurally preserved fossil plants from Antarctica. 1. Antarcticycas, gen. nov., a Triassic cycad stem from the Beardmore Glacier area. *American Journal of Botany* **72**, 1410–1423.

Soot-Ryen T (1952) Laternula elliptica (King and Broderip 1831) from the Pecten-Conglomerate, Cockburn Island. *Arkiv fur Zoologi* **4**(9), 163–164.

Spath LF (1953) The Upper Cretaceous cephalopod fauna of Graham Land. *Scientific Reports of the Falkland Islands Dependencies Survey* **3**, 1–60.

Specht RL, Dettmann ME and Jarzen DM (1992) Community associations and structure in the Late Cretaceous vegetation of southeast Australasia and Antarctica. *Palaeogeography, Palaeoclimatology, Palaeoecology* **94**, 283–309.

Speden IG (1962) Fossiliferous Quaternary marine deposists in the McMurdo Sound region, Antarctica. *New Zealand Journal of Geology and Geophysics* **5**, 746–777.

Splettstoesser JF (1980) Coal in Antarctica. *Economic Geology* **75**(6), 936–942.

Spörli B (1966) *The Ellsworth Mountains of Antarctica: An American Geological Expedition, in Mountain World 1964/65.* Allen & Unwin, Chicago.

Spörli KB (1992) Stratigraphy of the Crashsite Group, Ellsworth Mountains, West Antarctica; Geology and paleontology of the Ellsworth Mountains, West Antarctica. *Memoir – Geological Society of America* **170**, 21–35.

Spörli KB and Craddock C (1992) Stratigraphy and structure of the Marble, Independence, and Patriot Hills, Heritage Range, Ellsworth Mountains, West Antarctica; Geology and paleontology of the Ellsworth Mountains, West Antarctica. *Memoir – Geological Society of America* **170**, 351–364.

Spörli KB and Craddock C (1992) Structure of the Heritage Range, Ellsworth Mountains, West Antarctica; Geology and paleontology of the Ellsworth Mountains, West Antarctica. *Memoir – Geological Society of America* **170**, 375–392.

Spörli KB, Craddock C, Rutford RH and Craddock JP (1992) Breccia bodies in deformed Cambrian limestones, Heritage Range, Ellsworth Mountains, West Antarctica; Geology and paleontology of the Ellsworth Mountains, West Antarctica. *Memoir – Geological Society of America* **170**, 365–374.

Spörli B and Craddock C (1968) Analysis of Ellsworth mountains and Ruppert coast geologic data. *Antarctic Journal of the United States* **3**(5), 179.

Stahl BJ and Chatterjee S (1999) A late Cretaceous chimaerid (Chondrichthyes, Holocephali) from Seymour Island, Antactica. *Paleontology* **42**, 979–989.

Stahl BJ and Chattejee S (2002) A Late Cretaceous callorhynchid (Chondrichthyes, Holocephali) from Seymour Island, Antarctica. *Journal of Vertebrate Paleontology* **22**, 848–850.

Stephenson A and Fleming WLS (1940) King George the Sixth Sound. *Geographical Journal* **96**(3), 153–164.

Stevens GR (1967) Upper Jurassic fossils from Ellsworth Land, west Antarctica, and notes on Upper Jurassic biogeography of the South Pacific region. *New Zealand Journal of Geology and Geophysics* **10**(2), 345–393.

Stilwell JD (1988) The biostratigraphy of early Tertiary macroinvertebrates from the La Meseta Formation, Seymour Island, Antarctic Peninsula. Unpublished Master's thesis, The Purdue University, West Lafayette, Indiana.

Stilwell JD (1994) Latest Cretaceous to earliest Paleogene molluscan faunas of New Zealand: changes in composition as a consequence of the break-up of Gondwana and extinction. Unpublished PhD dissertation, The University of Otago, Dunedin, New Zealand.

Stilwell JD (2000) Eocene Mollusca (Bivalvia, Gastropoda and Scaphopoda) from McMurdo Sound: systematics and paleoecologic significance. In: *Paleobiology and Paleoenvironments of Eocene Rocks, McMurdo Sound, East Antarctica.* (Eds JD Stilwell and RM Feldmann) *American Geophysical Union Antarctic Research Series* **76**, 261–320.

Stilwell JD (2002) History of the geological exploration of Cockburn Island, Antarctic Peninsula. *Polish Polar Research* **23**(1), 47–73.

Stilwell JD (2003a) Patterns of biodiversity and faunal rebound following the K-T boundary extinction event in Palaeocene Austral molluscan faunas. *Palaeogeography, Paeoclimatology, Palaeoecology* **195**(3–4), 319–356.

Stilwell JD (2003b) Diversity, distribution and evolutionary history of the Austral, Late Cretaceous to Eocene, gastropod *Struthioptera* (Stromboidea: Aporrhaidae). *Journal of the Royal Society of New Zealand* **33**(1), 315–340. [Special issue – JD Campbell Memorial Volume.]

Stilwell JD (2005) A new species of *Calliotropis* (Mollusca: Gastropoda: Vetigastropoda: Trochidae; Eucyclinae): from the Eocene of Antarctica. *Molluscan Research* **25**(1), 9–13.

Stilwell JD, Consoli CP, Sutherland R, Salisbury S, Rich TH, Vickers-Rich PA, Currie PJ and Wilson GJ (2006) Dinosaur sanctuary on the Chatham Island, Southwest Pacific: first record of theropods from the K-T boundary Takatika AGrit. *Palaeogeography, Palaeoclimatology, Palaeoecology* **230**(3–4), 243–250.

Stilwell JD and Feldmann RM (Eds) (2000) Paleobiology and paleoenvironments of Eocene rocks, McMurdo Sound, East Antarctica. *American Geophysical Union Antarctic Research Series* **67**, 1–373.

Stilwell JD and Gaździcki A (1998) A new limid bivalve from the La Meseta Formation (Eocene) of Seymour Island, Antarctic Peninsula. *Acta Geologica Polonica* **48**, 149–155.

Stilwell JD, Harwood DM and Whitehead JM (2002) Mid-Tertiary macroinvertebrate-rich clasts from the Battye Glacier Formation, Prince Charles Mountains, East Antarctica. *Antarctic Science* **14**(1), 69–73.

Stilwell JD, Jones CM, Levy RH and Harwood DM (1998) First fossil bird from East Antarctica. *Antarctic Journal of the United States – Review 1998* **33**(1), 1–7.

Stilwell JD, Levy RH, Feldmann RM and Harwood DM (1997) On the rare occurrence of Eocene callianassid decapods (Arthropoda) in their burrows from Mount Discovery, East Antarctica. *Journal of Paleontology* **71**(2), 284–287.

Stilwell JD, Levy RH and Harwood DM (1993) Preliminary paleontological investigation of Tertiary glacial erratics from the McMurdo Sound region, East Antarctica. *Antarctic Journal of the United States – Review 1993* **28**(5), 16–19.

Stilwell JD and Zinsmeister WJ (1987a) Late Cretaceous faunal assemblage of Humps Island collected during the 1986–87 expedition to the Antarctic Peninsula. *Antarctic Journal of the United States* **22**(5), 9–10.

Stilwell JD and Zinsmeister WJ (1987b) Late Cretaceous fossils from Ula Point, James Ross Island, collected during the 1986–87 expedition to the Antarctic Peninsula. *Antarctic Journal of the United States* **22**(5), 8–9.

Stilwell JD and Zinsmeister WJ (1987c) Late Cretaceous fossils from Cockburn Island collected during the 1986–87 expedition to the Antarctic Peninsula. *Antarctic Journal of the United States* **22**(5), 5–6.

Stilwell JD and Zinsmeister WJ (1992) Molluscan systematics and biostratigraphy of the Lower Tertiary La Meseta Formation, Seymour Island, Antarctic Peninsula. *American Geophysical Union Antarctic Research Series* **55**, 1–192.

Stilwell JD and Zinsmeister WJ (2002a) Paleobiogeographic synthesis of Eocene fauna and flora, McMurdo Sound. In: *Paleobiology and Paleoenvironments of Eocene Rocks, McMurdo Sound, East Antarctica.* (Eds JD Stilwell and RM Feldmann) *American Geophysical Union Antarctic Research Series* **67**, 365–372.

Stilwell JD and Zinsmeister WJ (2002b) A new, large acteonid gastropod (Mollusca) from the latest Cretaceous of Antarctica. *Journal of Paleontology* **76**(6), 1098–1101.

Stilwell JD and Zinsmeister WJ (2003) A new gastropod mollusk, *Antarctissitys austrodema* gen. et sp. Nov., from the late Maastrichtian of Antarctica and its evolutionary significance. *Cretaceous Research* **24**(2), 111–118.

Stilwell JD, Zinsmeister WJ and Oleinik AE (2004) Early Paleocene Mollusks of Antarctica: systematics, Paleoecology and Paleobiogeographic significance. *Bulletins of American Paleontology* **367**, 1–89.

Stipanicic PN and Bonetti MIR (1970) Posiciones estratigáficas y edades de las principales flora jurásicas argentines. 11. Floras doggerianas y málmicas. *Ameghiniana* **7**(2), 101–118.

Stock JM and Cande SC (2002) Tectonic history of Antarctic seafloor in the Australia–New Zealand–South pacific sector, implications for Antarctic continental tectonics. *Bulletin of the Royal Society of New Zealand* **35**, 251–259.

Stolarski J (1996) Paleogene corals from Seymour Island, Antarctic Peninsula. *Palaeontologica Polonica* **55**, 51–63.

Stolarski J (1998) *Conopora* (Stylasteridae, Hydrozoa) from the Eocene of Seymour Island. *Antarctic Science* **10**, 487–492.

Stolarski J and Taviani M (2001) Oligocene Scleractinia corals from CRP-3 Drillhole, McMurdo Sound (Victoria Land Basin, Antarctica). *Terra Antartica* **8**(3), 1–4.

Stone P and Willey LE (1973) Belemnite fragments from the Cumberland Bay type sediments of South Georgia. *Bulletin of the British Antarctic Survey* **36**, 129–132.

Stuiver M, Denton GH and Borns HW (1976) Carbon-14 dates of *Adamussium colbecki* (Mollusca) in marine deposits at New Harbor, Taylor Valley. *Antarctic Journal of the United States* **11**(2), 86–88.

Sturm A and Carryer SJ (1970) Geology of the region between Matusevich and Tucker Glaciers, north Victoria Land, Antarctica. *New Zealand Journal of Geology and Geophysics* **13**(2), 408–435.

Suess E (1883) *Das Antlitz der Erde.* 2 vols. F. Tempsky, Prague.

Summerhayes C (2005) Antarctic Science. *EOS, Transactions, American Geophysical Union* **86**(32), 293, 298.

Talent JA, Mawson R, Aitchison JC, Becker RT, Bell KN, Bradshaw MA, Burrow CJ, Cook AG, Dargan GM, Douglas JG, Edgecombe GD, Feist RM, Jones PJ, Long JA, Phillips-Ross JR, Pickett JW, Playford G, Rickards RB, Webby BD, Winchester-Seeto T, Wright AJ, Young GC and ZhenY-Y (2000) Devonian palaeobiogeography of Australia and adjoining regions. *Memoirs of the Association of Australasian Palaeontologists* **23**, 209–219, figs 13–17. Vertebrata: Young, Long & Burrow (authors).

Tambussi CP, Acosta Hospitaleche CI, Reguero MA and Marenssi SA (2006) Late Eocene penguins from West Antarctica: systematics and biostratigraphy. In: *Cretaceous-Tertiary High-Latitude Palaeoenvironments: James Ross Basin, Antarctica.* (Eds JE Francis, D Pirrie and JA Crame) *Geological Society Special Publication London* **258**, 145–162.

Tambussi CP, Noriega JI, Gazdzicky A, Reguero MA, Tatur A and Vizcaino SF (1994) The first occurrence of Ratite bird in the Paleogene of Antarctica. In: *XXI Polar Symposium Proceedings*. pp. 45–48. Warszawa, Poland.

Tambussi CP, Noriega JI, Gazdzicky A, Tatur A, Reguero MA and Vizcaino SF (1994c) Ratite bird from the Paleogene La Meseta Formation, Seymour Island, Antarctica. *Polish Polar Research* **15**(1–2), 15–20.

Tambussi CP, Noriega JJ, Santillana SN and Marenssi SA (1995) Falconid Bird from the Middle Eocene La Meseta Formation, Seymour Island, West Antarctica. In: *55th Annual Meeting of the Society of Vertebrate Paleontology, November 1–4, 1995. Abstracts* **15**, 55A. Pittsburg, Pennsylvania.

Tambussi CP and Tonni EP (1988) A diomedeid (Aves, Procellariformes) from the Late Eocene of Antarctica. In: *5th Jornadas Argentiniana de Paleontologia de Vertebrados.* (Eds JC Quiroga and AL Cione) La Plata, Universidad de La Plata, p. 4. [In Spanish]

Tasch P (1968) Trace fossils from the Permian Polarstar Formation, Sentinel Mountains, Antarctica. *Transactions Kansas Acadamy Science* **71**(2), 184–194.

Tasch P (1969) Antarctic paleobiology: new fossil data and their significance. *Antarctic Journal of the United States* **4**(5), 198–199.

Tasch P (1970) Antarctic and other Gondwana conchostracans and insects: new data; significance for the Drift theory. *Proceedings and Papers Second Gondwana Symposium, South Africa, 1970*, pp. 589–592.

Tasch P (1973) Jurassic beetle from southern Victoria Land, Antarctica. *Journal of Paleontology* **47**(3), 590–591.

Tasch P and Riek EE (1969) Permian insect wing from Antarctic Sentinel Mountains. *Science* **164**(3887), 1529–1530.

Tavera J (1970) Fauna Titoniana-Neocomiana de Isla Livingston, Islas Shetland del Sur, Antártica. *Serie Cientifico, Instituto Antártico Chileno* **1**(2), 175–186 *(Contrnes Instituto Antártico Chileno 25.)*

Taviani M and Beu AG (2001) Palaeogene macrofossils from CRP-3 Drillhole, Victoria Land Basin, Antarctica. *Terra Antartica* **8**(4), 423–434.

Taviani M and Beu AG (2003) The paleoclimatic significance of Cenozoic marine macrofossil assemblages from Cape Roberts Project drillholes, McMurdo Sound, Victoria Land Basin, East Antarctica. *Palaeo3* **198**, 131–143.

Taviani M, Beu AG and Jonkers HA (2000) Macrofossils from CRP-2/2A, Victoria Land Basin, Antarctica. *Terra Antartica* **7**, 513–526.

Taviani M, Beu AG and Lombardo C (1998) Pleistocene macrofossils from CRP-1 Drillhole, Victoria Land Basin, Antarctica. *Terra Antartica* **5**(3), 485–491.

Taviani M and Claps M (1998) Biogenic Quaternary carbonates in the CRP-1 drillhole, Victoria Land Basin, Antarctica. *Terra Antartica* **5**, 411–418.

Taviani M and Zahn R (1998) The stable oxygen isotope record of Pleistocene and Miocene bivalves in the CRP-1 drillhole, Victoria Land Basin, Antarctica. *Terra Antartica* **4**(3), 419–424.

Taylor BJ (1965) Aptian cirripedies from Alexander Island. *Bulletin of the British Antarctic Survey* **7**, 37–42.

Taylor BJ (1966) Taxonomy and morphology of Echinodermata from the Aptian of Alexander Island. *Bulletin of the British Antarctic Survey* **8**, 1–18.

Taylor BJ (1967) Trace fossils from the Fossil Bluff Series of Alexander Island. *Bulletin of the British Antarctic Survey* **13**, 1–30.

Taylor BJ (1969) Small tubiform fossils from the Lower Cretaceous of Alexander Island. *Bulletin of the British Antarctic Survey* **21**, 71–78.

Taylor BJ (1971) Thallophyte borings in phosphatic fossils from the Lower Cretaceous of south-east Alexander Island, Antarctica. *Palaeontology* **14**(2), 294–302, pls. 52–53.

Taylor BJ (1972a) Stratigraphical correlation in south-eastern Alexander Island. In: *Antarctic Geology and Geophysics.* (Ed. RJ Adie) pp. 149–153. Universitetsforlaget, Oslo, Norway.

Taylor BJ (1972b) An urdidid isopod from the Lower Cretaceous of south-east Alexander Island. *Bulletin of the British Antarctic Survey* **27**, 97–103.

Taylor BJ (1973) The cuticle of Cretaceous macrurous Decapoda from Alexander and James Ross Islands. *Bulletin of the British Antarctic Survey* **35**, 91–100.

Taylor BJ (1979) Macrurous Decapoda from the Lower Cretaceous of south-eastern Alexander Island. *British Antarctic Survey, Scientific Reports* **81**, 1–39.

Taylor EL (1996) Enigmatic gymnosperms? Structurally preserved Permian and Triassic seed ferns from Antarctica (6th Argentine Congress on Paleontology and Biostratigraphy). *Review of Palaeobotany and Palynology* **90**, 303–318.

Taylor EL and Taylor TN (1990) Structurally preserved Permian and Triassic floras from Antarctica. In: *Antarctic Paleobiology.* (Eds EL Taylor and TN Taylor) pp. 149–163. Springer, New York.

Taylor EL, Taylor TN, Isbell JL and Cúneo NR (1989a) Fossil floras of southern Victoria Land 1.Aztec Mountain. *Antarctic Journal of the united States* **24**(5), 24–25.

Taylor EL, Taylor TN, Isbell JL and Cúneo NR (1989b) Fossil floras of southern Victoria Land. 2. Kennar Valley. *Antarctic Journal of the United States* **24**(5), 26–27.

Taylor EL, Taylor TN and Cúneo NR (1992) The present is not the key to the past: a Permian petrified forest in Antarctica. *Science* **257**, 1675–1677.

Taylor F, Whitehead JM and Domack EW (2001) Holocene paleoclimate change in the Antarctic Peninsula: evidence from the diatom, sedimentary and geochemical record. *Marine Micropaleontology* **41**(1–2), 25–44.

Taylor TG and Goddard EJ (1914) Short notes on palaeontology. In: *Glaciology, Physiography, Stratigraphy, and Tectonic Geology of South Victoria Land.* (Eds TWE David and RE Priestley), Scientific Reports Series A, **3**(1), London, W Heinemann.

Taylor TN (1989) Fossil fungi (Endogonaceae) from the Triassic of Antarctica. *American Journal of Botany* **76**, 389–396.

Taylor TN, Taylor EL and Isbell JL (1989) Glossopterid sreproductive organs from Mount Achernar, anstarctica. *Antarctic Journal of the United States* **24**(5), 28–29.

Thomson JA (1918) The genus *Bouchardia* (Brachiopoda) and the age of the younger beds of Seymour Island, West Antarctic. *Geological Magazine (N.S., Decade 6)* **5**, 258–263.

Thomson JW (1971) The geology of Matthews Island, South Orkney Islands. *Bulletin of the British Antarctic Survesy* **26**, 51–57.

Thomson JW (1973) The geology of Powell, Christoffersen and Michelsen Islands, South Orkney Islands. *Bulletin of the British Antarctic Survey* **33 & 34**, 137–167.

Thomson JW (1974) The geology of the South Orkney Islands. 111. Coronation Island. *Scientific Report of the British Antarctic Survey* **86**, 1–39.

Thomson MRA (1967) A probable Cretaceous invertebrate fauna from Crabeater Point, Bowman Coast, Graham Land. *Bulletin of the British Antarctic Survey* **14**, 1–14.

Thomson MRA (1969) The marine origin of water-Iain volvanic sediments of south-west Adelaide Island. *Bulletin of the British Antarctic Survey* **19**, 83–88.

Thomson MRA (1971a) Gastropoda from the Lower Cretaceous sediments of south-eastern Alexander Island. *Bulletin of the British Antarctic Survey* **25**, 45–58.

Thomson MRA (1971b) Inarticulate Brachiopoda from the Lower Cretaceous of south-eastern Alexander Island. *Bulletin of the British Antarctic Survey* **25**, 85–94.

Thomson MRA (1972a) [1971] Ammonite faunas of south-eastern Alexander Island and their stratigraphical significance. In *Antarctic Geology and Geophysics.* (Ed. RJ Adie) pp. 155–160. Universitetsforlaget, Oslo, Norway.

Thomson MRA (1972b) Lower Cretaceous *Lamellaptychus* (Aptychi, Ammonoidea) from south-eastern Alexander Island. *Bulletin of the British Antarctic Survey* **30**, 35–40.

Thomson MRA (1972c) New discoveries of fossils in the Upper Jurassic Volcanic Group of Adelaide Island. *Bulletin of the British Antarctic Survey* **30**, 95–101.

Thomson MRA (1974) Ammonite faunas of the Lower Cretaceous of south-eastern Alexander Island. *ScientificReports British Antarctic Survey* **80**, 1–44.

Thomson MRA (1975a) First marine Triassic fauna from the Antarctic Peninsula. *Nature, London* **257**(5527), 577–578.

Thomson MRA (1975b) Fossils from the South Orkney Islands: 11 Matthews Island. *Bulletin of the British Antarctic Survey* **40**, 75–79.

Thomson MRA (1975c) Upper Jurassic Mollusca from Carse Point, Palmer Land. *Bulletin of the British Antarctic Survey* **41 & 42**, 31–42.

Thomson MRA (1975d) New palaeontological and lithological observations on the Legoupil Formation, north-west Antarctic Peninsula. *Bulletin British Antarctic Survey* **41** & **42**, 169–185.

Thomson MRA (1982) A comparison of the ammonite fauna of the Antarctic Peninsula and Magellanes Basin. *Journal of the Geological Society, London* **139**, 763–770.

Thomson MRA (1984) Cretaceous ammonite biostratigraphy of western James Ross Island, Antarctica. In: *Memoria III Congreso Latinoamericano de Paleontologia, Mexico, 1984.* pp. 308–314. Universidad Nacional Autonoma de Mexicom, Instituto de Geología, Mexico City.

Thomson MRA (1989) Antarctic invertebrate fossils: the Mesozoic-Cainozoic record. In: *The Geology of Antarctica.* (Ed. RJ Tingey) pp. 487–498. Oxford University Press, Oxford.

Thomson MRA and Tranter TH (1986) Early Jurassic fossils from Central Alexander Island and their geological setting. *British Antarctic Bulletin* **70**, 23–39.

Thomson MRA and Willey LE (1975) Fossils from the South Orkney Islands. I. Coronation Island. *Bulletin British Antarctic Survey* **40**, 15–21.

Tingey RJ (Ed.) (1991) *The Geology of Antarctica.* Monographs on Geology and Geophysics Volume 17. Oxford University Press, Oxford.

Tonni EP (1980) A large pseudodontornithid (Pelecaniformes, Odontopterygia) from the early Tertiary of Antarctica. *Ameghiniana* **17**(3), 273–276. [In Spanish]

Tonni EP and Tambussi CP (1985) New remains of odontopterygia (Aves: Pelecaniformes) from the early Tertiary of Antarctica. *Ameghiniana* **21**(2), 121–124. [In Spanish]

Torres T, Barale G, Méon H, Phillippe M and Thévenard F (1997) Cretaceous floras from Snow Island (South Shetland Islands, Antarctica) and their biostratigraphic significance. In: *The Antarctic Region: Geological Evolution and Processes.* (Ed. CA Ricci) pp. 1023–1028. Terra Antartica Publication, Siena.

Torres T, Barale G, Thévernard F, Philippe M and Galleguillos H (1997) Morfologia y sistemática de la flora del Cretácico Inferior de President Head, isla Snow, archipelago de las Shetland del Sur, Antártica. *Sesrie Cientifica del Instituto Antártico Chileno* **2**, 15–50.

Torres T and Lemoigne Y (1988) Maderas fósiles terciarias de la Formacíon Caleta Arctowski, Isla Rey Jorge, Antártica. *Instituto Antártico Chileno Serie Cientifica* **37**, 69–107.

Torres T and Lemoigne Y (1989) Fossil wood findings of angiosperms and gymnosperms of the Upper Cretaceous at Williams Point, Livingston Island, South Shetland Islands, Antarctica. *Instituto Antártico Chileno Serie Cientifica* **39**, 9–26.

Torres T, Marrenssi S and Santillana S (1994) Fossil wood of La Meseta Formation, isla Seymour, Antártica. *Serie Cientifica del Instituto Antártico Chileno* **44**, 17–38.

Torres T, Marenssi S and Santillana S (1994) Maderas fósiles de la isla Seymour, formacíon La Meseta, Antártica. *Instituto Antártico Chileno Serie Científica* **39**, 43–58.

Townrow JA (1967a) Fossil plants from the Allan and Carapace Nunataks, and from the Upper Mill and Shackleton Glacier, Antarctica. *New Zealand Journal of Geology and Geophysics* **10**, 456–473.

Townrow JA (1967b) On a conifer from the Jurassic of East Antarctica. *Papers and Proceedings of the Royal Society of Tasmania* **101**, 137–147.

Trendall AF (1959) The geology of South Georgia – II. *Scientific Reports of the Falkland Islands Dependencies Survey* **19**, 1–48.

Trewin NH and McNamara KJ (1995) Arthropods invade the land: trace fossils and palaeoenvironments of the Tumblagooda Sandstone (?late Silurian) of Kalbarri, Western Australia. *Transactions of the Royal Society of Edinburgh: Earth Sciences* **85**, 177–210.

Tröger KA and Weber W (1985) Description of a Cambrian fauna from the Neptune Range, Pensacola Mountains, Antarctica. *Zeitschrift Geologischer Wissenschaft, Berlin* **13**, 359–367.

Truswell EM (1980) Permo-Carboniferous palynology of Gondwanaland: progress and problems in the decade to 1980. *BMR Journal of Australian Geology and Geophysics* **5**, 95–111.

Truswell EM (1989) Cretaceous and Tertiary vegetation of Antarctica: a palynological perspective. In: *Antarctic Paleobiology.* (Eds TN Taylor and EL Taylor) pp. 82–87. Springer Verlag.

Truswell EM (1991) Antarctica: a history of terrestrial vegetation. In: *The Geology of Antarctica.* (Ed. RJ Tingey) pp. 499–537. Cambridge University Press, Cambridge.

Truswell EM, Dettmann ME and O'Brien PE (1999) Mesozoic palynofloras from the Mac.Robertson shelf, East Antarctica: geological and phytogeographic implications. *Antarctic Science* **11**(2), 239–255.

Truswell EM, Quilty PG, McMinn A, Macphail MK and Wheller G (2005) Late Miocene vegetation and palaeoenvironments of the Drygalski Formation, Heard Island, Indian Ocean: evidence from palynology. *Antarctic Science* **17**, 427–442.

Turner RD (1967) A new species of fossil *Chlamys* from Wright Velley, McMurdo Sound, Antarctica. *New Zealand Journal of Geology and Geophysics* **10**, 446–455.

Turner S and Young GC (1992) Thelodont scales from the Middle-Late Devonian Aztec Siltstone, southern Victoria Land, Antarctica. *Antarctic Science* **4**, 89–105.

Urreta MBA (1989) The Cretaceous decapod Crustacea of Argentina and the Antarctic Peninsula. *Palaeontology* **32**, 499–552.

Valenzuela E and Herve F 1972 [1971]. Geology of Byers Peninsula, Livingston Island, South Shetland Islands. In *Antarctic Geology and Geophysis.* (Ed. RJ Adie) pp. 83–89. Universitetsforlaget, Oslo, Norway.

Varne R, Gee RD and Quilty PG (1969) Macquarie Island and the cause of linear magnetic anomalies. *Science* **164**, 230–233.

Vialov OS (1962) Problematica of the Beacon Sandstone at Beacon Height West, Antarctica. *New Zealand Journal of Geology and Geophysics* **5**, 718–732.

Vickers-Rich P and Rich TH (1997) The polar dinosaurs of southeastern Australia. In: *Dinofest International.* (Eds DL Wolberg, E Stump and GD Rosenberg) pp. 253–258. Academy of Natural Sciences, Philadelphia.

Vickers-Rich P and Rich TH (2000) Dinosaurs of the Antarctic. *Scientific American Special Edition* **14**, 40–47.

Vizcaino SF, Bond M, Reguero MA and Pascual R (1997) The youngest record of fossil land mammals from Antarctica; its significances on the evolution of the terrestrial environment of the Antarctic Peninsula during the late Eocene. *Journal of Paleontology* **71**, 348–350.

Vizcaino SF, Carlini AA and Reguero MA (1988) Primer registro de un marsupial Didelphimorphia en Antartida. Su implicancia biogeografica. In: *V Jornadas Argentinas de Paleontologia de Vertebrados. Resumenes.* pp. 30–31. La Plata, Argentina.

Vizcaíno SF, Pascual R, Reguero MA and Gooin FJ (1998) Antarctica as a background for mammalian evolution. In: *Paleógeno de América del Sur y de la Península Anártica.* (Ed. S Casadio) *Asociación Paleontológica Argentina, Publicación Especial* **5**, 199–209.

Vizcaino SF, Reguero MA, Goin FJ and Pascual R (1996) The Antarctic continent as background for mammalian evolution. In: *Congreso del Paleogeno de America del Sur y XII Jornadas Argentinas de Paleontologia Vertebrados. Resumenes.* p. 38. La Pampa, Argentina.

Vizcaino SF, Reguero MA, Goin FJ and Pascual R (1998) The Antarctic continent as background for mammalian evolution. In: *Asociacion Paleontologica Argentina, Publicacion Especial 5, Paleogeno de America del Sur y de la Peninsula Antartica.* (Ed. S Casadio) **5**, 199–209.

Vizcaino SF, Reguero MA, Goin FJ, Tambussi CP and Noriega JI (1995a) An approach to the structure of the Eocene terrestrial vertebrate community from Antarctica. *Journal of Vertebrate Paleontology, Abstracts* **15**, 58A.

Vizcaino SF, Reguero MA, Goin FJ, Tambussi CP and Noriega JI (1995b) Eocene terrestrial vertebrate assemblage from Antarctica: an essay community structure and paleozoogeography. In *55th Annual Meeting of the Society of Vertebrate Paleontology, November 1–4, 1995. Abstracts.* Pittsburgh, USA.

Vizcaino SF, Reguero MA, Goin FJ, Tambussi CP and Noriega JI (1996) An approach to the structure of the Eocene terrestrial vertebrate community from Antarctic Peninsula. In: *Congreso del Paleogeno de America del Sur y XII Jornadas Argentinas de Paleontologia Vertebrados. Resumenes.* p. 39. La Pampa, Argentina.

Vizcaino SF, Reguero MA, Goin FJ, Tambussi CP and Noriega JI (1998) An approach to the structure of the Eocene terrestrial vertebrate community from Antarctic Peninsula. In: *Paleógeno de América del Sur y de la Península Antártica.* (Ed. S Casadio) *Asociación Paleontológica Argentina, Publicación Especial* **5**, 177–183.

Vizcaino SF, Reguero MA, Marenssi SA and Santillana SN (1994) The fossol record of land mammals from Antarctica. In: *XXI Polar Symposium (Warszawa, Poland, September 22–24, 1994) Proceedings,* pp. 49–54.

Vizcaino SF, Reguero MA, Marenssi SA and Santillana SN (1995) New land mammal-bearing localities from the Eocene La Meseta Formation, Seymour Island, Antarctica. In: *VII International Symposium on Antarctic Earth Sciences. Abstracts.* p. 397. Siena, Italy.

Vizcaino SF, Reguero MA, Marenssi SA and Santillana SN (1996) Eocene terrestrial environment and biota of the northern Antarctic Peninsula. *56th Annual Meeting of the Society of Vertebrate Paleontology.* New York, October 16–19, 1996. Abstracts, 35A.

Vizcaino SF, Reguero MA, Marenssi SA and Santillana SN (1996d) Eocene terrestrial environment and biota of the northern Antarctic Peninsula. *Journal of Vertebrate Paleontology, Abstracts* **16**, 70A.

Vizcaino S, Reguero M, Marenssi S and Santillana S (1997) New land mammal-bearing localities from the Eocene La Meseta Formation, Seymour Island, Antarctica. In: *The Antarctic Region: Geological Evolution and Processes.* (Ed. CA Ricci) pp. 997–1000. Terra Antartica Publication, Siena, Italy.

Vizcaino SF and Scillato-Yane GJ (1994) Un Tardigrada (Mammalia Xenarthra) del Eoceno medio de la Isla Marambio (Seymour), Antartida. In: *3ras Jornadas Antarticas. Actas.* pp. 63–64. Buenos Aires, Argentina.

Vizcaino SF and Scillato-Yane GJ (1995) An Eocene tardigrade (Mammalia, Xenarthra) from Seymour Island, West Antarctica. *Antarctic Science* **7**, 407–408.

Vodrážka R and Crame JA (2011) First fossil sponge from Antarctica and its paleobiogeographical significance. *Journal of Paleontology* **85**(1), 48–57.

Wade FA (1972) Geologic survey of Marie Byrd Land. *Antarctic Journal of the United States* **7**(5), 144–145.

Walker SM (illustrated by J Bindon) (2008) *The Search for Antarctic Dinosaurs.* Millbrook Press, Minneapolis.

Wanntorp L, Vajda V and Raine JI (2011) Past diversity of Proteaceae on subantarctic Campbell Island, a remote outpost of Gondwana. *Cretaceous Research* **32**(3), 357–367.

Ward DJ and Grande L (1991) Chimaeroid fish remains from Seymour Island, Antarctic Peninsula. *Antarctic Science* **3**, 323–330.

Warnke DA, Richter C, Florindo F, Damuth JE, Balsam WL, Strand K, Ruikka M, Juntila J, Theissen K and Quilty PG (2004) Data report: HiRISC (High-Resolution Integrated Stratigraphy Committee) Plio-Pleistocene interval, 0–50 m mbsf, at ODP Leg 188 Site 1165, Prydz Bay, Antarctica: a data report. *Proceedings ODP, Scientific Results.*

Watkins D (1992) Upper Cretaceous nannofossils from ODP Leg 120, Kerguelen Plateau, Southern Ocean. *Proceedings of the ODP Scientific Results* **120**, 343–370.

Watkins D, Quilty PG, Mohr BAR, Mao S, Francis JE, Gee CT and Coffin MF (1992) Paleontology of the Cretaceous of the central Kerguelen Plateau. *Proceedings ODP Scientific Results* **120**: College Station TX (Ocean Drilling Program), pp. 951–959.

Weaver L, McLoughlin S and Drinnan A (1997) Fossil woods from the Upper Permian Bainmedart Coal Measures, northern Prince Charles Mountains, East Antarctica. *AGSO Journal of Australian Geology & Geophysics* **16**, 655.

Webb P (1972) Comments on the reported occurrence of ?*Globotruncana contusa* (Cushman) from the Upper Cretaceous of James Ross Island, Graham Land. *New Zealand Journal of Geology and Geophysics* **15**(1), 183.

Webb PN (1991) A review of the Cenozoic stratigraphy and paleontology of Antarctica. In: *Geological Evolution of Antaractica.* (Eds MRA Thomson, JA Crame and JW Thomson) pp. 599–607. Cambridge University Press, Cambridge.

Webb PN and Harwood DM (1993) Pliocene fossil Nothofagus (Southern Beech) from Antarctica: phytogeography, dispersal strategies and survival in high latitude glacial-deglacial environments. In: *Forest Development in Cold Climates.* (Ed. J Alden) p. 135. Plenum Press, New York.

Webb PN and Wrenn JH (1975) Foraminifera from DVDP holes 8.9. and 10, Taylor Valley. *Antarctic Journal of the United States* **10**(4), 168–169.

Webers GF (1966) Upper Cambrian archaeocyathid from Antarctica. [Abstract]. *Geological Society of America, Special Papers* **87**, 183.

Webers GF (1970) Paleontological investigations in the Ellsworth Mountains, west Antarctica. *Antarctic Journal of the United States* **5**(5), 162–163.

Webers GF 1972 [1971]. Unusual Upper Cambrian fauna from west Antarctica. In: *Antarctic Geology and Geophysics.* (Ed. RJ Adie) pp. 235–237. Universitetsforlaget, Oslo, Norway.

Webers GF, Craddock C and Splettstoesser JF (1992) Geology and palaeontology of the Ellsworth Mountains, west Antarctica. *Geological Society of America Memoirs* **170**, 1–459.

Weller S (1903) The Stokes collection of Antarctic fossils. *Journal of Geology* **11**(4), 413–419.

Welton B and Zinsmeister WJ (1980) Eocene neoselachians from the La Meseta Formation, Seymour Island, Antarctic Peninsula. *Contributions in Science, Natural History Museum of Los Angeles County* **329**, 1–10.

White EI (1968) Devonian fishes of the Mawson-Mulock area, Victoria Land, Antarctica. . *Trans-Antarctic Expedition 1955–58. Scientific Reports. Geology* **16**, 1–26.

White Jr. JF and Taylor TN (1989) A trichomycete-like fossil from the Triassic of Antarctica. *Mycologia* **81**, 643–646.

White ME (1962) Permian plant remains from Mount Rymill, Antarctica. In: *Report for 1961. Plant fossil collection. Records of the Bureau of Minerals and Resources, Geology and Geophysics, Australia* **114**, 13–18 [unpublished].

White ME (1973) Permian flora from the Beaver Lake area, Prince Charles Mountains, Antarctica. *Bureau of Mineral Resources, Geology and Geophysics. Bulletin* **126**, 13–18.

Whitehead JM, Ehrmann W, Harwood DM, Hillenbrand, Quilty PG, Hart C, Taviani M, Thorn V and McMinn A (2006) Late Miocene paleoenvironment of the Lambert Graben embayment, East Antarctica, evident from: mollusc paleontology, sedimentology and geochemistry. *Global and Planetary Change* **50**, 127–147.

Whitehead JM, Harwood DM, McKelvey BC, Hambrey MJ and McMinn A (2004) Diatom biostratigraphy of the Cenozoic fjordal Pagodroma Group, Northern Prince Charles Mountains, East Antarctica. *Australian Journal of Earth Sciences* **51**, 521–547.

Whitehead JM, Quilty PG, Harwood DM and McMinn A (2001) Early Pliocene paleoenvironment of the Sørsdal Formation, Vestfold Hills, based on diatoms. *Marine Micropaleontology* **41**, 125–152.

Whitehead JM, Quilty PG, McKelvey B and O'Brien PE (2006) A review of the Cenozoic stratigraphy and glacial history of the Lambert Graben-Prydz Bay region, East Antarctica. *Antarctic Science* **18**, 83–99.

Wiedman LA, Feldmann RM, Lee DE and Zinsmeister WJ (1988) Brachiopoda from the La Meseta Formation (Eocene), Seymour Island, Antarctica. In: *Geology and Paleontology of Seymour Island.* (Eds MO Woodburne and RM Feldmann) *Geological Society of America Memoir* **169**, pp. 449–457.

Wilckens O (1905–07) Die Lamellibranchiaten, Gastropoden & c., der oberen Kreide Südpatagoniens. *Berichte der Naturforschenden Gesellschaft zu Freiburg* **15**, 97–166.

Wilckens O (1910) Die Anneliden, Bivalven und Gastropoden der antarktischen Kreideformation. *Wissenschaftliche Ergebnisse der Schwedischen Südpolar-Expedition* **3**(12), 1–132.

Wilckens O (1911) Die Mollusken der Antarktischen Tertiärformation. *Wissenschaftliche Ergebnisse der Schwedischen Südpolar-Expedition 1901–1903* **3**(13), 1–62.

Wilckens O (1924) Die tertiäre Fauna der Cockburn-Inssel (Westantarktika). *Further Zoological Results of the Swedish Antarctic Expedition* **1**(1), 1–18.

Wilckens O (1932) Fossilien und Gesteinne von Süd-Georgien. *Scientific Results Norway Antarctic Expedition* **8**, 1–28.

Wilckens O (1947) Paläontologische und geologische Ergebnisse der Reise von Kohl-Larsen (1928–29) nach Süd-Georgien. *Abhandlungen der Senckenbergischen Naturforschenden* **474**, 1–66.

Willey LE (1972) Belemnites from south-eastern Alexander Island: I. The occurrence of the family Dimitobelidae in the Lower Cretaceous. *Bulletin of the British Antarctic Survey* **28**, 29–42.

Willey LE (1973) Belemnites from south-eastern Alexander Island: II. The occurrence of the family Belemnopseidae in the Upper Jurassic and Lower Cretaceous. *Bulletin of the British Antarctic Survey* **36**, 35–59.

Willey LE (1975a) Upper Jurassic and Lower Cretaceous Grammatodontinae (Bivalvia) from southern Alexander Island. *Bulletin of the British Antarctic Survey* **41** & **42**, 1–22. .

Willey LE (1975b) Upper Jurassic and lowest Cretaceous Trigoniidae (Bivalvia) from south-eastern Alexander Island. *Bulletin of the British Antarctic Survey* **41 & 42**, 77–85.

Willey LE (1975c) Upper Jurassic and Lower Cretaceous Pinnidae (Bivalvia) from southern Alexander Island. *Bulletin of the British Antarctic Survey* **41 & 42**, 121–131.

Williams M, Siveter DJ, Asworth AC, Wilby PR, Horne DJ, Lewis AR and Marchant DR (2008) Exceptionally well preserved lacustrine ostracods from the Middle Miocene of Antarctica: implications for high-latitude palaeoenvironments at 77 degrees south. *Proceedings of Biological Sciences* **275**(1650), 2449–2454.

Williams PL (1970) Geology of the Lassiter Coast. *Antarctic Journal of the United States* **5**(4), 98–99.

Williams PL and Rowley PD (1971) Geologic studies of the Lassiter Coast. *Antarctic Journal of the United States* **6**(4), 120.

Willis PMA and Stilwell JD (2000) A probable piscivorous crocodile from Eocene deposits of McMurdo Sound, East Antarctica. In: *Paleobiology and Paleoenvironments of Eocene Rocks, McMurdo Sound East Antarctica* (Eds JD Stilwell and RM Feldmann) *American Geophysical Union Antarctic Research Series Volume* **67**, 355–358.

Wiman C (1905a) Über die alttertiären Vertebraten der Seymourinsel. In: *Wissenschaftliche Ergebnisse der Schwedischen Südpolar-Expedition 1901–1903.* **3**(1), 1–37. Stockholm, Sweden.

Wiman C (1905b) Vorläufige Mitteilung über die alttertiären Vertebraten der Seymourinsel. *Bulletin of the Geological Institute of the University of Upsala* **6**, 247–252.

Wolfe KJ, Long JA, Bradshaw MA, Harmsen F and Kirkbride M (1990) Fish-bearing Aztec Siltstone (Devonian) in the Cook Mountains, Antarctica. *New Zealand Journal of Geology and Geophysics* **33**, 511–514.

Woodburne MO and Case JA (1996) Dispersal, vicariance, and the Late Cretaceous to Early Tertiary land mammal biogeography from South America to Australia. *Journal of Mammalian Evolution* **3**, 121–161.

Woodburne MO and Zinsmeister WJ (1982) Fossil land mammal from Antarctica. *Science* **218**, 284–285.

Woodburne MO and Zinsmeister WJ (1984) The first land mammal from Antarctica and its biogeographic implications. *Journal of Paleontology* **58**, 913–948.

Woolfe KJ (1990) Trace fossils as paleoenvironmental indicators in the Taylor Group (Devonian) of Antarctica. *Palaeogeography, Palaeoclimatology, Palaeoecology* **80**, 301–310.

Woolfe KJ, Long JA, Bradshaw M, Harmsen F and Kirkbride M (1990) Discovery of fish-bearing Aztec Siltstone (Devonian) in the Cook Mountains, Antarctica. *New Zealand Journal of Geology and Geophysics* **33**, 511–514.

Wrenn JH and Hart GF (1988) Paleogene dinoflagelate cyst biostratigraphy of Seymour Island, Antarctica. In: *Geology and Paleontology of Seymour Island, Antarctic Peninsula.* (Eds RM Feldmann and MO Woodburne) *Geological Society of America. Memoirs* **169**, 303–320.

Wright TO, Ross RJ and Repetski JE (1984) Newly discovered youngest Cambrian or oldest Ordovician fossils from the Robertson Bay terrane (formerly Precambrian), northern Victoria Land, Antarctica. *Geology* **12**, 301–305.

Wrona R (2004) Cambrian microfossils from glacial erratics of King George Island, Antarctica. *Acta Palaeontologica Polonica* **49**, 13–56.

Wrona R and Zhuravlev AY (1996) Early Cambrian archaeocyaths from glacial erratics of King George Island (South Shetland Islands), Antarctica. *Palaeontologica Polonica* **55**, 10–36.

Xu H-H and Berry CM (2008) The Middle Devonian lycopsid *Haskinsia* Grierson et Banks from the Ruppert Coast, Marie Byrd Land, West Antarctica. *Review of Palaeobotany and Palynology* **150** (1–4), 1–4.

Yochelson EL, Flower RH and Webers GF (1973) The bearing of the new Late Cambrian monoplacophoran genus *Knightoconus* upon the origin of the Cephalopoda. *Lethaia* **6**(3), 275–309.

Young GC (1981) Biogeography of Devonian vertebrates. *Alcheringa* **5**, 225–243.

Young GC (1982) Devonian sharks from south-eastern Australia and Antarctica. *Palaeontology* **25**, 817–843.

Young GC (1988) Antiarchs (placoderm fishes) from the Devonian Aztec Siltstone, southern Victoria Land, Antarctica. *Palaeontographica A* **202**, 1–125.

Young GC (1989a) The Aztec fish fauna of southern Victoria Land evolutionary and biogeographic significance. In: *Origins and Evolution of the Antarctic Biota.* (Ed. JA Crame) *Geological Society of London Special Publication* **47**, 43–62. London, UK.

Young GC (1989b) New occurrences of culmacanthid acanthodians (Pisces, Devonian) from Antarctica and southeastern Australia. *Proceedings of the Linnean Society of New South Wales* **111**, 11–24.

Young GC (1990) Devonian vertebrate distribution patterns, and cladistic analysis of palaeogeographic hypotheses. In: *Palaeozoic Palaeogeography and Biogeography, Geological Society Memoirs, London* (Eds WS McKerrow and CR Scotese) **12**, 243–255.

Young GC (1991) Fossil fishes from Antarctica. In: *The Geology of Antarctica.* (Ed. RJ Tingey) pp. 538–567. Oxford University Press, Oxford, UK.

Young GC (1993) Middle Palaeozoic macrovertebrate biostratigraphy of Eastern Gondwana. In: *Palaeozoic Vertebrate Biostratigraphy and Biogeography.* (Ed. JA Long) pp. 208–251. Belhaven Press, London, UK.

Young GC and Laurie JR (Eds) (1996) Devonian (Chapter 4). In: *An Australian Phanerozoic Timescale.* pp. 96–109. Oxford University Press, Melbourne.

Young GC and Burrow C (2004) Diplacanthid acanthodians from the Aztec Siltstone (late Middle Devonian) of southern Victoria Land, Antarctica. In: *Lower Vertebrates from the Palaeozoic.* (Ed. GC Young) First International Palaeontological Congress (IPC 2002) Sydney, Australia. July 2002. *Proceedings of Symposium 6 (Palaeoazoic Vertebrates) Fossils & Strata* **50**, 23–43.

Young GC, Burrow CJ, Long JA, Turner S and Choo B (2010) Devonian macrovertebrate assemblages and biogeography of East Gondwana (Australia, Antarctica). *Palaeoworld* **19**, 55–74.

Young GC and Long JA (2005) Phyllolepid placoderm fish remains from the Devonian Aztec Siltstone, southern Victoria Land, Antarctica. *Antarctic Science* **17**(3), 387–408.

Young GC, Long JA and Ritchie A (1992) Crossopterygian fishes from the Devonian Aztec Siltstone of Antarctica: systematics, relationships and biogeographic significance. *Records of the Australian Museum, Supplement* **14**, 1–77.

Young GC and Goujet D (2003) Devonian fish remains from the Dulcie Sandstone and Craven Peak Beds, Georgina Basin, central Australia. *Records of the Western Australian Museum Supplement* **65**, 1–85.

Young GC and Turner S (2000) Devonian microvertebrates and marine – nonmarine correlation in East Gondwana: Overview. *Courier Forschungsinstitut Senckenberg* **223**, 453–470.

Young VT (1986) Early Devonian fish material from the Horlick Formation, Ohio Range, Antarctica. *Alcheringa* **10**, 35–44.

Zastawaniak E (1981) Tertiary leaf flora from the Point Hennequin Group of King George Island (South Shetland Islands, Antarctica) Preliminary report. *Studia Geologica Polonica* **72**, 97–108.

Zastawniak E (1990) Late Cretaceous leaf flora of King George Island, West Antarctica. In: *Proceedings of the Symposium: Palaeofloristic and Palaeoclimatic Changes in the Cretaceous and Tertiary.* (Eds E Konbloch and Z Kvaček) pp. 81–86. Geological Survey, Prague.

Zastawniak E (1993) Macroscopic plant remains from Upper Cretaceous and Tertiary of King George Island, (South Shetland Islands, West Antarctica). *Wiadomości Botaniczne* **37**, 217–219.

Zastawniak E (1994) Upper Cretaceous leaf flora from the Blaszyk Moraine (Zamek Formation), King George Island, South Sheltand Islands, West Antarctica. *Acta Palaeobotanica* **34**, 119–163.

Zastawaniak E, Wrona R, Gazdzick A and Birkenmajer K (1985) Plant remains from the top of the Point Hennequin Group (Upper Oligocene), King George Island (South Shetland Islands, Antarctica) Preliminary report. *Studia Geologica Polonica* **81**, 143–164.

Zeuner FE (1959) Jurassic beetles from Grahamland, Antarctica. *Palaeontology* **1**(4), 407–409, pl. 72.

Zhou Z and Li H (1994) Early Tertiary ferns from Fildes Peninsula, King George Island, Antarctica. In: *Stratigraphy and Palaeontology of Fildes Peninsula, King George Island.* (Ed. Y Shen) *State Antarctic Committee Monograph* **3**, 173–189.

Zhuravlev AY and Wood RA (1996) Axoxia as the cause of the mid-Early Cambrian (Botomian) extinction event. *Geology* **29**, 311–314.

Zinsmeister WJ (1976a) A new genus and species of the family Struthiolariidae, *Antarctodarwinella ellioti,* from Seymour Island, Antarctica. *The Ohio Journal of Science* **76**(3), 111–114.

Zinsmeister WJ (1976b) The Struthiolariidae (Gastropoda) fauna from Seymour Island, Antarctic Peninsula. *Actas VI Congreso Geologico Argentina, Sept. 21–27, 1975* **1**, 609–618.

Zinsmeister WJ (1977) Note on a new occurrence of the southern hemisphere aporrhaid gastropod *Struthioptera* Finlay and Marwick on Seymour Island, Antarctica. *Journal of Paleontology* **51**(2), 399–404.

Zinsmeister WJ (1978) Eocene nautiloid fauna from the La Meseta Formation of Seymour Island, Antarctic Peninsula. *Antarctic Journal of the United States* **13**(4), 24–25.

Zinsmeister WJ (1979) Biogeographic significance of the late Mesozoic and early Tertiary molluscan faunas of Seymour Island (Antarctic Peninsula) to the final breakup of Gondwanaland. In: *Historical Biogeography, Plate Tectonics and the Changing Environment.* (Eds J Gray and AJ Boucot) pp. 349–355. Oregon State University Press, Eugene, Oregon.

Zinsmeister WJ (1981) Middle to late Eocene invertebrate fauna from the San Julian Formation at Punta Casamayor, Santa Cruz Province, southern Argentina. *Journal of Paleontology* **55**(5), 1083–1102.

Zinsmeister WJ (1982a) Review of the Upper Cretaceous-lower Tertiary sequence on Seymour Island, Antarctica. *Journal of the Geological Society of London* **139**(6), 779–786.

Zinsmeister WJ (1982b) Late Cretaceous-early Tertiary molluscan biogeography of the southern circum-Pacific. *Journal of Paleontology* **56**, 84–102.

Zinsmeister WJ (1983c) First U.S. expedition to the James Ross Island area, Antarctic Peninsula. *Antarctic Journal of the United States* **17**(5), 63–64.

Zinsmeister WJ (1984a) Geology and paleontology of Seymour Island, Antarctic Peninsula. *Antarctic Journal of the United States* **19**(2), 1–5.

Zinsmeister WJ (1984b) Late Eocene bivalves (Mollusca) from the La Meseta Formation, collected during the 1974–75 joint Argentine-American expedition to Seymour Island, Antarctic Peninsula. *Journal of Palaeontology* **58**, 1497–1527.

Zinsmeister WJ (1985) Unusual occurrence of nautiloids from the Upper Eocene La Meseta Formation, Seymour Island. *Antarctic Journal of the United States* **20**(5), 43–44.

Zinsmeister WJ (1986) Fossil Windfall at Antarctica's Edge. *Natural History* **95**(5), 61–66.

Zinsmeister WJ (1987) Unusual nautilid occurrence in the upper Eocene La Meseta Formation, Seymour Island, Antarctica. *Journal of Paleontology* **61**(4), 724–726.

Zinsmeister (1998a) Discovery of fish mortality horizon at the K-T boundary on Seymour Island: re-evaluation of events at the end of the Cretaceous. *Journal of Paleontology* **72**, 556–571.

Zinsmeister (1998b) Event plane analysis: a technique for the spatial analysis of biostratigraphic data. *Geological Society of America, Program with Abstracts* **28**(6), 249.

Zinsmeister WJ (1990) Note on the occurrence of a limpet in living position from the Late Cretaceous, Seymour Island, Antarctica. *Journal of Paleontology* **64**, 477–478.

Zinsmeister WJ (1998) Discovery of fish mortality horizon at the K-T boundary on Seymour Island: re-evalutation of events at the end of the Cretaceous. *Journal of Paleontology* **72**, 556–571.

Zinsmeister WJ (2001) Late Maastrichtian short-term events on Seymour Island, Antarctic Peninsula. *Geology* **109**, 213–229.

Zinsmeister WJ and Camacho HH (1980) Late Eocene Struthiolariidae (Mollusca: Gastropoda) from Seymour Island, Antarctic Peninsula and their significance to the biogeography of early Tertiary shallow-water faunas of the Southern Hemisphere. *Journal of Paleontology* **54**, 1–14.

Zinsmeister WJ and Feldmann RM (1984) Cenozoic high latitude heterochroneity of southern hemisphere marine faunas. *Science* **224**, 281–283.

Zinsmeister WJ, Feldmann RM, Woodburne MO and Elliot DH (1989) Late Cretaceous/earliest Tertiary transition of Seymour Island, Antarctica. *Journal of Paleontology* **63**, 731–738.

Zinsmeister WJ and Griffin M (1995) Late Cretaceous and early Tertiary aporrhaid gastropods from the southern rim of the Pacific Ocean. *Journal of Paleontology* **69**, 692–702.

Zinsmeister WJ and Macellari CE (1988) Bivalvia (Mollusca) from Seymour Island, Antarctic Peninsula. In: *Geology and Paleontology of Seymour Island, Antaractic Peninsula.* (Eds RM Feldmann and MO Woodburne) *Geological Society of America Memoir* **169**, 253–284.

Zinsmeister WJ and Stilwell JD (1990) First record of the family Ringiculidae (Gastropoda) from the Middle Tertiary of Antarctica. *Journal of Paleontology* **64**, 373–376.

Zullo VA, Feldmann RM and Wiedman LA (1988) Balanomorph Cirripedia from the Eocene La Meseta Formation, Seymour Island, Antarctica. In: *Geology and Paleontology of Seymour Island, Antarctic Peninsula.* (Eds RM Feldmann and MO Woodburne), *Geological Society of America Memoir* **169**, 459–464.

INDEX

Figures in **bold** refer to illustrations.

www.ingramcontent.com/pod-product-compliance
Ingram Content Group UK Ltd.
Pitfield, Milton Keynes, MK11 3LW, UK
UKHW060221061026
15625UKWH00007B/433

9 780643 096356